DESIGN OF EXPERIMENTS USING THE TAGUCHI APPROACH

DESIGN OF EXPERIMENTS USING THE TAGUCHI APPROACH

16 STEPS TO PRODUCT AND PROCESS IMPROVEMENT

Ranjit K. Roy, Ph.D., P.E.

A WILEY-INTERSCIENCE PUBLICATION

JOHN WILEY & SONS, INC.

New York • Chichester • Weinheim • Brisbane • Singapore • Toronto

This publication is designed to provide accurate and authoritative information in regard to the subject matter covered. It is sold with the understanding that the publisher is not engaged in rendering professional services. If professional advice or other expert assistance is required, the services of a competent professional person should be sought.

Library of Congress Cataloguing-in-Publication Data:

Roy, Ranjit K.
   Design of experiments using the Taguchi approach : 16 steps to product and process improvement / Ranjit K. Roy.
     p. cm.
     "A Wiley-Interscience publication."
     ISBN 0-471-36101-1 (cloth : alk. paper)
     1. Quality control—Statistical methods. 2. Taguchi methods (Quality control) 3. Experimental design. I. Title
     TS156.R688 2001
     658.5'62—dc21   00-036841

Printed in the United States of America

10 9 8 7 6 5 4 3

CONTENTS

■■■■■ PREFACE

Learning takes place when ideas and concepts are exchanged between an instructor and students or readers. The author's involvement in promoting and training use of Taguchi's experimental design technique dates back to the early 1980s, when it was introduced in the United States. The early attendees of seminars on the subject were there more through professional curiosity than application interest. The attendees seemed satisfied that they got what they expected from the seminars, but over the years, as demands for quality-related skills grew, so did the demand for application knowledge from industrial and academic professionals at all levels. Instructional seminars with attendees of all sorts of backgrounds, from those with advanced university degrees to others with no university instruction, always poses a challenge to the instructor. As a trainer wishing to satisfy the most attendees, the author has always felt a continuing need to make the instructional method easier. Having conducted several hundred classes and workshops, published one textbook (*A Primer on the Taguchi Method*, 1990), and developed one software package (Qualitek-4) on the subject during the past few years, it appears that the task of making the learning process simpler is never-ending. *Design of Experiments Using the Taguchi Approach: 16 Steps to Product and Process Improvement* is an attempt to offer another learning approach that allows the experimenter to understand the technique on his or her own with detailed examples, exercises, and preprogrammed solutions.

Although the design of experiments using the Taguchi approach is statistical in nature, no formal background in statistical science is necessary to master the subject. Over a dozen reference texts on the technique have been published by experimenters in the industry. A number of organizations also routinely offer courses and workshops on the subject. But there are not many hands-on seminars with application workshops to be found anywhere, nor is there any textbook that offers readers the option to learn in their own home or workplace. This book is written using the language most common on the manufacturing floor so that it will be easier for technicians, working engineers, and professional managers in the manufacturing industries to read and understand.

For developing application expertise, a sound knowledge of the theory and a working understanding of the application principles and computational aids used in processing the experimental results are necessary ingredients. To learn the theory, one would generally need to read one or more of the books available on the subject. Application tips are more easily understood from an experienced instructor than by reading books on application examples. When experiments are carried out and there is a need for analysis of results, it is a good idea to obtain one of the available packages of analysis software. The software assures the accuracy of analysis and saves time. This book provides resources in all three elements of the learning process. It contains applicable theories in brief and step-by-step application of the concepts. It also contains a fully functional software CD-ROM that readers can use to review example experiments. By using the software, the readers learn each application step by experiencing hands-on how to enter data for experiments, analyze results, and draw conclusions from the analysis.

The author sincerely hopes that practicing engineers in the industry and academia will find this book helpful in learning how to apply the Taguchi experimental design technique and promote applications of the technique within their own work environment. The author would greatly appreciate receiving your questions, comments, and suggestions.

RANJIT K. ROY

E-mail: rkroy@rkroy.com

ACKNOWLEDGMENTS

I would like to thank Bob Argentieri of John Wiley & Sons, Inc. for expressing an interest in the book and working with me throughout the publication. Bob worked with my last book, *A Primer on the Taguchi Method*, while he was with Van Nostrand Reinhold. My thanks also go to Bob Hilbert, Stacy Rympa, and other staff members of John Wiley who worked on the production of this book.

I'm also indebted to my seminar attendees and hosts. It is their collective feedback and comments that motivated me to write this book. Over the last several years, I devoted much of my time to publishing easy-to-use literature, developing software for calculations with experimental results, and setting up effective training classes. Now you will be able to use this book–software combination to learn the techniques at your own pace. Thanks for your reviews and suggestions to all my public seminar attendees, students of my class at Oakland University (Rochester, Michigan), attendees at my seminars at company sites, readers of my earlier book, and users of Qualitek-4 software.

The source for much of the application experience I shared in this book came from seminars and workshops conducted for my corporate sponsors. I would like to take this opportunity to thank Larry Smith, Mark Slagale, and Rod Munro of Ford Motor Company in Dearborn, Michigan for their unwavering sponsorship of my training in many parts of the organization. I'm also much obliged to Probir Guha at Budd Plastic of Troy, Michigan, Hossein Nikoui of the Oakwood Group, Dearborn, Michigan, and Brian Rauch of John Deere Dubuque Works in Dubuque, Iowa for providing me with the opportunity to train their people. My special thanks also go to Dean Stamitis of Contemporary Consultants, Jayanta Chandra of Eaton Corporation, and Pavel Blecharz of Quality Management Institute for reviewing my manuscript materials and providing valuable feedback. Finally, I express my sincere thanks to those who are always in mind: my daughter Purba and her husband, Walter; my younger daughter, Paula; and my wife, Krishna, for allowing me to be absent from countless family activities so that I could put the manuscript together.

The publisher and I would like take this opportunity to express our gratitude to Peggy Jennings, Vice President–Operations of the American Supplier Institute, Inc., Center for Taguchi Methods, for granting permission to reproduce the orthogonal arrays, triangular tables, and related linear graphs. These materials are contained in the *System of Experimental Designs* by Genechi Taguchi (1987).

SYMBOLS AND ABBREVIATIONS

A, B, C	Notation used for factors under experimental study
a_1, b_2 or A_1, B_2	Descriptions of level 1 of factor A, level 2 of factor B, etc.
C_p, C_{pk}	Capability indices
DOF, f	Degrees of freedom
F	F-ratio
L-4, L-8, etc.	Orthogonal arrays
m	$y_{avg} - y_o$
MSD	Mean-squared deviation
n	Number of results
QC	Quality characteristic
S	Sums of squares
S'	Pure sums of squares
S/N	Signal-to-noise ratio
S_t	Total sums of squares
T	Total of all results
V	Variance
y	Result
y_a, y_{avg}	Average performance
y_{min}	Minimum result
y_0	Target/nominal
σ	Standard deviation
σ_n	Standard deviation (population)
σ_{n-1}	Standard deviation (classical definition/sample)
Σ	Summation sign

◼◼◼◼ INTRODUCTION

Welcome to *Design of Experiments Using the Taguchi Approach*, designed to provide you with a comprehensive background in the design of experiments (DOE)/Taguchi technique. You will not only learn the details of application methods, you will also learn how to set up experiments and accomplish all analysis tasks using any of several available software packages. *A Full-Featured Working Model* of the Qualitek-4 software is included with this book for your learning experience. All example experiments used in the book are included with the software.

The book is laid out to prepare you for immediate application of the approach in 16 steps. Each step covers necessary minimum skills and background you will need to apply the technique confidently in your own projects. In the first step you will get an overview and understand what design of experiment technique is all about and how Dr. Genechi Taguchi applied the technique to improve the quality of products. In the next several steps you will learn the method for designing simple experiments, followed by steps applying the same method to study interactions between factors and mixed-level factor designs. Advanced techniques to analyze experimental results and Taguchi robust design strategy are covered in the last several steps. Experiment planning and strategies for effective experiments are discussed in Step 15.

Each step starts with a learning objective and ends with a summary of what you should have learned in the step. The steps also contain exercises for readers to use to practice what they learn. The solutions to the exercises are included at the end of each step. The overall goal is to keep the discussions simple and teach the application method quickly. The computer software included will be used to carry out calculations whenever the capability is available. The use of complicated equations is restricted unless necessary to communicate a point under discussion. Similarly, a detailed discussion of background and theory is included only when it enhances understanding, as applications are the main focus. Discussion of experiments with dynamic characteristics, which is one of the more sophisticated concepts proposed by Dr. Taguchi, is left for further study.

DOE APPLICATION SKILLS

The benefits derived from using DOE/Taguchi depend largely on the manner in which it is used. Although the main purpose of this book is to take you through all the steps involved in the design and analysis of experiments, much emphasis is placed on proper preparation and planning of the experiment. After completing all the steps, you will

1

be ready to apply the technique confidently to all of your projects and continue to become effective as you become more and more confident in your applications.

Who Should Read This Book

Anyone who intends to learn and practice the DOE/Taguchi technique will find this book helpful. If you envision getting involved in the detailed levels of planning, designing, conducting, and analyzing results of experiments, you should read this book. The book is written at a level understandable for readers with or without a college degree. You need not have a preconceived background or experience in quality education. If you work in industry, you should read this book. If you are a quality assurance manager/engineer/technician, design engineer, manufacturing engineer, product assurance engineer, scientist, researcher, or manager involved in quality improvement or problem-solving efforts, this book is for you.

Background You Need to Learn and Practice DOE

Design of experiments is a statistical technique used to study multiple *variables* (also called *factors* or *parameters*) simultaneously. A formal course or background in statistics is not required. Readers interested in application of the technique will grasp the methods easily, regardless of their background.

Where the DOE/Taguchi Approach Is Applicable

DOE is highly effective wherever and whenever it is suspected that the performance of a part or process is controlled by more than one factor. Contrary to what many people may believe, most planned experiments do not always involve a large number of test samples. A simple design of experiment may require as few as four separate experiments or trials. More sophisticated experiments require a larger number of trials but always produce a wealth of information about the project. Although DOE is most often applied using experimental hardware, it is equally effective when applied to analytical simulation. When used for product design optimization, analytical simulation is the common approach, because hardware is not often available.

Learning Strategy

The materials in this book have been presented with applications in mind. Your goal should be to learn how to apply all steps. Like any other technique, DOE/Taguchi has math and an application process (method). The mathematical operations necessary for analysis of results are quite involved but are fixed and follow strict rules. Calculators handle this part. The methods, on the other hand, vary depending on the applications and are difficult to define by rules. If you want to be an expert practitioner, you would eventually need to learn both the math and the method well. But you need not become an expert right away to benefit from DOE. The approach in this book is to give you all the methods, teach you how to interpret the results of calculations, show some sample calculations, and teach you to rely on the computer programs to carry out most routine calculations. As you gain application experience, you will find it necessary to refer to other books to develop clearer concepts of the mathematical treatments involved.

About the Experiment Design and Analysis Software You Need

Included with this book is a CD-ROM containing a *Full-Featured Working Model* (DEMO) of the Qualitek-4 software. The DEMO program of Qualitek-4 contains all example experiments used in the book. All screen figures and graphs were also made using this program. You will be able to expedite your learning process if you study this book while running the program alongside. The copy of the Qualitek-4 program allows you to review all example experiments regardless of size. As long as you are working with the existing files, you will be able to utilize all capabilities of the program, just like the regular version. The program also allows you to create your own experiments utilizing the L-8 array. By the end of this text, you will find that the Taguchi L-8 array can be used to design over 15 different experiments. If for some reason the software package you received is not compatible with your system, you may try to procure the latest version of the software from the manufacturer, Nutek, Inc. The DEMO copy (the latest version of the software included with the book) is available free from the Nutek Web site. Download the software and try to install it. If this does not work, try another system.

What You Will Need to Run the Qualitek-4 Software

You will need an IBM PC, Compaq 486, or better-quality computer equipped with MS DOS–based Windows 3.1+, 95, or later software. Your computer should also have at least 8 megabytes of RAM and 10 megabytes of available disk space. As a word processor, you should have Microsoft Word Version 7.0 or higher loaded in your system. Before you begin the steps, install the Qualitek-4 software and make sure that it is working fully in your system. The computer system used by the author is a Pentium with Windows 95, and the word-processing software is Microsoft Word Version 7.0. If you are using Windows 3.1 with other versions or another brand of word-processing program, you may need to determine the appropriate commands equivalent to those indicated in the book.

QUICK STEP SUMMARY

Following are brief reviews of what each step includes and what you will learn.

Step 1:   Design of Experiments and the Taguchi Approach

This step gives an overview of the technique. The main purpose is to give a clear understanding of what Dr. Taguchi has done to make the design of experiments (DOE) technique more effective in applications, and how he related the outcome of the technique to improve the quality of products and processes. At end of this session you will have a clear understanding of the philosophy behind the Taguchi style of DOE and its intended application objectives.

Step 2: Definition and Measurement of Quality

The main focus of application of DOE is to improve quality. But what is quality? How do we measure quality? The definition of quality varies widely depending on the applications, but it must be defined before any experimental technique can be produced with meaningful results. Taguchi offers a generalized definition for quality of performance. He regards performance as the major component of product or process quality. A reduced variation results in a reduction in scraps, less rejection of product, and fewer warranty returns, consequently reducing costs and improving customer satisfaction. By the end of this step you will have a clear picture of how to measure quality in terms of variation and how to compare two population performances.

Step 3: Common Experiments and Methods of Analysis

A direct way to determine cause and effect is to run experiments. The number of experiments necessary to study the influencing factors depends on the nature of the trend in influence of the factor and its levels. Experiments generally bring to mind a picture of hardware, but it need not be so. When hardware or a physical representation of a product or process is not available, mathematical simulation of the performance can serve as a vehicle for experimentation. Investigation of the influence of factors, one at a time, is a common practice among scientific professionals and requires the use of common sense. When several factors are studied simultaneously, pursuing an economical experimental plan unfortunately requires a little more than common sense— it requires DOE. After completing this step you will understand the logic behind experimental design, appreciate the rational for a statistical approach to determining the trend of influence of various causes, and learn how to express and compare population performances.

Step 4: Experimental Design Using Orthogonal Arrays

Depending on the project, that is, the number of factors and levels included, there are many possible ways in which an experiment can be laid out. A number of standard orthogonal arrays (tables of numbers) have been constructed to facilitate experimental design. Each of these arrays can be used to design experiments to suit several experimental situations. By the end of this step, you will understand how easy it is to design experiments using orthogonal arrays and the simplicity of the process.

Step 5: Experimental Design with Two-Level Factors Only

Experiments with all factors at two levels are quite common in the industrial environment. A number of orthogonal arrays, such as L-4, L-8, L-12, L-16, L-32, L-64, and so on, are created specifically for two-level factors. Using these arrays, experiments of all sizes can be designed easily as long as all factors involved are tested at two levels. By completing this step, you will learn how quickly experiments involving two-level factors can be designed and analyzed using standard orthogonal arrays.

Step 6: Experimental Design with Three- and Four-Level Factors

When only two levels of factors are studied, the behavior of the factors is necessarily assumed to be linear. When nonlinear effects are suspected, more than two levels of the factors are better. Although many larger two-level orthogonal arrays can be modified to accommodate three- and four-level factors, a set of standard arrays, such as L-9, L-18, L-27, modified L-16, and modified L-32, are also available. This step covers design and analysis of common experiments with three- and four-level factors.

Step 7: Analysis of Variance

Calculations of result averages and averages of factor-level effects, which involve only simple arithmetic operations, produce answers to major questions that were unconfirmed in earlier steps of the project. However, questions concerning the influence of factors in the variation of results in terms of discrete proportions can be obtained only by performing analysis of variance (ANOVA). This step shows basic steps in calculation of all terms of an ANOVA table. A number of examples are illustrated, and the common use and interpretation of ANOVA results are also discussed.

Step 8: Experimental Design for Studying Factor Interaction

Interaction among factors, which is the effect of one factor on another, is quite prevalent in industrial experiments. When experiments with factors alone do not produce satisfactory results, or when interactions among factors are suspected, the experiment must accommodate interaction studies. In this step you learn how to design experiments to include interaction and how to analyze the results to determine if interaction is present. You also learn how to determine the most desirable condition when interaction is found to be significant.

Step 9: Experimental Design with Mixed-Level Factors

Designing experiments with all factors at one level is easily accomplished using the standard arrays available. But there are not many orthogonal arrays that readily accommodate many situations with mixed-level factors that we commonly find in industrial settings. Most mixed-level designs, however, can be accomplished by altering the standard orthogonal arrays. This step includes methods of upgrading and downgrading columns of the orthogonal array to accommodate factors at two, three, and four levels.

Step 10: Combination Designs

For some applications, the factors and levels are such that standard use of an orthogonal array does not produce an economical experimental strategy. In such situations a special experimental design technique such as a combination design may offer a significant savings in number of samples. This step makes you familiar with the necessary assumptions that must be made to lay out the experiments using a combination design technique.

Step 11: Strategies for Robust Design

Variations among parts manufactured to the same specifications are common even when all factors are properly controlled. Reduction of variation is our ultimate goal. It will favorably affect one aspect of quality by establishing consistency in performance. The most common variations are considered to be caused by factors that are not controllable or are too expensive to control. These are called *noise factors*. In robust design strategy, the approach is not to control the noise factors but to minimize their influence by adjusting the controllable factors that are included in the study.

Step 12: Analysis Using Signal-to-Noise Ratios

The traditional method of calculating the average effects of factors, thereby determining the desirable factor levels (optimum condition), is to look at the simple averages of the results (performance). Although average calculation is simpler, it does not capture the variability of data within the group. A better way to compare population behavior is to use the mean-squared deviation (MSD) of the results. For convenience of linearity and to accommodate wide-ranging data, a logarithmic transformation of MSD (called the *S/N* ratio) is recommended for the analysis of experimental results. In this step you will learn how the MSD is calculated for various quality characteristics and how analysis using *S/N* ratios differs from the standard practice of using averages of results to compute the effects of the factors.

Step 13: Results Comprising Multiple Criteria of Evaluation

Although analyzing results to satisfy one objective (criterion) at a time is a common practice, most product or process applications involve more than one performance objective. Generally, different objectives are also evaluated by different criteria of evaluation, each of which has different units of measure and different relative weighting. Combining the different evaluation criteria into a single quantity is desirable for determining the optimum factor combination that best satisfies all the objectives. This requires devising a special scheme. This step introduces the concept of overall evaluation criterion (OEC) and shows you how to create an OEC formula appropriate for your own applications.

Step 14: Quantification of Variation Reduction and Performance Improvement

Most DOE applications allow you to determine the optimum design that is expected to produce overall better performance. *Improvement in performance often means that either or both average and variations have improved.* When the new design is put into place, it is expected to reduce the number of scraps and warranty returns. This reduction, in turn, offsets the cost of the new design. The savings expected from the improved design in terms of money can be calculated by using the loss function proposed by Taguchi. In this step you learn how to estimate the savings expected from the improvement predicted by the experimental results. Further, you will also learn how the expected improvement in performance from the new design is expressed in terms of popular capability indices such as C_p and C_{pk}.

Step 15:   Effective Experiment Preparation and Planning

Planning is the first and most important step in the application process. Planning for DOE/Taguchi requires structured brainstorming with project team members. The nature of discussions in the planning session is likely to vary from project to project and is best facilitated by one who is expert in the technique. Because good planning draws on in-depth knowledge of the subject, it is introduced as one of the last steps. A recommended agenda for a typical project application is discussed.

Step 16:   Case Studies

The knowledge you gathered so far can be your lifelong asset if you put it to practice. Find your own applications or be involved in a team project. Build your confidence by doing a few small experiments first. Use the example case studies to guide you along the way. Case studies presented in this step contain all or many of the following items:

- Project description and participants (Last names and first initials)
- Objectives and evaluation criteria
- Factors and levels
- Experiment design
- Collection of data and preparation of results
- Analysis of results (main effects, ANOVA, and optimum condition)
- Conclusions and recommendations
- Variation reduction (reduction of standard deviation) and savings (relative to trial 1 performance)
- Confirmation results (if performed)

Appendix

The appendix contains a glossary; some commonly used statistical tables, orthogonal arrays, and triangular tables, references; and a practice session with Qualitek-4 software.

Design of Experiments and the Taguchi Approach

What You Will Learn in This Step

- The Taguchi approach
- How the Taguchi approach relates to design of experiments (DOE)
- What Taguchi has done to make DOE more applicable
- The philosophy behind the Taguchi approach
- A new discipline of working together for the greatest benefits

You will not need to use the Qualitek-4 computer software in this step.

Thought for the Day

If you have your sight, you are blessed. If you have insight, you are a thousand times blessed.

—(Author unknown)

OVERVIEW OF DESIGN OF EXPERIMENTS AND THE TAGUCHI APPROACH

The lessons in this step present an overview of the design of experiments (DOE) technique and are intended to give you a basic understanding of what Taguchi has done to make the technique more effective in applications. By the end of this step you will have a clear understanding of the philosophy behind the Taguchi style of DOE and its intended application objectives to improve the quality of products and processes. If you are a first-time reader, you should read this section carefully.

There are no numerical exercises in this step. You don't need the computer or to do any calculations. Your objective in this step will be to get insight into the topic. The best way to do that is to remove yourself temporarily from being a practitioner and think of being the head of your own organization. Be the owner or top manager of your company, and try and get answers to such questions as: What is DOE, how is the

Taguchi's name related to it, what is new, what can it do for my company's product, why should I know about it, who should learn it, and why should I spend money training our people in this technique? If you could get answers to these questions, your learning process would be much easier.

For detailed statistical theory and background, you should read Chapters 1 through 4 of *A Primer on the Taguchi Method* by the author. You may also use the book to develop better understanding of experimental design principles when discussion in this book is not sufficient. Read this step in a leisurely fashion so as to get a clear picture in your mind of what it is and what it can do for you.

What Is Design of Experiments?

Design of experiments (DOE) is a statistical technique introduced by Sir R. A. Fisher in England in the early 1920s. His primary goal was to determine the optimum water, rain, sunshine, fertilizer, and soil conditions needed to produce the best crop. Using the DOE technique, Fisher was able to lay out all combinations (also called *treatments* or *trial conditions*) of the factors included in experimental study. The conditions were created using a matrix, which allowed each factor an equal number of test conditions. Methods for analyzing the results of such experiments were also introduced. When the number of combinations possible became too large, schemes were devised to carry out a fraction of the total possibilities such that all factors would be evenly present. Fisher devised the first method that made it possible to analyze the effect of more than one factor at a time. We discuss DOE in more detail in Step 3.

After Fisher introduced the technique and demonstrated its use in agricultural experiments, much more research and development followed. Unfortunately, most of the work remained in the academic environment. Although the need to study multivariable effects is widespread in the industrial environment, not many industries other than a few segments of the chemical and fertilizer industries have applied the DOE technique in their production processes. In fact, as academic knowledge grew, the further it got from a method that industry could absorb and apply. The more sophisticated the theory supporting DOE became, the less appealing it looked to practicing engineers.

Things You Already Know

- *DOE* is a statistical technique used to study the effects of multiple variables simultaneously.
- *Factor* is synonymous with such words as *variable, factor, input,* and *ingredient.* Just as input and output are relative to the subject under discussion, so are factors unique to the system under investigation.

Who Is Taguchi?

Dr. Genechi Taguchi is a Japanese scientist who spent much of his professional life researching ways to improve the quality of manufactured products. After World War II, the Japanese telephone system was badly damaged and dysfunctional. Taguchi was appointed as head of Japan's newly formed Electrical Communications Laboratories

(ECL) of Nippon Telephone and Telegraph Company. Much of his research at ECL involved developing a comprehensive quality improvement methodology that included use of the DOE technique. Practiced and perfected at the Nippon Telephone and Telegraph Company, Taguchi's concept was adopted by many companies, including Toyota, Nippon Denso, Fuji Film, and other Japanese firms. During 1960s and 1970s, he made frequent trips outside Japan to teach his concepts, but his technique was not introduced in the United States until the early 1980s. As an executive director of the American Supplier Institute in Allen Park, Michigan, Taguchi spends a few months each year consulting with various industries in the United States.

Why Is Taguchi's Name Associated with DOE Today?

The quality engineering method that Taguchi proposed is commonly known as the *Taguchi method* or *Taguchi approach*. His approach is a new experimental strategy in which he utilizes a modified and standardized form of DOE. In other words, the Taguchi approach is a form of DOE with special application principles. For most experiments practiced in the industry today, the difference between DOE and the Taguchi approach is in the method of application, as will be clear in the next few steps.

WHAT'S NEW?

Since the introduction of DOE by Fisher in the 1920s, and the 1940s when Taguchi started to research with it, there was much development with this statistical technique, but its use in industry was rare. To make it more useful, Taguchi first proposed a way to define quality in general terms. He showed that DOE could be used not only to improve quality, but also to quantify the improvements made in terms of savings in dollars. To make the technique easier and friendlier to apply, he standardized the application method. For laying out experiments, he created a number of special orthogonal arrays, each of which is used for a number of experimental situations. To analyze the results of the experiments for the purpose of determining the design solution that produces the best quality; he introduced a new way to analyze the results. His use of the signal-to-noise ratio for analysis of repeated results helps experimenters easily assure a design that is immune (robust) to the influence of uncontrollable factors.

New Philosophy and Attitude Toward Building Quality

Taguchi is a strong proponent of designing quality into products. He contended that the only way to improve quality permanently is to design it into the product. During his time, and even today, quality improvement effort in most companies only involves inspection of a part after production. Most often, it is too late and too expensive to do anything about quality once the parts are produced. Instead, if quality concerns are addressed upstream in design, or often further up-front in concept design, many improvements can be achieved for a smaller amount of expense.

Typically, a manufactured product will have phases in its life cycle such as concept design, development, validation, and production. In concept design, many ideas are examined, a few are selected, and engineering drawings are created. The development phase is the time to build a few prototypes of the concepts, try them out, and select the best-functioning part. The part is then validated in the validation phase by building a few samples and trying them for functional objectives. When validation is satisfied, production can begin.

Traditionally, quality activities in a company were restricted to the production floor. Typical quality activities were to inspect parts at the end of the production line. This was the common structure of things for most companies. For years we became accustomed to this order of business. The emphasis on up-front quality by Taguchi made us think for the first time about quality in all phases of engineering activities. Quality activities in organizations in areas other than production should not cause us to abandon the quality effort after items are produced; it simply means that we should not forget to ask about quality in activities leading to production. At each stage of engineering, a number of quality building skills are available, as shown in Figure 1.1. Not all are applicable to one business, but certainly one or more are appropriate for each type of business and can be applied to its activities.

Experience has shown that quality improvement efforts are more successful when done in activities before production. The return on investment for every dollar spent is much greater in design than in production, and even greater when applied further up-front in the concept design. Experienced quality professionals are quite comfortable with numbers such ratios of gain as 10:1 and 100:1. This relative merit of quality activities or return on investment for money spent at different phases of engineering is demonstrated by the model shown in Figure 1.2.

Things to Remember

- *What is the new quality philosophy?* Do it up-front. Design quality into a product.
- *When and where does quality start?* Quality is a factor in every phase of engineering: The sooner, the better.

Building quality in a product is preferable to inspecting for quality.

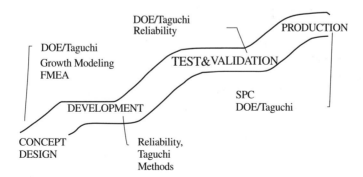

Figure 1.1   Road map to production and the applicable quality improvement skills.

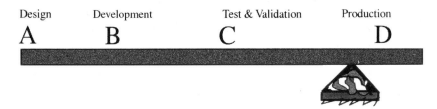

BIGGEST BANG FOR YOUR BUCK

Design	Development	Test & Validation	Production
A	B	C	D

Figure 1.2 Return on investment in quality activities: much more in design than in production.

Quality can be perceived differently by different people. It is also different for different products, processes, or services under discussion. Even in strictly technical terms, quality can be performance, durability, reliability, delivery, shape, or size. For some products it might only be appearance, or service after delivery. Indeed, product quality, as perceived by the customer, can be one or more of the criteria mentioned. How can we have a general definition of quality? Do we need a common and easy-to-understand definition of quality?

A clearly defined measure of quality is essential before any effort to improve quality is pursued. Such measures of quality must also be defined in measurable and numerical terms. Taguchi has given us a practical definition of quality that can be applied to most products and processes around us. First, understand that quality has many elements and that not one person or activity in your company is responsible for it. Performance is a strong element of the total quality. It is, of course, something that most engineering, quality assurance, and scientific professionals in the company have more influence on than any other aspect of quality. Improvement of performance therefore adds undeniably to the overall quality of products.

What about other elements of quality? Let's assume that everybody in their respective roles is doing their part in improving quality. Elements of quality are like the links in a chain: For overall quality improvement, all the links must be equally strong. While performance is improved, let's assume that efforts can be made to improve, say, reliability, delivery, or customer service.

How can we measure improvement in performance? How could we measure when a slight improvement has been made? Taguchi defines quality as consistency of performance around the target (nominal or average value when available). Of course, many products and processes do not have a target. The target in many situations also is desired to be zero or a higher value. No matter how performance is measured (units of measure), if the performance is consistent, variation around the target is reduced. For example, if the product is a 9-volt transistor battery, the goal will be to make batteries that are all close to 9 volts (target value) most of the time. If, on the other hand, the object is to achieve higher performance—say, the working life of a particular light bulb—the life values desired will be for all bulbs to last a longer time, not one a long time and another a short time. So it is quite possible that based on consistency, your preference will be for a design that performs with an average value less than that of another design that produces a higher performance level. When variation is reduced (i.e., performance is consistently closer to the target; Figure 1.3), fewer parts are produced out of specification. When more parts are produced within specification limits, there are fewer scraps, reworks, and warranty returns. This is the basis for achieving quality improvement in the Taguchi approach.

In high-volume production environments, what quality objective do we want to achieve? How can we measure it in quantitative terms? Improving quality by achieving consistency in performance does not happen accidentally. It is a result of planned action for a desired objective. Under normal conditions the performance may be such that the *mean* of the population performance is away from the target and the distribution is shallower. What is desired is to be near the target more often. When most performances are on the target, the distance of the mean of the population will be closer to the target and the distribution becomes narrower. The distribution becomes narrower because for consistent performance, the standard deviation becomes smaller. So what specific measurable statistics will tell us whether or not we have improved quality?

Improving quality will require that we reduce variations around the target by achieving consistency of performance. To achieve consistency will require that both the distance of the population *mean* from the *target* and the standard deviation of the population be reduced. What can be done to bring the mean closer to the target? How can the standard deviation be reduced? Design of experiments is a technique that can be applied to affect both these performance characteristics.

Things to Remember

- *How is quality measured?* By consistency of performance.
- *How can consistency of performance be achieved?* By reducing the distance of the population mean to the target and by reducing the standard deviation of the population performance.

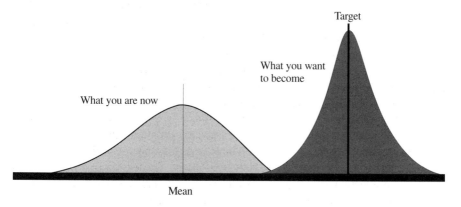

Figure 1.3   Improving quality by reducing variation around the target.

New Way to Measure Cost of Quality

The focus in measuring expense for quality improvement is not to count the cost of time and material spent in quality improvement efforts; rather, it is to calculate the financial loss suffered when quality is not what it can be. When quality is poor, rejects, rework, and warranty costs are higher. How much do they mean in terms of reduced profit or additional expenses? The traditional practice has been to calculate the cost by multiplying the number of rejects and the cost of production. This method of cost estimation looks at the product while it is still in the production facility. But what about what the parts will do after they leave the plant? Taguchi pointed out that subquality products might have harmful effects throughout their functional life in the hands of the customer. He suggested that poor quality can cause unnecessary service cost, waste manpower, and even have ill effects on the community. His way of looking at the cost of poor quality is to estimate the cost of societal loss.

The concept of loss to the society can be understood through a few simple examples. Consider an automobile manufacturer utilizing a part of poor quality. Perhaps the part is made to specifications but has a wider variation. This means that some customers will never have problems but that many others will. Those that experience the problem will need to take time off from their work to take their vehicle to the dealer for repair. In anticipation of possible defects, dealers as a rule maintain a large pool of mechanics, and thus make a small profit. Dealers would probably be happier making more money by selling better products than by fixing something that should have been right the first time. In addition to the cost of the repair, because consumers lose hours off their own work, there is a loss of productivity to the society as a whole. How can we put a dollar value on such a loss?

Imagine a company producing subquality underground gasoline tanks for neighborhood gas stations. If these products are not made to meet quality requirements, some may fail prematurely. In addition to cost of repair by the owner of such installations, leaky tanks may result in contamination of underground water supplies, lakes, and rivers. Such pollution may damage natural habitats and render the water sources unfit for use by the community. How do you calculate the harm done to the society?

Taguchi proposed a mathematical formula called the *loss function* for estimating the monetary loss caused by lack of quality. The loss function estimates loss even if parts are made within specification limits. This is necessary to allow for the fact that a company that makes all parts within specification limits still has warranty and customer complaints. That is, there is some loss associated with a population of parts no matter how well they are produced. As long as any parts differ from the target specifications, there is some loss. The shape of the Taguchi loss function is shown in Figure 1.4.

Taguchi defines loss as a quadratic expression in terms of measured quality characteristics (y) of the part that ranges between the target value and the specification limits, that is, upper and lower specification limits. The loss function is defined such that when the part is made on the target, the loss is absent. The loss becomes the same as the cost of production of a single part, which is the cost of rejection, when all parts are made outside specification limits. The loss between the target and the specification limits is of parabolic shape, symmetrical about the target.

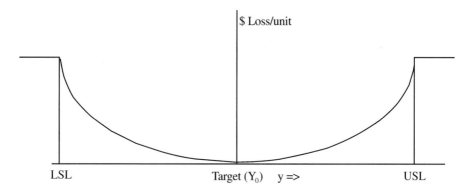

Figure 1.4 Taguchi loss function.

In this step you need only be familiar with the concept of loss function. A detailed description and how the function is used to quantify the improvement achieved are discussed in later steps. A history of the development of the Taguchi loss function is given in *A Primer on the Taguchi Method* by the author.

Things to Remember

- *Cost of poor quality.* The actual cost of poor quality is higher than what can be calculated at the time of production. The Taguchi loss function can estimate the cost of quality at any performance level.
- *Loss within specification limits.* It is possible to have a loss even if parts are made within specification limits. Loss is absent only when a part is made exactly on target.

New Disciplines

The Taguchi approach to quality improvement, which utilizes DOE and which you will learn in subsequent sessions, is most effective when the experiments are planned and carried out as a team project. It requires that the experiments be a group effort rather than an individual effort. It sponsors an approach where consensus is reached before launching an experiment rather than having to spend time convincing others about the results after the experiment is completed. This philosophy is particularly appropriate in industrial settings, where projects involve a larger number of people in the success and implementation of the outcome. The new discipline demands that all decisions regarding the experimental studies be reached by consensus in a formally convened meeting of project team members. It requires that democratic decisions be preferred to individual expertise or experience.

There are major differences between doing things in the old ways and this new approach. The two approaches are shown in Figures 1.5 and 1.6. The old ways of doing things when confronted with a problem or a project to improve design, followed a series of research and try out, and if that did not work, trying something else. Before a

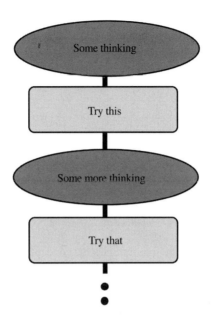

Figure 1.5 Typical old approach (series process).

solution is found, whether or not the best, a large number of experiments or studies will have been carried out.

The new way is a parallel process which calls for thinking through an entire project from the beginning. The project goals are discussed together in a planning session. Here the extent of the investigations is decided carefully by consensus. The number of experiments, the cost, and individual roles and responsibilities are all decided and agreed upon before doing a single experiment. These are decided on in the experiment planning session, which is the first and most important step in the entire DOE process using the Taguchi approach for quality improvement. The points of difference between the two approaches are listed in Table 1.1.

There are five basic phases (Figure 1.6) in applying the Taguchi experimental design technique to a project. For projects in industrial settings, it is important to follow these steps closely. Although the tasks in most phases are defined clearly, phase I, experiment planning, is the most valuable. If the project involves people from different disciplines of an organization, the success of the entire project will depend on the planning process. Following is a brief description of each phase.

I. Planning. Formally, this is the experiment planning session, preferably facilitated by a person who is not involved in the project. All decisions about the project, such as the objectives, their measurement method, and the factors that may influence the results, are made by the participating members in a democratic process

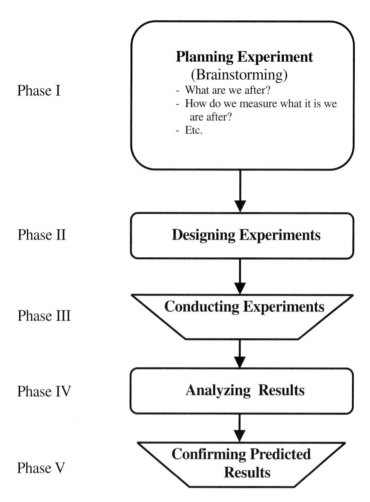

Figure 1.6 DOE application phases: the new discipline (parallel process).

TABLE 1.1 Typical Old Approach and the New Discipline of Working Together

Old Approach	New Discipline
• Carried out alone or with fewer people.	• Carried out with more people.
• Things are decided by the leader or experts.	• Things are decided by group consensus.
• Scope of experiment changes with time.	• Scope of experiment is known after the planning session.
• Project is usually concluded as soon as an improvement is found.	• All planned experiments are carried out regardless of the outcome.

in which each person has one vote. Generally, this will be a whole-day session. The details of the content of the planning process are discussed in Step 15.

II. ***Designing.***   The planning sessions conclude with all necessary information in place regarding the experimental strategy. Based on the factors and levels identified, an experiment is designed that readily specifies the number of experiments and the manner in which each experiment will be carried out.

III. ***Conducting.***   Once the experiments are designed, they are conducted following the exact design combination (recipe) prescribed and in the order required by the layout for statistical validity.

IV. ***Analyzing.***   In this stage the results collected from the experiments are analyzed. Analysis procedures are standardized to determine information about the project under study. Depending on the number of samples tested in each experimental condition, different analysis techniques are suggested. The primary goal of analysis is to obtain information about the new design condition and an estimate of the improvement expected.

V. ***Confirming.***   Generally, the best design condition predicted by the analysis is not one that currently exists or one that has been tested. Therefore, to verify if improvement is really achievable or to determine how close the estimate matches actual performance, a number of samples made to the new design specifications are tested.

SIMPLER AND STANDARDIZED EXPERIMENT DESIGN TECHNIQUE

The manner in which Taguchi used the then-existing DOE technique was also new. By the late 1940s when Taguchi began his work, about 25 years after R. A. Fisher introduced it, the technique was well developed and was recognized as a sophisticated statistical tool. But its use remained primarily in the research environment. As a vehicle for accomplishing his product and process quality improvement objective, Taguchi adopted the DOE technique. Recognizing that DOE in its current state was not popular, he researched ways to standardize its application methods. For experiment designs, Taguchi created a set of tables of numbers. These tables, known as *orthogonal arrays*, are used to lay out experiments of particular factor constituents. For analyses of results, Taguchi follows basic statistical calculations such as average and analysis of variance (ANOVA), but blends with it a new approach to analyze results based on the deviation from the target instead of absolute values. The study of results based on their deviation from the target (or the average when a target is absent) allows selection of the design condition that is most consistent and yields reduced variation, which leads to improved quality.

As you recognize by now, DOE using the Taguchi approach is not just another type of DOE or an easier way to make use of it. It is an entirely new process or system of strategies for improving quality where DOE is just one necessary tool. The other elements in the process are his definition of quality, the loss function, and the new discipline of working together, all of which are discussed above. There are several other concepts in his total quality improvement process, which are discussed later as appropriate. Henceforth the term *DOE* will be used to refer to the entire process of quality improvement following Taguchi's approach.

The Taguchi version of the DOE technique involves mainly learning how to design experiments, how to conduct them, and how to analyze the results. The philosophy and the discipline of how the technique leads to quality improvement is all very important, but discussion of them is kept to a minimum in later sessions. Although most subsequent sessions involve learning the details of using the technique, the brief description below presents an overview of the subject.

What Is DOE All About?

The DOE technique helps us study many factors (variables) simultaneously and most economically. By studying the effects of individual factors on the results, the best factor combination can be determined. When applied to product or process design, the technique helps to seek out the best design among the many alternatives. The technique can also be used to solve scientifically problems whose solution lies in the proper combination of ingredients (factors or variables) rather than innovations or a single identifiable cause.

Consider an application of DOE technique to a common process such as baking bread using an automatic breadmaker. The recipe for bread in my Panasonic (model SD-BT10P) Automatic Bread Bakery prescribes the recipe shown under level 1 in Table 1.2. This recipe worked well most often, but a few of the consumers within the household desired sweeter and softer bread. Alternatives for each ingredient (level 2 in Table 1.2) were suggested. Since more sweetness was desired from the alternative recipe, a little extra sugar was used as the level 2 value. A special type of automatic breadmaker flour was used in the second recipe. Margarine was used to replace butter

TABLE 1.2   Ingredients for Baking Milk Bread Using Panasonic Bread Bakery

	Level	
Factor	1	2
1: Flour	$2\frac{1}{3}$ cups all-purpose	$2\frac{1}{3}$ cups special brand
2: Sugar	2 tablespoons	$2\frac{1}{2}$ tablespoons
3: Salt	1 teaspoon	$\frac{1}{2}$ teaspoon
4: Butter	1 tablespoon	1 tablespoon margarine
5: Milk	1 cup homogenized	1 cup low-fat
6: Dry yeast	1 teaspoon brand 1	1 teaspoon brand 2

and the salt was cut to half of the original measure. Alternatives for milk and dry yeast were also used. Which of the other five ingredients will make a better-tasting cake in addition to the little extra sugar (the sixth factor)? How many loaves do we need to bake to find that out?

Perhaps it is clear by now that the levels of a factor are alternatives, values, or measures of the same factor. Thus one alternative for each ingredient creates a situation where there are six factors with two levels for each factor, 12 levels (ingredients) all together. For a complete recipe, only 6 of the 12 ingredients are needed. A common-sense layout or combination formula gives $2^6 = 64$ possible combinations of available ingredients. This can be seen by supposing recipe 1 to be comprised of all ingredients under level 1, recipe 2 with all ingredients under level 2, recipe 4 with one ingredient from level 1 and five ingredients from level 2, and so on.

A direct way to determine which of the 64 different recipes will give the most desirable bread is to bake all the recipes. But there are too many. Should we lower the number of factors? Is six too many factors? A number of factors in the range 6 to 15 is quite common in industrial projects. Seven factors each with two levels requires 128 separate experiments, whereas 15 two-level factors creates $2^{15} = 32,768$ distinct test conditions. In the Taguchi approach, only a small fraction of all possible conditions need to be tested. The number of conditions necessary depends on the number of factors and their level. The experimental layout is accomplished by use of the special tables of numbers called orthogonal arrays. An L-16 orthogonal array is used to design an experiment with 15 factors all at two levels, which requires only 16 experiments instead of the 32,768 combinations possible. Similarly, when there are three factors each on two levels, only four experiments are needed instead of eight. Thus experiments designed using orthogonal arrays reduce the number of experiments to a much more practical and affordable size. Generally, the larger the number of factors, the greater is the reduction from the total possibilities. Common experimental situations and the corresponding experimental configurations are shown in Table 1.3.

The results of such a planned DOE are generally analyzed to determine a number of things about the process performance. In the case of bread, the influence of the various ingredients will tell us whether to select homogenized or low-fat milk, and whether margarine or butter provide better taste. In the end, of course, such information will help us decide the best conditions for baking bread in the future. As part of

TABLE 1.3 Various Experimental Situations and Corresponding Experiment Size

Experimental Situation	Maximum Possible Conditions	Size of Taguchi Experiments
3 two-level factors	8	4
7 two-level factors	128	8
11 two-level factors	2,048	12
15 two-level factors	32,768	16
4 three-level factors	81	9
7 three-level factors	2,187	18

the simplification, Taguchi has also prescribed a few standard procedures for analysis purposes. The details of these standardized design and analysis techniques are the primary subject of discussion in subsequent steps.

Where Should DOE Be Applied?

In general, DOE (Taguchi or otherwise) is applicable to any situation that depends on many influencing factors (i.e., variables, inputs, parameters, or ingredients). It is a technique that lets you scientifically select the best option when you are confronted with many possibilities. We encounter many options and possibilities in all aspects of our daily lives. There are many ways to cook chicken soup, just as there are several ways to go about preparing the ground for a vegetable garden. Every day, creative financial experts come up with ways to economize in companies' financial affairs. In traveling the 25 miles to an airport, a person might take different routes depending on the time of day or the season. Which is the best way to go? What process should be followed? Any time you can ask a question like this, be aware that there are objective ways to find an answer to such a question.

Even when many factors influence an outcome, why couldn't we just try testing one factor at a time and determine the effects of the factors? Why do we need to follow rigid rules to determine how to test a recipe? Experience has shown that the results found by testing one factor at a time often do not hold true when all factors are in action in a real-life application. The only way to study real behavior is when the influences of all factors have an equal opportunity to be present. Only designed experiments can capture such effects.

What Types of Industries Can Benefit from DOE?

All kinds of industries can utilize DOE. Where there are products and processes, DOE can be applied. Even a service industry can use DOE when there a valid model is available. If the options can be studied, whether through actual hardware, models, or analytical simulations, the DOE technique can be applicable. Therefore, common analytical simulations such as finite element analysis make possible highly cost-effective design optimization studies using DOE.

Who Should Benefit Most from DOE?

DOE can produce maximum returns when applied in research, concept design, and product development. Engineers and scientific personnel in these activities should learn and apply DOE just as they do basic principles of physics, chemistry, and mechanics. The next areas to benefit from DOE are manufacturing and production processes. DOE should be an essential skill for all manufacturing and process specialists. Every process, including machining, heat treating, casting, molding, soldering, brazing, bonding, painting, welding, gluing, and coating, is controlled by a number of factors. They can all be fine-tuned by appropriately designed experiments. Engineers at a manufacturing plant should routinely use DOE techniques to improve production processes. Third in terms of return on investment, but first in number of applications, is the use of DOE in problem solving. Most DOE applications today are in problem solving. Plant problem solution specialists can greatly enhance their ability to tackle difficult problems by adding DOE application skills to their quality toolbox.

NOTE: Taguchi breaks down his quality engineering strategies into three phases, which he calls *off-line quality control. Off-line* refers to the fact that it is practiced away from or parallel to production processes. It is not intended to be applied while actual production is in progress.

- *System design.* This phase deals with innovative research. Here one looks for what each factor and its level should be rather than how to combine many factors to get the best result.
- *Parameter design.* This phase comprises what are commonly known as Taguchi methods and is the subject matter of this book. In this phase it is assumed that the factors are already known and production is in progress but there is room for improvement. The strategy here is to improve performance by adjusting the levels of the factors rather than looking for new factors. Quality improvement is achievable without incurring much additional cost. This strategy is obviously well suited for the production floor.
- *Tolerance Design.* This phase must be proceeded by parameter design activities. The objective here is to determine the control characteristics for each factor level identified in the earlier studies.

System design and tolerance design are not discussed in this book.

SUMMARY

In this step you have learned Taguchi's strategy for quality improvement and how he proposed using DOE to improve consistency of performance. You have also learned how working as a team and deciding things by consensus can lead to the greatest benefits for a project. You should be aware of the following basic principles:

- *Do it up front.* The best way to build quality is to design it into the product. This requires that we apply quality improvement tools as far ahead in the design process as possible.
- *Measure cost of quality.* A more realistic way to compare performance is by examining variation around the target (average in the absence of a target). Reduced variation is reflected in reduced rejection and warranty claims, which generally translate into cost savings. The cost estimate due to quality must include the effects of a product throughout its entire life. The loss function may be used to estimate dollar savings resulting from design improvement.

- *Devise a simpler design of experiment.*   Taguchi simplified the use of DOE by all practitioners by incorporating the following in the application process:
  - A standardized analysis procedure
  - Clear guidelines for interpretation of results
  - Special data transformation to achieve reduced variation
  - Formal study of uncontrollable factors by the robust design technique

EXERCISES

1.1 What does Taguchi mean by *quality*?

1.2 In the Taguchi approach how is quality measured?

1.3 Which of the following statistical terms do you affect when you improve quality, and how? Select all correct answers.
  (a) Move the population mean closer to the target.
  (b) Reduce the standard deviation.
  (c) Reduce the variation around the target.

1.4 Looking from a project engineering point of view, determine which of the following correspond to the Taguchi method or to conventional practice.
  (a) Carry it out alone or with a smaller group.
  (b) Carry it out with a larger group and plan experiments together.
  (c) Decide what to do by judgment.
  (d) Evaluate results after completion of all experiments.
  (e) Evaluate experiments as you go and alter plans as you learn.
  (f) Determine the best design by "hunt and peck."
  (g) Follow a standard technique to analyze results.

1.5 In general, which of the following types of businesses or activities can benefit from the Taguchi approach? Select all appropriate answers.
Areas:
  (a) Engineering design
  (b) Analysis/simulation
  (c) Manufacturing
Projects:
  (d) To optimize design
  (e) To optimize process parameters
  (f)To solve production problems

1.6 The first step in use of the Taguchi method is the planning session, generally known as *brainstorming*. Brainstorming in the Taguchi method is different from conventional brainstorming in several ways. Which of the following are favorable in such a planning session? Select all appropriate answers.

    (a) It requires the project leader to be open to the group's input and be willing to implement consensus decisions.

    (b) It works well when the group members work as a team.

    (c) It is more productive when the session is carried out in an open and democratic environment.

EXERCISE ANSWERS

1.1 Quality is defined by consistency of performance.

1.2 By reduced variation around the target (or mean).

1.3 All are affected.

1.4 Answer depends on the current practices in the organization.

1.5 All types could benefit.

1.6 All are favorable.

Definition and Measurement of Quality

What You Will Learn in This Step

- How to measure the quality of products or processes
- How to express the performance of a single sample in quantitative terms
- The definition of quality characteristic (QC) and how it relates to the measured results
- How to measure performance when there are multiple objectives and express the overall performance in terms of a single number
- How to express common population performance indices
- How to compare the performance of two population samples with multiple objectives

You will install the Qualitek-4 software that accompanies this book and use it to carry out a number of examples and exercises. Make sure that you have the Qualitek-4 CD-ROM and that your computer is ready for use.

Thought for the Day

You can't cross the sea merely by standing and staring at the water. Don't let yourself indulge in vain wishes.

—Rabindranath Tagore

PERFORMANCE EVALUATION AND MEASUREMENT

The main purpose of experimental studies is to find out how we are doing with our present design, how our competition is doing, or whether a design is better or worse than an alternative design. But the key to being able to compare such results is to know how to measure the performance we are after. Do we always know how to measure where we are or what improvement we've achieved? In this step you will take a closer look at the technique of measurement and evaluation of performance. Clear and accurate definition and measure of performance lead to better comparison of individual sample and group performances. In analyzing DOE results, a few new terms and a new method of data reduction are introduced. You will learn what we mean when we say *quality characteristic* (QC) instead of *response* or *result*, how the performances of two groups are compared using *mean-squared deviation* (MSD) instead of *average*, and how to combine multiple criteria of evaluation in a single index called the *overall evaluation criterion* (OEC).

- *Quality Characteristic (QC).* The DOE technique is used to select the best option from among many. Depending on the product, process, or system to which it is applied, the objective is measured by different units of measure. The measured value of the objective, which is commonly referred to as response, results, or output, will also be called the quality characteristic (QC). The QC is more than the result, however, because it not only represents the magnitude of the result in terms of a measured unit, it also indicates the sense of desirability of the result.

- *Mean-squared deviation (MSD).* There are two types of performance comparison. In the first we compare the performance of two individual samples. In this case, direct comparison of the measured value of the results works satisfactorily. In the second situation we need to compare the performance of two groups. Traditionally, average performance values of the groups have been used. The new approach is to compare the mean-squared deviation (MSD) of group performances. The MSD includes both the average and standard deviation of group performance. [Consistent with the definition of quality, when comparing the MSD values of the groups, the one with the more consistent performance (i.e., the one that has less variation around the target) is preferred.]

- *Overall evaluation criterion (OEC).* For most industrial projects, we are after more than one objective. If the design is adjusted to maximize the result based on one objective, it may not necessarily produce favorable results for the other objectives. A compromise might have to be obtained by analysis of the overall evaluation criterion (OEC), which is obtained by combining into one number evaluations under various criteria.

Process View of System under Study

Improving product and process designs, which depends on many factors, is the primary project in using DOE. For general reference, a product in its functional state can be viewed as a process. So it is a process whose performance we want to improve. Obviously, questions such as what performance is, what factors are in the process, and what the input and output are occupy our thoughts. Measurement of performance depends on the process under consideration. The specifics differ, but all processes can be viewed under a general scheme as shown in Figure 2.1. The input and output of a process are strictly related to the process. In other words, what is input for one process is not necessarily the same for another process. The best way to determine what an item is to the process is to take a view from within the system.

Everything we encounter in an industrial environment—a product whose performance we care about or a process whose outcome we wish to improve—can be viewed in terms of a process model. Typically, the process will have a system whose function is to transform the input into the desired objective with the help of factors. In this case

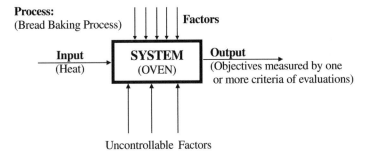

Figure 2.1 Elements of a typical process.

the system acts like a transformation machine, which can be as simple as a single element or as complex as a giant machine. In the bread-baking project, the system is the dough, made up of the factors butter, flour, sugar, and so on. The uncontrollable factors are those parts of the system that are not easily controlled, such as the kitchen temperature or the type of oven used to bake the bread. Input consists of one or more things that make the system function and produce the output intended. In this case the heat or electricity is the input, and the taste of the bread is the output desired.

For many industrial applications, the various system elements, such as input, output, factors, and uncontrollable factors, can be quite confusing. Trying to fit your application into a process model can give a better understanding of your system and help develop consensus among project team members. The system elements are always strictly relative to the system and sometimes subject to interpretation by the project team. Common project and probable process elements are listed in Table 2.1.

- *System.* This is the heart of the process that models your application. It consists of mechanism, parts, containers, machines, or assemblies required to make the process operational. In the bread-baking process, it is the container that holds the dough. The baking machine or oven is the system. In case of an injection molding process, the system is the molding machine itself.

- *Input.* This is what will be needed to start the system and to sustain its function. In the bread-baking process, heat or electricity, which does the cooking, is the input. Generally, it is not the subject of our DOE study, as in most cases we do not have a choice about its condition. (Studies of systems where input varies, known as dynamic system are beyond the scopes of this book.)

- *Factors.* These are the ingredients that the system requires to produce the intended objectives. Only those factors that are considered to have direct influence on the output and those that are included in the investigation are considered as factors in a DOE study. Those that are not included in the study remain as fixed components in the system. In bread-baking process, the factors are flour, sugar, butter, yeast, and so on. The primary focus of the DOE study is the search for a combination of levels (value or condition) of factors that best satisfy the desired objective.

TABLE 2.1   Example Applications and the Process Elements

Description of Application (System)	Input to the Process	Factors (Influencing the Process)	Uncontrollable Factors	Output Objectives
Coil spring design, used for automobile shock absorbers	Winding head motion	Type of steel, number of turns, wire diameter, etc.	Carbon content, machine setup	Deflection (stiffness)
Printed circuit board, used for common electronic instruments	Soldering head, electricity, etc.	Solder type, bead width, temperature, flux density, etc.	Operator skill, time of the day, etc.	Continuity, solder uniformity, etc.
Plastic injection molding of bracket assembly, used for supporting a cooling fan in a power supply unit	Electricity and other machine settings	Raw material, pack time, mold temperature, closing pressure, etc.	Operator, mold cleanliness, etc.	Stiffness, presence of voids, etc.
Transmission gear shaft machining process, used to mount gears, which transmits torque	Power to the turning machine	Type of steel, feed rate, depth of cut, turning speed, type of tool, etc.	Condition of tool wear, blank surface roughness, etc.	Surface finish, diameter, shaft straightness, etc.
Automobile generator fabrication, used to produce electricity from rotational motion of the engine	Engine rotation	Casement structure, air gap, impregnation, control brush, stator structure, type of winding, etc.	Engine stability, operating environment, etc.	Electrical power output, durability life, operating noise, etc.

- *Output.* This includes things that we are after. It shows how well we have achieved the objective (purpose). Usually, we are after more than one objective. For comparison purposes, each objective may be evaluated by separate evaluation criteria. In the bread-baking process, the output (performance of the process) can be measured in terms of how well it did in taste (an evaluation criterion) and shelf life (another criterion of evaluation). Since most determinations from DOE studies are based on the experimental results, which measure how well the objective is reached, it is very important that the output desired be clearly defined and measured.

- *Uncontrollable factors.* These are variables that are known to have an influence on the output but are either unidentifiable, difficult to control, or are not economically controlled. In Taguchi DOE, uncontrollable factors (called *noise factors*) are incorporated into the experimental process in a special method known as *robust design*. This is discussed in several of the last steps.

Much of the discussion in this book (and happens to be the subject of experimental design) deals with how to lay out the experimental strategy suited to the number of factors in the process. The technique of designing an experiment for a given number of factors and their levels has been standardized, which makes it easier to plan the experiments and identify the factors applicable to the study. The factors are identified during the planning session, which is the first step in the application process. Although the planning session is the first step, the content of discussions in this meeting requires full knowledge of the DOE technique. Therefore, detailed discussions of the planning session will have to follow sessions on the design methodology. Instead, some basic output evaluation principles, which form part of the planning session's discussion and which are essential for identification of factors to be included in the study, are discussed in this step.

Types of Results

Terms such as *output*, *response*, or *results* are used to indicate the level of performance of the process in some distinguishable form. The level of performance can be measured and expressed as two broad types (Figure 2.2). The first is where performance

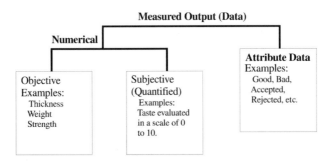

Figure 2.2 Type of measured output.

objective is measurable in terms of a number. For example, when the objective is weight, the measured value could be 2 pounds, 300 grams, or another value. Objectives, such as strength, thickness, horsepower, and voltage, will have numerical quantities representing the level of performance in experiments. The second type of performance evaluation is done using descriptions such as good, average, and unacceptable. Called *attribute data*, these classify the performance of different sample experiments into a fixed number of categories.

EVALUATIONS FOR COMPARISON

Generally, the term *result* indicates a measured value of output. It will take different values and units of measurement, depending on the objective desired. For example, in the bread-baking case, suppose that the objectives desired are taste and softness, both of which are evaluated by some agreed upon range of numbers. Then the result would refer to the number assigned under each evaluation. When there are multiple objectives, the results will indeed mean different things depending on the objective of interest. The issue about multiple objectives and how they are handled is discussed later in this lesson. For now, let us assume that we are concerned with only one objective.

The term *evaluation* refers to a process in which the performance (output, or status of product) is measured using an objective method and expressed in quantitative terms. In a simple evaluation, just the level of achievement is expressed. If you want to know the length of a rod, the rod is measured. If the length is found to be 10.25 inches, if nothing else is said or known, we make and convey no other meaning. Suppose that you were looking for a long object and all you find is this 10.25-inch-long rod. Now you have an expectation attached to the measured value. Your sense of desirability is that *bigger is better*. Now consider scores in the game of golf. When you ask a friend how his game was today and your friend replies that his score was 85 (in 18 holes), you know whether it is better than his score of 78 yesterday. You know that in the game of golf, a lower score is more desirable; in other words, the sense is that *smaller is better*. Suppose that in a basketball match between two teams, team A scored 95 and team B scored 105. It is clear in your mind who won, as you know that in this case the higher scorer is the winner (*bigger is better*). Here the individual evaluation is always compared with what the standard is, or what the more desirable values are. Because in DOE studies available alternative situations are compared with each other, all evaluations are performed with a sense of desirability. The sense of desirability associated with evaluations and measured values, in addition to the units of measure, is called the *quality characteristic*.

Loosely speaking, *quality characteristic* (QC) is the new term for *result*. But it is more than the result; it is also the sense of desirability of the result. More often than not, QC will refer to the sense of desirability. This sense is the direction in which the result is expected to go. QC is, therefore, a result with a direction. In this regard, QC is analogous to *velocity*, which is *speed* with direction.

No matter the applications, the method of measurement, or the units in which the results are expressed, there could be three different types of quality characteristics:

bigger is better (QC = B), *smaller is better* (QC = S), and *nominal is best* (QC = N). Many QCs have industry-wide accepted standards. Most others are unique to the application. They also depend on the units of measure and are subject to interpretation. Consider the term *efficiency*, used to measure, say, the performance of an engine. This evaluation has a standard range of measure (0 to 100%, no units) and a standard sense of desirability (*bigger is better*). Another common practice is the measurement of sound or noise. The unit of measure is decibels (e.g., 85 dB), which is something we want less of. Other measurements commonly used in engineering and scientific practices are stress, measured in pounds per square inch, which has QC = S; stiffness, measured in pounds per inch, which has QC = B; and warranty, expressed in dollars, which has QC = S. Examples of other quality characteristics are described below.

Bigger Is Better (QC = B).   In this type of measurement, the larger magnitude of evaluation will be preferred over smaller ones. Theoretically, there is no upper limit on the results. In practice, some upper limit is needed for numerical correctness. To achieve consistency, the average performance can be considered as the target value.

Examples of QC = B

- Taste of bread measured on a scale of 0 to 10 such that number 10 is assigned to the best bread
- Flow rate of a water pump, measured in cubic feet of discharge per minute
- Size of garden-grown tomato, measured in pounds of weight (assuming that bigger tomatoes are desirable)
- Miles per gallon obtainable from a brand of gasoline
- Time interval between tool changes in a machining process, measured in minutes
- Hardness of a part, measured as Rockwell hardness numbers
- Strength of a material, expressed in pounds per square inch
- Annual salary that one makes
- Savings, expressed in dollars
- Scores in games such as soccer, basketball, football, and tennis
- Reliability and durability life of a component, such as the life of household light bulbs, measured in hours

Smaller Is Better (QC = S).   Here the smaller magnitude of the results is always preferred over the others. The theoretical target is zero. The practical value of the lowest achievable value can be set to some appropriate number.

Examples of QC = S

- Number of voids in a cake (on a scale of 0 to 20)
- Noise of an automobile generator, measured in decibels (dB)

- Rejection rate in a production process
- Loss, calculated in dollars for a new part
- Stress of a structural member, measured in pounds per square inch
- Taxes that a company has to pay for doing business in a state
- Amount of gasoline that a car needs to travel 100 miles
- Average repair cost that an automobile needs during its service life
- Average number of defects of a particular model of vehicle in 12,000 miles
- Number of unpopped kernels when popping corn in a microwave oven

Nominal Is Best (QC = N). In this type of measurement, a fixed value is always desired. The fixed level of achievement desired is called the *target* or *nominal value*. For comparison purposes, the smaller deviation of the results from the target value is credited. Generally, plus or minus deviations are treated equally by considering the magnitudes only.

Examples of QC = N

- Outside diameter of a piston designed to fit into a 5-inch-diameter cylinder (target = 5)
- Distance of golf ball from the hole (target = 0)
- Weight of a tennis or Ping-Pong ball (target = fixed weight)
- Measured dimension of a part made to a specification (target = specified with ± tolerances)
- Voltage of a 9-volt transistor battery
- Idle speed of an automobile engine (target = fixed revolutions per minute)

A number of common terms have been used to describe and measure output. You will need to understand and define these terms as applicable to your project. The term *quality characteristic* is used mostly for analysis of DOE results. These output measurement terms are explained in relation to the example bread-baking process in Figure 2.3.

Example 2.1 A grinding machine is used to create the finished mounting surface of a transmission housing. An instrument is used to measure the surface finish in terms of reflection of a narrow beam of light. The amount of reflected light, expressed as a percentage (80%, 90%, etc.) of the incident light, is known to be directly proportional to surface smoothness. Describe the quality characteristic of the process performance being measured.

Solution: Since a better surface finish will result in more reflected light, a higher number is desirable in this case. Hence the criterion of evaluation is surface finish, the evaluation is, say, 85%, and the sense of desirability is *bigger is better*. Thus the qual-

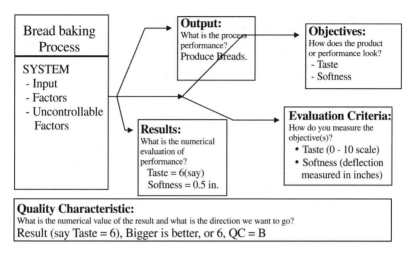

Figure 2.3   Relationships among output measuring terms.

ity characteristic in this case is surface finish measured by the percentage of reflected light: *bigger is better* (QC = B).

Example 2.2   To determine the moistness of cakes from a baking process, a standard-size cake was measured using a gram scale. The expected range of weight is between 30 and 60 grams. The 30-gram weight is considered too dry, whereas 60 grams makes the cake very moist. A weight of 45 grams is known to make the cake most desirable from the point of view of moistness. What is the quality characteristic of measurement of moistness of the cake?

Solution: QC = weight of standard piece of cake in grams; *nominal is best* (target = 45 grams).

Example 2.3   In an aluminum engine block casting process, among several criteria of evaluation of the process, the number of voids is considered important to the customer. As part of the routine inspection, cast engine block samples are sawed at predefined locations and examined for voids. The number of voids above a certain dimension are counted and recorded for quality audits. Describe the quality characteristic for evaluation of the casting process.

Solution: QC = number of voids; *smaller is better*.

SETTING UP QUALITEK-4 SOFTWARE IN YOUR COMPUTER

This is a good time to set up the Qualitek-4 (QT4) software in your personal computer. When you study this and future steps in this book, you will need to use QT4. So it

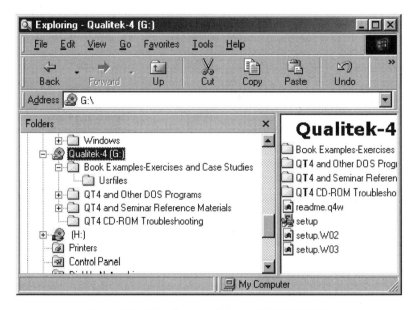

Figure 2.4   Directories and files in QT4 CD-ROM.

would be a good idea for you to install the software on the computer that you will use most often. Remove the CD-ROM from the back cover of the book. Turn your computer on and make sure that no other programs are running in the background. If there are, close all the programs. Now insert the QT4 CD-ROM in your CD drive (Figure 2.4).

From the *Start* menu of your Windows (lower-left corner of your monitor), select the *Run* option and type *D* or *?:\Setup.exe*. (The ? indicates any other character designations as appropriate for your computer.) Now click *OK* to proceed with the setup shown in Figure 2.5. ("Click" would always refer to pressing the left mouse button unless stated otherwise.) As you proceed with installation by clicking the *Next* buttons, you will be prompted to enter your name and company. Enter the information and

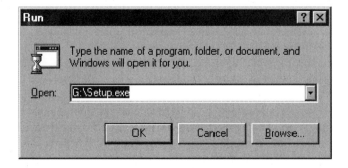

Figure 2.5   Windows *Run* screen used for installation of QT4.

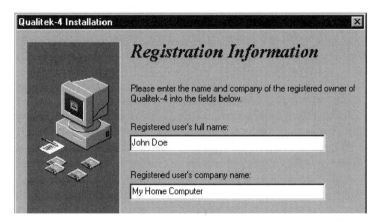

Figure 2.6   Qualitek-4 registration information screen: user's name and company.

click *Next* to go to the registration screen (Figure 2.6). The program on your CD-ROM installs a *Full-Featured Working Model* (DEMO) of the QT4 software when you enter *Demo* in place of a registration number (Figure 2.7). If the word *Demo* already appears in the box, you need not type anything. Click *Next* to proceed.

The installation program will allow options to select or name a directory of your choice for QT4. By default, QT4 will be placed in the Q4W subdirectory under the *Program Files* directory of your C: drive. If you prefer to place QT4 under a different directory, you may do so by clicking on the *Browse* button (Figure 2.8). Otherwise, return to the default directory and click *Next* to proceed. QT4 makes use of three other directories, which it utilizes to place example experiment files, to store data files, and to create backup files. By default it will create them as subdirectories under \Q4W directories or any other you name. Unless you have some special reason, click *Next* to these prompts. Check *Yes* to back up the file creation prompt and click *Next* to finish installing QT4. The screen (Figure 2.9) displays the installation in progress.

Upon completion of installation, the QT4 icon (Figure 2.10) is placed in a group window called *QT4*. The program name QT4 is also listed under the *Program* group

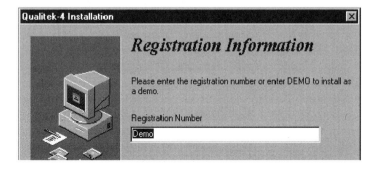

Figure 2.7   QT4 registration information screen: register number.

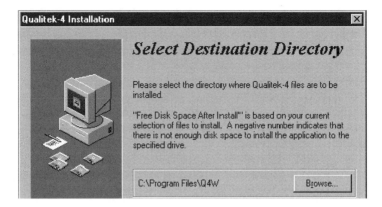

Figure 2.8   QT4 directory selection.

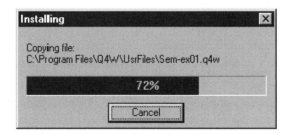

Figure 2.9   QT4 installation in progress.

Figure 2.10   Windows icon for Qualitek-4.

in the *Start* menu. You may now run QT4 from either the *Start* menu or by double-clicking the QT4 icon. For more frequent use, you may find it convenient to place a copy of the QT4 icon of Figure 2.10 in your Windows desktop. To do so, click the QT4 icon in the QT4 group window. (If the icon is not visible, you can display it by clicking the right mouse button on the *Start* menu, then select *Open*. Double-click *Program*, then double-click *Qualitek-4*.) While the QT4 icon is highlighted, click the right mouse button and select the *Copy* option. Click the right button of the mouse on any empty spot of the Windows desktop (this is your computer screen with the Windows background and several program icons), then select *Paste* from the option list.

Copying Example Experiment Files to the QT4 Program Directory

The examples and exercise experiment files are saved under the *USRFILES* subdirectory in *Book Examples-Exercises* and the reference files directory in the program CD. You need to copy the entire content of this subdirectory to your QT4 program directory. (The default directory is *PROGRA~1\Q4W*.) If you are using Windows Explorer, you may encounter the prompt shown in Figure 2.11. Click the *Yes to all* button at this prompt.

Running the QT4 Program

You can run the QT4 program alongside your other Windows program. Windows multitasking capability allows you to work with several programs at a time. So while you are in the middle of writing a report using, say Microsoft Word, you can click on *Start* and select *Qualitek-4* from the *Program* menu. On the other hand, if you have no other program running or the programs that are running have been minimized (click on the _ box in the top-right corner of the program screen), you can double-click on the QT4 icon. This will start the QT4 program and you will see the screen shown in Figure 2.12. Click *OK* to proceed.

The *Notice* screen shown in Figure 2.13 explains the limitation of the QT4 demonstration program. Even though the capacity of this program is limited to new use of experiments with an L-8 array, all examples are done using this program. You would also be able to use the program to solve most exercises included in this book. Click *OK* to proceed.

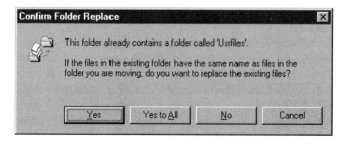

Figure 2.11 Windows Explorer prompt for copying experiment files.

Figure 2.12   QT4 front screen.

The registration screen of QT4 (Figure 2.14) indicates information about the licensed user and the registration number. Click *OK* to proceed as you did for the previous two screens.

The QT4 screen shown in Figure 2.15 is called the *experiment configuration screen*. It is commonly referred as the *main screen* and is the starting point of all capabilities for which you will utilize QT4. This is where you will start to design experiments, describe trial conditions, analyze results, and use the software to carry out many other calculations in the later steps. In this step you will use QT4 to do some calculations of the type for which we commonly use scientific calculators.

When you run QT4 for the first time, it will always show the *PISTON.Q4W* experiment as the default file. If you have been able to run the program and come to this point, you have done well. Soon we will explore some capabilities of the program.

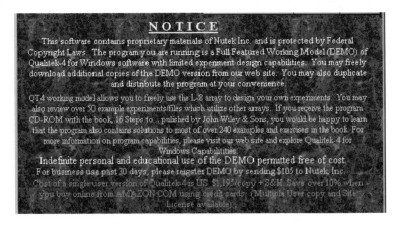

Figure 2.13   QT4 demo notice.

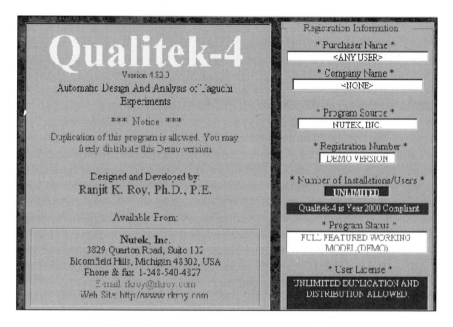

Figure 2.14   QT4 screen registration information.

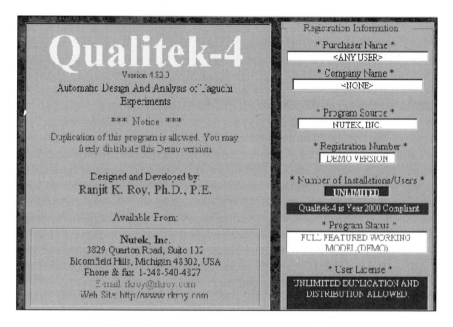

Figure 2.15   Experiment configuration screen.

You may now exit the program (using the *Exit from file* menu of the program) or stay at the main screen (Figure 2.15). For all our future use of the software, you will always start from this screen.

COMPARISON OF INDIVIDUAL PERFORMANCES

Comparing the performances of individual sample performances is made much easier when the QC is defined. Who is the better golf player: player 1, who scored 76, or player 2, who did 82? You can answer this question easily, because you know the QC (player 1 is better because the QC is *smaller is better*). If you were to decide the better among two candidates, one weighing 170 pounds and the other weighing 210 pounds, you cannot determine who is better, because the QC is not defined or known to you yet. You don't know the rules of the evaluation or the purpose of selection. If it were for an athletic event at which a lighter person were desired, your answer would be clear. If, on the other hand, it was for a football game and heavier weight was desirable, the answer would be quite the opposite.

When we are comparing data from a single performance with those of another, there is nothing new other than the fact that the QC plays an important role in helping us decide which one is measurable and which is not. The idea of consistency or reduced variation does not come with a single data point. Just as we cannot speak of standard deviation when there is only one number, the concept of consistency does not apply to a single sample performance. Thus the new way of measuring quality in terms of consistency of performance, which looks at the reduced variation around the target (or mean in the absence of a target), can be satisfied only when there are multiple samples in a group.

COMPARISON OF GROUP PERFORMANCES

In a modern mass production process, parts are manufactured in large volumes. In high-volume production using machines, part-to-part variations are common even when attempts are made to keep the equipment within specifications. For reasons not always known, when measured carefully, one part will differ from another. The dimension specified on the drawing or which is intended cannot be assumed to be true for all samples. The need to compare a batch produced by one machine with that of another, or a batch made yesterday with one made today, is very common. What is a good scheme to compare one group with another? What are the ground rules to follow?

Just as rules of the game are essential for a competitive sport, comparisons of performances of groups can greatly benefit from clearly defined procedures. There are many possible ways to compare two groups of numbers. Comparison of averages is one. Finding the single best performer in a group is another. Comparing variation within a group is yet another. Each of these approaches has a particular rationale and suitability. Which one should we follow? Consistent with the definition of quality, and as quality improvement is the purpose for use of the DOE technique; variations within groups are what we want to compare.

Consider two machines producing 9-volt transistor batteries. Three samples from machine A measure voltages of 7, 9, and 11, and those from machine B measure 8, 9, and 10. Which machine is better? Considering the averages, they are equal. If the rules were established such that the group with the highest voltage would be the winner, machine A would be better, as it has one sample of 11 volts. But if we are committed to prefer the group with better consistency, machine B will be preferred since, by inspection, 8, 9, and 10 have less variation than 7, 9, and 11. It is obvious from this simple example that comparison of group performance will differ depending on the approach we follow. Once again, our approach will prefer the group with reduced variation.

Measuring Variations with Mean-Squared Deviation

Variation within a group of numbers that represent a set of measured values, or sample performance, can be calculated by adding deviations of all individual numbers from a fixed reference value. For the *nominal is best* quality characteristic ($QC = N$), the reference value is the target value (same as the nominal value). Suppose that in the case of the two machines producing 9-volt batteries discussed above, the target is 9 volts. Let's see how we can calculate the variations.

Machine A: 7, 9, and 11

Machine B: 8, 9, and 10

Since the target is 9, deviations of samples from machine A are:

Deviation of sample that reads 7 volts = $7 - 9 = -2$

Deviation of sample that reads 9 volts = $9 - 9 = 0$

Deviation of sample that reads 11 volts = $11 - 9 = 2$

By adding the above, the total deviation becomes $-2 + 0 + 2 = 0$. This cannot be true, as it does not reflect the real variation situation. As a matter of fact, when it comes to variation, data short of the target or past the target get an equal variation value. This can be done by ignoring the sign of the variation (i.e., by considering the absolute value of the data for all calculation purposes). To avoid complications caused by the signs (+ and −) of the deviations, an option commonly practiced is to square the deviation. Thus the squares of the deviations of the sample data from machine A are recalculated as follows:

Square of deviation of sample that reads 7 volts = $(7 - 9)^2 = 4$

Square of deviation of sample that reads 9 volts = $(9 - 9)^2 = 0$

Square of deviation of sample that reads 11 volts = $(11 - 9)^2 = 4$

Adding yields the total squared deviation: $4 + 0 + 4 = 8$.

The total of the squares of deviations is now divided by the number of samples in a group that are used to produce the mean. This is done to minimize the effect of number of samples in a group, such that the number of samples need not be equal for the sake of comparison. Thus, for machine A,

$$\text{mean-squared deviation (MSD)} = \frac{8}{3} = 2.667$$

Similarly, the MSD for machine B is calculated to be

$$\text{MSD} = \frac{(8-9)^2 + (9-9)^2 + (10-9)^2}{3}$$

$$= \frac{1+0+1}{3}$$

$$= 0.667$$

The better of machines A and B now can be determined easily by considering the quality characteristic (QC) of the measuring device. In this case we are comparing the MSD of the output voltages (QC = N) of the samples. So what should the QC be when the MSD of the results is considered?

Just as many measurements in science and technologies have well-established connotations, the word *deviation* has always been used for something we want as small as possible. In other words, the deviation or mean of deviation carries the *smaller is better* QC. Now since machine B has MSD = 0.667 as compared to MSD = 2.667 for machine A, machine B is considered to perform much better than machine A. What about the QCs for other kinds of results? What do we do when there is no target value?

MSD is a yardstick that can be used for all types of data. Whether we are comparing several engines looking for the most powerful (QC = B), or selecting the machine that produce most batteries on a target value (QC = N), when we calculate the variations within the data in terms of the MSD, the QC for the comparison using the MSD will always be *smaller is better*. To assure that the MSD will always have the same *smaller is better* QC for data types with all three QCs, its formulas for the three data QCs are defined separately.

MSD for *"Nominal Is Best"* Data. In this type of data, there is a nominal (i.e., target) value: for example, measured values of samples from a lathe turning 2 ± 0.005-inch-diameter pins. For a general definition of the MSD, assume that the sample readings (QC = N) are expressed by using the notation Y_1, Y_2, Y_3, and so on, and that the target value is represented by Y_0 and the number of samples by n.

$$\text{Sample data: } Y_1, Y_2, Y_3, Y_4, Y_5, \cdots, Y_n, \qquad \text{QC} = \text{N}, \qquad \text{target} = Y_0$$

Then

$$\text{MSD} = \frac{(Y_1 - Y_0)^2 + (Y_2 - Y_0)^2 + (Y_3 - Y_0)^2 + \cdots + (Y_n - Y_0)^2}{n}$$

Example 2.4 The measured door-closing loads from two stations of a particular model of automobiles were found to be:

Station 1: 11, 13, 16, 14, 9, 13, 15, 12, and 14.5 pounds (pull force)

Station 2: 15, 11, 12, 14, 10, 9, and 15 pounds (pull force)

If the desired closing force is 12 pounds, which of the two stations produces better assemblies?

Solution: Calculate and compare the MSD for both stations. The station with the smaller MSD is considered to have done a better job. For station 1:

$$\text{MSD}_1 = \frac{(11 - 12)^2 + (13 - 12)^2 + (16 - 12)^2 + (14 - 12)^2 + (9 - 12)^2}{9}$$

$$+ \frac{(13 - 12)^2 + (15 - 12)^2 + (12 - 12)^2 + (14.5 - 12)^2}{9}$$

$$= \frac{1 + 1 + 16 + 4 + 9 + 1 + 9 + 0 + 6.25}{9}$$

$$= 5.25$$

For station 2:

$$\text{MSD}_2 = \frac{(15 - 12)^2 + (11 - 12)^2 + (12 - 12)^2 + (14 - 12)^2}{7}$$

$$+ \frac{(10 - 12)^2 + (9 - 12)^2 + (15 - 12)^2}{7}$$

$$= \frac{9 + 1 + 0 + 4 + 4 + 9 + 9}{7}$$

$$= 5.142$$

Since MSD_2 is less than MSD_1, station 2 is considered better.

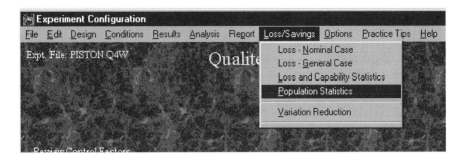

Figure 2.16 Menu option for MSD calculation.

Using QT4 for MSD Calculation

Now let us do the MSD calculation using the QT4 program, which you installed earlier. As the starting point, you should be at the main screen (or *Experiment configuration* screen) of QT4 shown partially in Figure 2.16. If you are running the program but not at the main screen, return there by clicking *Cancel*, *Return*, or *OK* a few times. Otherwise, run QT4 by clicking the QT4 icon from Windows Desktop or from the *Start* and *Program* menus. Click *OK* three times to be at the main screen. When you are at the main screen, click on the *Loss/savings* menu items of the QT4 program, then select (click once on) *Population statistics* (Figure 2.16). This will take you to the screen, where you can enter any arbitrary data and have the program calculate the MSD and many other statistics.

In the *Population statistics* screen, enter the data for station 1 of Example 2.4 in the *User data* field (Figure 2.17). You must enter one number (11, 13, etc.) at a time. To

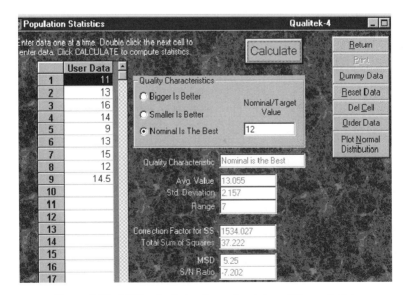

Figure 2.17 MSD calculation for station 1 data of Example 2.4.

begin to key in data, you may have to double-click the field, then use the keyboard/arrow keys to move within the field. Once you entered the data, you may want to order the data by clicking the *Order data* button. Ordering data is not necessary for MSD calculation and has not been done for this example. Before you can ask QT4 to calculate the MSD, you need to tell it about the QC for the data. You should do this by checking (click on one of the circular option boxes). In this example, nominal QC is checked, and since you checked nominal QC, you need to enter the nominal value of the data, which is 12. QT4 allows you to enter a nominal value only when you check the nominal option. You may check that you cannot enter data in nominal value box by checking any of the other two QC options.

Once you have completed entering data and selecting the QC option, click the *Calculate* button to carry out the calculation for MSD and a few other things that you are not interested in yet. The MSD value, the one we are interested in, is the second from the bottom (5.25) in the list of items shown in Figure 2.17. You will later know about items such as *S/N, correction factor*, and *Total sum of squares*, and, of course, you are already familiar with such terms as *average, standard deviation*, and *range*, all of which are calculated at the same time. Now that you know about the capabilities, whether or not you are carrying out experiments, you can always use this screen to calculate some of the statistics noted above.

While in the *Population statistics* screen, click on the *Reset* button and enter data for station 2 of Example 2.4 and proceed as you did for station 1. The screen will display the calculated MSD value (Figure 2.18).

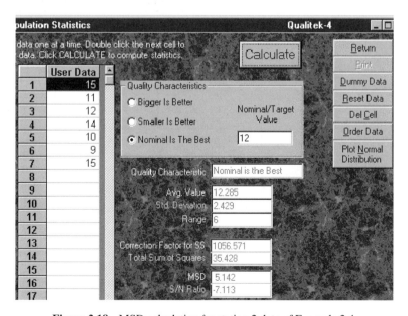

Figure 2.18 MSD calculation for station 2 data of Example 2.4.

There are times when you may want to see the graphical representation of the data under consideration. By *graphical representation* we generally mean the frequency distribution, which closely predicts the behavior of the population from which the data were selected. This is not a subject of discussion in this book, nor do you need a background in the theory of distribution. But we will assume that some of you are familiar with the frequency distribution or what the shape of this curve means. Therefore, we simply indicate that you have the option to plot your data set in graphical form by clicking on the *Plot Normal Distribution* button in this screen. Obviously, for plotting purposes, the assumption is made that the data you entered closely match a normal distribution. The distribution (generally referred to the shape of the curve) and the target value (when applicable) for the station 2 data set are shown in Figure 2.19.

MSD for *"Smaller Is Better"* Data. The smaller magnitude of data is considered preferable in this case. It is equivalent to the target being zero (i.e., $Y_0 = 0$).

$$\text{Sample data: } Y_1, Y_2, Y_3, Y_4, Y_5, \cdots, Y_n$$

$$\text{MSD} = \frac{Y_1^2 + Y_2^2 + Y_3^2 + \cdots + Y_n^2}{n}$$

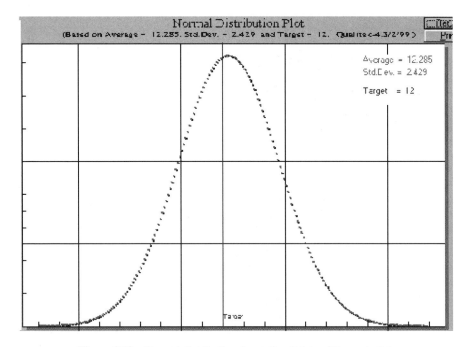

Figure 2.19 Normal distribution for station 2 data of Example 2.4.

Example 2.5 The cleaning-cycle times for nylon-coated brake linings using two process settings of Maxi-Blast metal cleaning machines are as follows:

Setting 1:   11, 9, 10, 7, 6, 13, 9, 8, 10.5, and 11.5 seconds (10 data points)
Setting 2:   12, 9, 8.5, 11, 13, 7, 9.5, and 10 seconds (8 data points)

Determine the setting that is likely to do the job in the least amount of time.

Solution: Since the least amount of time is preferred, the quality characteristic for the time data above becomes *smaller is better* (QC = S). The corresponding MSD is calculated for both stations using the appropriate formula. The setting with the smaller MSD is considered to have done a better job. For setting 2:

$$\mathrm{MSD} = \frac{12^2 + 9^2 + 8.5^2 + \cdots + 10^2}{8}$$

$$= \frac{826.5}{8}$$

$$= 103.312$$

Similarly, the MSD for setting 1 is found to be 94.349. Setting 1, with MSD = 94.349, is more desirable than setting 2, with MSD = 103.312.

The QT4 calculation of MSD for setting 1 is as shown in Figure 2.20. As in previous examples, you can calculate MSD in the *Population statistics* screen obtained from the *Loss/savings* menu in the main screen. Be sure to check the *smaller is better* QC before proceeding with the calculation. Similarly, you can calculate the MSD for setting 2 data of Example 2.5, as shown in Figure 2.21.

MSD for "Bigger Is Better" Data.   Larger values among the measured values are more desirable in this case. To maintain the same sense of desirability while using the MSD (smaller value desired), it is calculated by adding the squares of the inverse of the individual results. The expression for MSD in this case is:

Sample data: $Y_1, Y_2, Y_3, Y_4, Y_5, \cdots, Y_n$

$$\mathrm{MSD} = \frac{(1/Y_1)^2 + (1/Y_2)^2 + (1/Y_3)^2 + \cdots + (1/Y_n)^2}{n}$$

Example 2.6 The top speeds achieved by two separate bolt-on component designs in an experiment regarding jet ski performance are as indicated below.

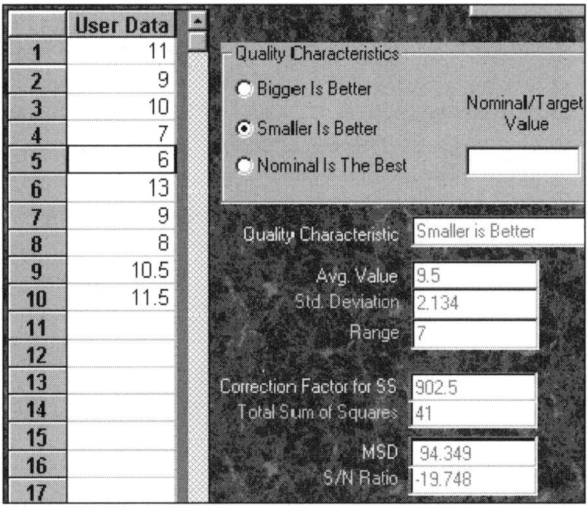

Figure 2.20 MSD calculation for setting 1 data of Example 2.5.

Design 1: 5.1, 5.2, 4.3, 4.5, 4.7, and 4.4 meters/second

Design 2: 4.3, 3.8, 5.1, 4.5, 4.2, 4.8, 5.0, and 4.6 meters/second

Compare and determine the design that is expected to produce the higher speed.

Solution: The MSD formula for QC = B must be used in this case, as the higher value of the speed is desired. The calculation procedure is demonstrated for design 1 data:

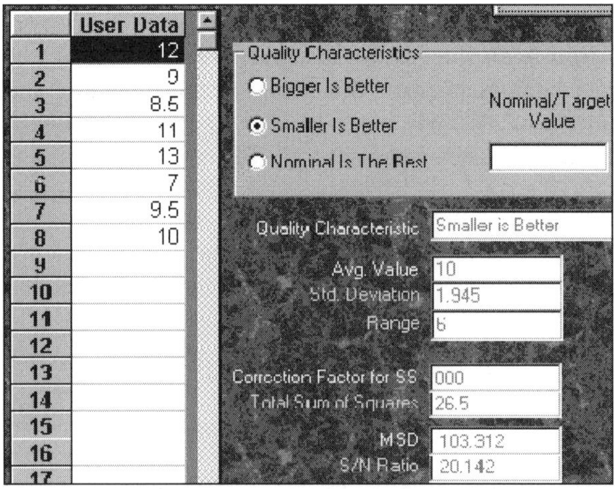

Figure 2.21 MSD calculation for setting 2 data of Example 2.5.

$$\text{MSD} = \frac{(1/5.1)^2 + (1/5.2)^2 + (1/4.3)^2 + \cdots + (1/4.4)^2}{6}$$

$$= \frac{(3.85 + 3.70 + 5.41 + 4.94 + 4.53 + 5.17) \times 10^{-2}}{6}$$

$$= \frac{27.6}{6 \times 10^{-2}}$$

$$= 0.045$$

For design 2, MSD = 0.049. Design 1 is more desirable, as its MSD is slightly smaller than that of design 2.

You can calculate the MSDs from the *Population statistics* screen of QT4, as before. Be sure to check the *bigger is better* option for QC before proceeding with the calculation. The MSD values for the design 1 and 2 data sets are as shown in Figures 2.22 and 2.23, respectively.

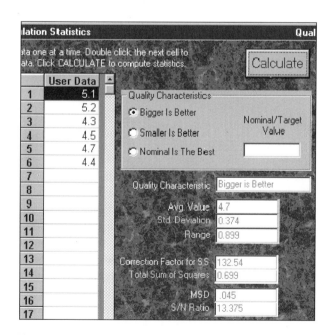

Figure 2.22   MSD calculation for design 1 data of Example 2.6.

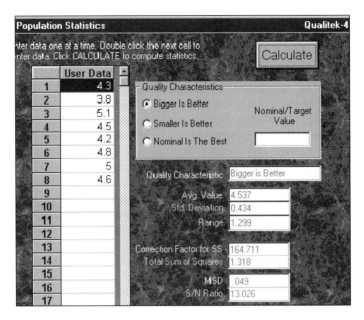

Figure 2.23 MSD calculation for design 2 data of Example 2.6.

RESULTS COMPRISING MULTIPLE CRITERIA OF EVALUATION

In DOE applications such as factor, level, and noise factor, *result* and associated terms such as *response, evaluation criteria, readings*, and *observations* have special meaning and are generally specific to the project under study. The same item can be a factor (input) in one application and an output in another. For example, for a process involving the manufacture of dairy products, butter is an output (measured and expressed as a result). When used in baking a cake, the same butter is a factor (an input to the process). What factors are and how they are identified and their values determined are discussed following this lesson. Let us first understand what is meant by *result*.

The *result* (synonymous with *response* or *performance*) is a measure of achievement of the objectives. Results are expressed in numerical or descriptive (attribute data: pass or fail, good or bad, etc.) terms and indicate the level of performance satisfying the objectives. But it is not just a number or attribute, it necessarily has a quality characteristic (QC) that is the sense of desirability, which allows it to be used for comparison purposes. For example, one of the objectives in a cake baking process could be to obtain the best-tasting cake. Suppose that the *taste*, which is evaluated subjectively (varies from person to person), is evaluated in terms of a number ranging from 0 to 10. If it is agreed that number 10 is assigned to the best-tasting cake, the *result* will be a *bigger is better* QC. With the QC value of the results known, one would be able to compare two cakes and determine which is better (say, cake number 8 is better than cake 7).

Evaluation and measurement of results are generally straightforward when there is only one objective for the project under study. But it can be quite involved when there are multiple objectives. Are there multiple objectives in most industrial projects? The answer is most often yes. To understand the complexity that multiple objectives create, consider these scenarios. In the cake-baking process considered earlier, suppose that we are not only after *taste*, but also *moistness* which is measured by weight in grams of a standard piece. Conventionally, the results of *Taste* and *Moistness* analyzed separately and the best cake recipe for the two objectives are determined. This approach does not guarantee that the two recipes will be the same. In most cases, only one recipe (design) can be used for production purposes (released design). What are we to do when the two recipes are different?

Consider a project to determine the best design parameters for an automotive water pump. The customer specifications indicate that *cost, durability life*, and *operating noise* are the criteria of evaluation. But not all three criteria are of equal importance. Samples tested at one design that cost $15.50 were found to last 2500 hours with a noise level of 82 decibels (dB). The corresponding numbers for another design are $17.40, 3,200 hours and 85 dB. Which design is better? How would you decide that objectively? Is there an overall evaluation scheme that you use to evaluate a number of such designs?

The concept of an overall evaluation criterion occurs when there are more than one objective that a product or process is expected to satisfy. Situations of this nature are quite common in many areas of our lives. Consider the educational system of rating students. Students are evaluated separately in each of the courses they take. But for comparison purposes, it is the grade-point average (GPA) that is used most of the time. Do you ever wonder where performance numbers such as 5.89, 4.92, and so on, come from in figure skating competitions? The numbers are averages of all the judges' scores on a scale of 0 to 6. Each judge uses the same range of evaluation numbers to evaluate different aspects of the performance: style of skating, how high they jump, how well they land, and so on. The scores that one judge assigns to a performer come from averaging his or her scores in all separate criteria of performance.

Given that use of an overall index is so common in sporting, educational, and social events, why is it rarely used in engineering and scientific studies? Primarily, because of three difficulties:

1. *Units of measurements.* Unlike GPA or figure skating scores, the criteria for evaluations in engineering and science are generally different (e.g., pounds per square inch for pressure and inches for length). When units are different, they cannot easily be added (or combined some way) to form a single quantity.

2. *Relative Weighting.* In formulating the GPA, all courses for students are weighted the same. This is generally not the case in scientific studies. Consider, for example, baking pound cakes. There might be three important objectives: taste, moistness, and smoothness. But these criteria of evaluation may not all be of equal importance. How do you decide which one is important? This is

generally done subjectively, by project team consensus during the experiment planning session.

3. *Sense of the quality characteristic.* The quality characteristic indicates the magnitude of the measured value in certain units of measure, along with the direction of desirability of the criterion being evaluated. Depending on the criterion and how it is measured, the QC can be *bigger is better, smaller is better*, or *nominal is best*.

For example, in the game of golf, a smaller (QC = S) score is better. In the game of basketball, on the other hand, a larger score is better (QC = B). Again, unlike sporting and educational events, the QC value for all criteria is always the same; the QC values for different criteria of evaluations are generally different. Even though two criteria may have the same units of measure (unitless numbers such as 42 and 38 for the two games of player 1 shown in Table 2.2). Unless the quality characteristics of different criteria are the same, the evaluation numbers cannot be combined readily. Are these two players (Table 2.2) of equal caliber? Is the addition of scores meaningful and logical? Obviously, the total scores of 80 for both players have no meaning, as these players do not perform equally. (Readers familiar with golf recognize that player 1's score of 42 for nine holes of golf is a much better score than the 52 for player 2). To make the total score meaningful, one of the numbers must be adjusted such that all QCs are aligned either up (QC = B) or down (QC = S). A logical and meaningful way to combine the two scores will be first to change the QC of the golf score to a *bigger is better* QC by subtracting the actual score from a fixed number, say 90 (a reference number), then adding it to the basketball score. The new total score using the modified QC for golf now becomes 86 for player 1 and 66 for player 2, as shown in Table 2.3. Perhaps the total scores now reflect the relative merit of the two players.

Multiple objectives that require evaluation by different criteria of evaluations are quite frequent in engineering projects. No matter the applications, be it product optimization, process studies, or problem solving, the desire to satisfy more than just one objective is most common. Because the criteria are different, it is a common practice to analyze one criterion at a time. This approach does not, of course, guarantee that the best design obtained for one criterion will also be desirable for the other criteria. What is needed is a properly formulated single index, call it an *overall evalu-*

TABLE 2.2   Comparison of Two Criteria with Different QC Value

| Sport | Player | | QC |
	1	2	
Golf (9 holes)	42	52	Smaller
Basketball	38	28	Bigger
Total score	80	80	

TABLE 2.3   Comparison of Two Criteria with Criteria Modified to Have the Same QC Value

	Player		
Sport	1	2	QC
Golf (9 holes)	48 (= 90 − 42)	38 (= 90 − 52)	Bigger (modified)
Basketball	38	28	Bigger
Total score	86	66	

ation criterion (OEC), which represents the performance of the test sample. Without a defined scheme such as OEC to generate a single quantity, evaluations of experiment samples with multiple criteria will continue to pose a major hurdle.

OEC FORMULATION

When a product or process under study is to satisfy more than one objective, performances of samples tested for each trial condition are evaluated by multiple criteria of evaluation. Such evaluations can be combined into a single quantity, the overall evaluation criterion (OEC), which is considered the result for the sample. But the evaluation of each individual criterion may have different units of measure, quality characteristics, and relative weighting. To combine the different criteria, they must first be normalized and weighted accordingly, as described for the cake baking process under discussion.

The evaluation criteria (Table 2.4) were defined such that taste was measured on a scale of 0 to 12 (QC = B); moistness was indicated by weight, with 40 grams considered as the best value (QC = N); and consistency was measured in terms of number of voids within a predefined section (QC = S). Evaluations for two test samples in the corresponding criteria are also shown (test 1: 9, 30, and 5; test 2: 6, 45, and 3). In Table 2.4, the worst reading is the number that represents the worst performance among all tests samples under consideration. Similarly, the best reading is the number that represents the best performance value among all samples. Understand that the data in Table 2.4 can be compiled with only when tests (planned experiments or comparative study) are carried out and you have evaluations for all test samples. Obviously, in the case of nominal QC, the nominal or target value is the best reading, and the worst deviant (from the target) value is the worst reading. The relative weighting numbers are subjective and are generally determined by consensus agreement by the members of the project group. We discuss these items in more details in later steps.

Example 2.7   Assume that the readings for cake samples 1 and 2 are as shown in the rightmost two columns of Table 2.4:

TABLE 2.4   Evaluation Criteria Description

Criterion	Worst Reading	Best Reading	QC	Weighting (%)	Test 1	Test 2
Taste	0	12	B	55	9	6
Moistness (grams)	25	40	N	20	30	45
Consistency	8	2	S	25	5	3

Test 1 readings: taste = 9, moistness = 30, and consistency = 5

Test 2 readings: taste = 6, moistness = 45, and consistency = 3

Determine which test sample is better based on the criteria of evaluation described in Table 2.4.

Solution: An overall evaluation criterion (OEC) is constructed such that it not only forces each criteria readings to have compatible units of measure and QC values, but also reflects proper relative weighting of each individual criterion. The OEC formula for test 1 evaluation is

$$\text{OEC} = \frac{9-0}{12-0} \times 55 + \left(1 - \frac{40-30}{40-25}\right) \times 20 + \left(1 - \frac{5-2}{8-2}\right) \times 25$$

$$= \frac{9}{12} \times 55 + \left(1 - \frac{10}{15}\right) \times 20 + \left(1 - \frac{3}{6}\right) \times 25$$

$$= 41.25 + 6.67 + 12.50$$

$$= 60.41$$

Rationale for the OEC Formula

In order for criteria of evaluations to be combined, each criterion must be adjusted to remove any units of measure and all QCs must be made the same. A common method of removing units of measure (normalization) is to divide by a reference number with the same units of measure (pounds/pounds = fraction, no units). The reference number is always the absolute value of the difference between the best and worst values. For example, the reference number for the taste criterion is 12 (12 − 0 = 12).

   For the bigger is better QC, the worst reading is subtracted from the test value. This number is then divided by the corresponding reference number.

   For the smaller is better QC, the best reading is subtracted from the test value. This number is then divided by the corresponding reference number. Since it was decided to change all QCs to align with that of the first criterion (taste, bigger is better) before adding them together, this new value is subtracted from 1 to change it to the bigger is better QC.

For the nominal is best case, the absolute difference between the test value and the best reading is divided by the corresponding reference number. Again, to align all QCs to that of the first criterion (taste, bigger is better), this value is then subtracted from 1 to change it to the bigger is better QC.

All of these values are then multiplied by the relative weighting to get a value that can be used to compare samples or tests. The percentage number representing the relative weighting (55%, 20%, etc.) is used as is, whole numbers (e.g., 55 for 55%), to assure that the OEC number calculated by the formula will always fall within 0 to 100 for the data set valid for the criterion.

Calculating OEC Using QT4

If calculation of the OEC value appears complicated, be assured that there is help. You will not need to calculate it by hand. The software can easily do it for you when you describe the nature of your readings to the software. Current QT4 capability includes manipulating criteria readings for a designed experiment (covered in later steps). There is no special capability to compute the OEC for one or two sample readings outside the designed experiment. Thus we will use an existing experiment file to accomplish our OEC calculation.

Run QT4 and go to the main screen. From the *File* menu, select the *Open* option. Then select *POUND.Q4W* (.Q4W is the common extension for all QT4 files) from the *Usrfiles* subdirectory in the QT4 program directory and click *OK* to open this file (Figure 2.24). This is a standard experiment that comes with QT4 that employs multiple

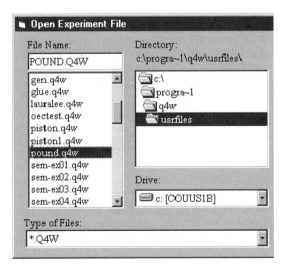

Figure 2.24 Experiment file for OEC calculation.

evaluation criteria for its eight experiments. For your use it would be a good idea to make a copy of this file and practice calculating the OEC for Example 2.7.

You can make a copy of the *pound* experiment file simply by saving it under a different name. While in the main screen, select *Save as* from the *File* menu. Name the file *BKEX-26.Q4W*, as indicated in Figure 2.25 and click *OK*. The extension .Q4W is optional. You now have a copy of the *Pound* file named *BKEX-26*, which already contains some OEC data. Notice that the main screen indicates (top left corner) your new file name. From the *Edit* menu, select the *Evaluation Criterion (OEC)* option. You will see the criterion description shown in Figure 2.26, which was used in *Pound* experiment. The screen shows OEC (results for *Pound*) values for eight trials (experiments). You will first need to modify the criterion description, then enter the evaluations for the two test samples in Example 2.7.

Examine the *Criteria description* part of the screen (lower left grids) and compare it with Table 2.4. The description in the first row (taste, 0, 12, B, and 55) is what you need for Example 2.7. For the second criterion, moistness, click on 45 and change it to 40. Click on 30 under relative weight and change it to 20. In the third row, replace 15 with 25. Now you have criteria descriptions (Figure 2.27) that match what you need for Example 2.7. Also make sure that the left column for all criteria descriptions is checked with an "X." If not, click on the empty box for an "X" to appear. (You don't need to change QCs in this example, but try clicking on the QC column and notice how B, S, and N can be adjusted.)

The numbers (11, 44, and 3) shown on the right side of the *Criteria description* box (Figure 2.26) are existing data for trial 1 under the sample 1 column. You will alter these data to reflect those (9, 30, and 5) for test 1 of Example 2.7. To do so, click on each number separately and replace it with the appropriate number (replace 11 with 9, 44 with 30, etc.). After entering the last evaluation (replacing 3 with 5), click on 9

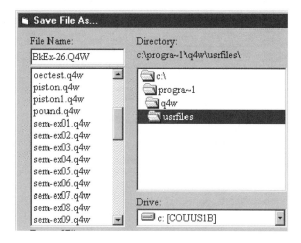

Figure 2.25   Saving an experiment file in QT4.

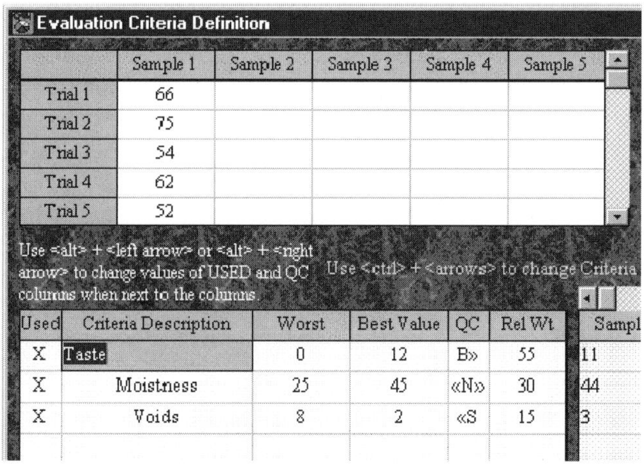

Figure 2.26 Evaluation criteria screen with sample description.

again. At this point you will see that the calculated OEC value (60.41) is shown at the bottom of the sample 1 column and also in the top left corner of the result grid (top rectangle: trial 1, sample 1) shown in Figure 2.27.

Things to Remember When Using QT4 OEC Capabilities

The items in the criteria description boxes have some special restrictions. The input options and ranges of values for each column are described briefly here.

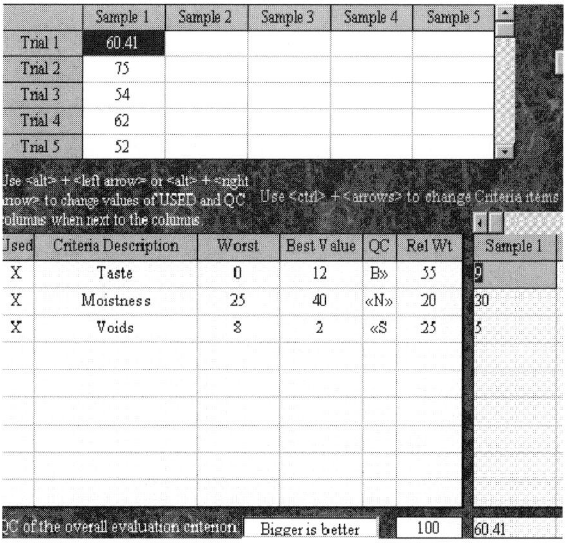

Figure 2.27 OEC value calculated for test 1 evaluations of Example 2.7.

- *Used.* This column must be checked (X) to indicate that it is included in the calculation. To check and uncheck, click on the appropriate space. Once unselected, the worst, best, and QC values for the criterion must be displayed before it can be reselected.

- *Criteria description.* This can be any alphanumeric description.

- *Worst.* This is the worst reading found among all test samples. This can be low- or high-magnitude number, depending on the quality characteristic of the criterion. In most application this number is selected by examining the performances of the samples already tested. In the absence of test data, it should be the expected value. For the nominal characteristic (QC = N), this value must be the worst deviant value from the target. (If the target is 40 for data ranging between 30 and 65, then worst = 65.)

- *Best value.* This is the value at the other end of the range. Like the worst value, its magnitude may be low or high, depending on the quality characteristic, and is determined by examining the sample performances tested. Before completion of the test, it is the expected best evaluation. In case QC = N, this value is the target/nominal value.

- *QC.* This is the quality characteristic of the criteria and is determined automatically (for bigger and smaller only) from the worst and best values supplied. To alter the QC status or to set the nominal characteristic, click on this space until the appropriate character (B, S, or N) is displayed.

- *Rel. Wt.* This stands for the relative weighting of the criterion of evaluation expressed in terms of percentage (0 to 100). These subjective numbers are usually determined by group consensus and must add to 100. To alter or enter a value, click on the space. (QT4 will not let you enter any evaluations until all criteria relative weights add up to 100.)

- *Sample 1.* Evaluations of tested samples are entered here. The data for up to 120 samples in each trial condition (in a designed experiment) can be entered. The numbers entered must be within the range specified by the worst and best values for the criteria. (If a certain evaluation is outside the range, you will need to adjust the worst and/or best value before entering it.)

General note: Sample evaluations are always for the trial condition indicated in the display box. To enter an evaluation for the desired trial condition, you will need to change it by clicking on the *Scroll* bar at the right of the top left OEC/result display grid.

The OEC screen will not let you exit until evaluations are entered for all samples. Either enter default data, then click *OK* to exit, or click *Cancel* to get back to the main screen. To keep the data entered, return often to the main screen and save the file from the *File* menu.

Use QT4 to calculate the OEC value for test 2 evaluations (6, 45, and 3) of Example 2.7 in the same manner. But instead of modifying the trial 1 data, you may want to modify the trial 2 data. To view the trial 2 evaluations under sample 1, click on the scroll bar on the right edge of the result grid. Make sure that the display box above the

	Sample 1	Sample 2	Sample 3	Sample 4	Sample 5
Trial 1	60.41				
Trial 2	61.66				
Trial 3	17.33				
Trial 4	36.16				
Trial 5	60.16				

Print	Ok
Tips	Cancel
Default Data	Reset Dat

Click Scroll Bars to Change
Sample# and Trial #.

Trial # Sample #
2   1

Use <alt> + <left arrow> or <alt> + <right arrow> to change values of USED and QC columns when next to the columns.   Use <ctrl> + <arrows> to change Criteria items and Sample #.

Used	Criteria Description	Worst	Best Value	QC	Rel Wt	Sample 1	Sample 2	Sample 3
X	Taste	0	12	B»	55	6		
X	Moistness	25	40	«N»	20	45		
X	Voids	8	2	«S	25	3		

Figure 2.28 OEC value calculated for test 2 evaluations of Example 2.7.

sample evaluations shows trial 2 and sample 1 as indicated in Figure 2.28. Now replace the evaluations under sample 1 by evaluations 6, 45, and 3 for test 2 of Example 2.7. When done, click on any part of sample 1 evaluations to see that the OEC value is calculated as 61.66.

You are now in a position to compare and determine which of the two test samples (tests 1 and 2 of Example 2.7) is better. Obviously, based on the OEC formula, which is defined such that it has a bigger QC, test 2 sample (OEC = 61.66) is slightly better than test 1 (OEC = 60.41).

In later steps you will have an opportunity to explore more options. For now, you may click *OK* to return to the main screen or try a few examples of your own to learn how QT4 lets you work with test samples as large as 120 samples per trial condition. (*Trial condition* and *sample number* have special meanings in DOE, as you will see later.)

SUMMARY

In this step you have learned how to measure performance by expressing it in quantitative terms and by associating a sense of desirability, called the quality characteristic, with such a measure. You have learned that for consistency of performance, mean-squared deviation (MSD) presents a better measure when two population performances are compared. You have also learned how to construct and use overall evaluation criteria (OECs) when performance includes meeting multiple objectives. You installed and used Qualitek-4 software to solve example problems.

REVIEW QUESTIONS

Q. What is the range of MSD values?

A. The MSD value will always be a positive number ranging between zero and infinity.

Q. How do you determine relative weighting for different criteria of evaluations?

A. Relative weighting is subjective. It is more meaningful to a project when these values are determined by group consensus. The total of all relative weightings must add up to 100%.

Q. Where should I use the OEC concept?

A. You can use the OEC concept in any situation that includes evaluation using multiple criteria. The concept can be applied in nonengineering applications with equal effectiveness. It can be used to express the performance of a single sample or can be used to compare performance among two or more test samples.

EXERCISES

2.1 Mr. Eveready has two assembly lines producing 9-volt batteries. Battery samples tested from the two lines produced the following data:

Line 1:   9.5, 8.3, 8.6, 9.2, 9.6, 10, 8.8, and 9.3 (8 samples)

Line 2:   8.5, 9.3, 9.8, 8.2, 8.4, 9.0, 9.4, 8.6, and 9.25 (9 samples)

To meet increased production demand, which line should Mr. Eveready put on overtime operation?

2.2 The mayor of Bloomfield Hills, Michigan wishes to determine which of two high schools in the city is better. The determination is to be made strictly on student scholastic achievement in the preceding year. The grade-point averages of 10 randomly selected students are as follows:

Andover High: 3.8, 2.7, 4.0, 2.1, 3.5, 4.2, 2.2, 3.9, 2.9, and 3.3

Lahser High: 2.8, 3.1, 4.3, 3.1, 3.5, 4.2, 2.1, 3.6, 2.7, and 3.1

Based on consistency of performance, which school is considered better?

2.3 An independent consulting company was contracted to determine the performance of two small cars based on defects reported by owners. A number of owners of each vehicle were randomly selected for interview. The number of total defects for the automobiles during the first 12 months of use were reported as follows:

Brand X: 2, 3, 5, 4, 3, 2, 1, 3, 0, and 2 (from 10 owners)

Brand Y: 1, 3, 2, 4, 6, 5, 3, 2, 0, 1, 1, and 2 (from 12 owners)

Considering the performance variation, which vehicle is expected to perform better?

2.4 Engineers in a plastic molding facility involved in determining machine settings for process parameters base their evaluations on satisfying three objectives: (1) a dimensional accuracy of mounting holes (15 ± 0.10 inches, QC = N), (2) stiffness at the center [300 to 500 pounds per inch (lb/in), QC = B], and (3) edge pops (number count 0 to 6, QC = S). Prior to evaluation of two

TABLE 2.5 Evaluation Criteria Description for Exercise 2.4

Criterion	Worst Reading	Best Reading	QC	Weighting (%)	Setting 1	Setting 2
Dimensional accuracy (in.)	15.10	15	N	45	15.05	14.93
Stiffness (lb/in.)	300	500	B	30	450	390
Edge pops	6	0	S	25	3	2

better settings, the group agreed to assign relative weightings of 45% to dimensional accuracy, 30% to stiffness, and 25% to edge pops. The criteria descriptions and the evaluations of parts made with the two settings are as shown in Table 2.5. Calculate the OEC for both settings, and based on its quality characteristic, determine the better setting.

2.5 Family physician Dr. Doshi is well known as a caring practitioner who takes a good amount of time explaining to his patients the meaning of test results. During consultations with patients he generally concentrates on five items: *blood pressure, cholesterol*–LDL (low density lipoprotein), *cholesterol*–HDL (high-density lipoprotein), *triglycerides*, and *blood sugar*. In his professional opinion, for example, a blood pressure (systolic) reading of about 115 mmHg is most desirable. Anything below or above this number (range: 90–115–240) would be a cause for concern. In addition to explaining the meaning of each test result separately to patients, he also offers a single number, which he prepares using the scheme shown in Table 2.6 and calls the Health Index

TABLE 2.6 Health Index Criteria Description for Exercise 2.5

Criterion	Worst Reading	Best Reading	QC	Weighting[a] (%)	Patient Test Results Current Year	Patient Test Results Previous Year
Blood pressure (mmHg)	240	115	N	30	110	135
Cholesterol–LDL[b]	160	90	S	15	128	100
Cholesterol–HDL[b]	15	80	B	15	22	34
Triglycerides[b]	800	100	S	20	280	130
Blood sugar[b]	120	100	N	20	105	112

[a]Relative weighting values are numbers representing a general consensus among the physicians employed in the clinic.
[b]These items are measured in milligrams per 100 milliliters

TABLE 2.7 Crashworthiness Criteria Description for Exercise 2.6

Criterion	Worst Reading	Best Reading	QC	Weighting (%)	Vehicle		
					A	B	C
HIC[a]	2,500	150	S	40	850	400	1300
Chest acceleration (grams)	90	20	S	20	42	55	64
Femur load (lb)	20,000	1,500	S	20	7,000	12,000	5,000
SCRD[b] (in)	12	2	S	20	4	6	3

[a]Head injury criterion (HIC) is a measure proportional to head acceleration (dummy's head).
[b]Steering column rearward displacement.

(like the OEC). The test results are also shown for a patient in the current and preceding years. Is the patient in better health now or last year?

2.6 Engineers involved in vehicle crashworthiness studies routinely carry out 30-mph front-barrier crash tests by placing instrumented dummies in prototype vehicles. For structural and safety considerations, a number of instrument readings are analyzed. A selected set of criteria of evaluation, their relative weighting, and evaluations for three test vehicles are shown in Table 2.7. Based on the defined criteria of evaluation, which among the three vehicles (A, B, and C) is expected to be safest?

EXERCISE ANSWERS

Use QT4 capability to solve Exercises 2.1 to 2.3: the *Population statistics* option of the *Loss/savings* menu from the main screen. Based on reduced variation (i.e., consistency of performance), a smaller MSD is always preferred, regardless of the quality characteristic.

2.1 Line 2 is preferred, as its MSD of 0.262 (QC = N, target = 9) is smaller than that for line 1 (MSD = 0.303).

2.2 Lahser High School is slightly better, as its MSD of 0.107 (QC = B, average = 3.25) is smaller than that of Andover (MSD = 0.111, QC = B, average = 3.26, which is better than Lahser's).

2.3 Brand X is expected to perform better as its MSD of 8.099 (QC = S, average = 2.5) is smaller than that of brand Y (MSD = 9.166), QC = S, average = 2.5).

2.4 You can use the same experiment file that you were working with earlier in QT4 (*BKEX-26.Q4W*). From the main screen open this file if it is not already loaded. From the *Edit* menu, select the *Evaluation criteria* option, modify the

Trial 1	42.5	52.16				Ti
Trial 2	37.66					Default I
Trial 3	23.86					Click Sc
Trial 4	36.98					Sam
Trial 5	14.63					Tri

Use <alt> + <left arrow> or <alt> + <right
arrow> to change values of USED and QC   Use <ctrl> + <arrows> to change Criteria items and San
columns when next to the columns.

Used	Criteria Description	Worst	Best Value	QC	Rel Wt	Sample 1	Sample
X	Dimensional Accurac	15.1	15	«N»	45	15.05	14.93
X	Stiffness	300	500	B»	30	450	390
X	Edge Pops	6	0	«S	25	3	1

QC of the overall evaluation criterion:	Smaller is better	100	42.5	52.16

Figure 2.29   OEC calculation for Exercise 2.4.

criteria description, and calculate OEC as done for Example 2.6. Click on the
Cancel button to exit this screen. Based on the OEC with the *smaller is better*
QC, setting 1 (OEC = 42.5) is much better than setting 2 (OEC = 52.16), as
shown in Figure 2.29.

Evaluation Criteria Definition   **Qualitek**

	Sample 1	Sample 2	Sample 3	Sample 4	Sample 5		Print
Trial 1	32.87	30.41					Tips
Trial 2							Default Dat
Trial 3							Click Scroll
Trial 4							SampleA
Trial 5							Trial #

Use <alt> + <left arrow> or <alt> + <right
arrow> to change values of USED and QC   Use <ctrl> + <arrows> to change Criteria items and Sample
columns when next to the columns.

Used	Criteria Description	Worst	Best Value	QC	Rel Wt	Sample 1	Sample 2
X	Blood Pressure	240	115	«N»	30	110	135
X	Cholst LDL	160	90	«S	15	128	100
X	Cholst HDL	15	80	B»	15	22	34
X	Triglycerides	800	100	«S	20	280	130
X	Blood Sugar	120	100	«N»	20	105	112

Figure 2.30   OEC calculations for patient data in Exercise 2.5.

Used	Criteria Description	Worst	Best Value	QC	Rel Wt	Sample 1	Sample 2	Sample 3
X	Head Injury HIC	2500	50	«S	40	850	400	1300
X	Chest Accl g	90	20	«S	20	42	55	64
X	Femur Load lb	20000	1500	«S	20	7000	12000	9000
X	SCRD in	12	2	«S	20	4	6	3
QC of the overall evaluation criterion:	Smaller is better				100	29.29	35.06	43.08

Figure 2.31   OEC calculation for the three-vehicle crash data in Exercise 2.6.

2.5   From the main screen of QT4, open file *BKEX-26.Q4W*. Select *OEC* from the *Edit* menu item and define criteria descriptions for the exercise. Click on the *Reset* button to clear data, enter evaluations for the patient in the current year (sample 1, OEC = 32.87) and the preceding year (sample 2, OEC = 30.41), and obtain the OECs shown in Figure 2.30. Based on the smaller QC value for the OEC, the patient is slightly worse now than last year.

2.6   From the main screen of QT4, open file *BKEX-26.Q4W*. Select *OEC* from the *Edit* menu item and define criteria descriptions for the exercise. Click on the *Reset* button to clear data, enter evaluations for vehicles A (sample 1, OEC = 29.29), B (sample 2, OEC = 35.06), and C (sample 3, OEC = 43.08) years, and obtain the OECs as shown in Figure 2.31. (Click *Cancel* to exit the OEC screen.) Based on the smallest QC for the OEC, vehicle A is the safest among the three vehicles tested.

Common Experiments and Methods of Analysis

What You Will Learn in This Step

In this step you will learn common experiments, methods of analyses of results, and some of the common terms used to describe the causes and effects of products and processes we experiment with. Before we start to discuss the design of experiments, the following are some common topics that will prepare you for planned experimental studies:

- Single-factor experiments for information about the nature of influence
- Common terms and terminology for experimental studies
- Types of factors: discrete and continuous
- Study of multiple factors with the least number of experiments
- Expression and comparison of population performance

In this step you will be using QT4 software to learn how to determine and express the performance status of test data. If you are familiar with the basic concepts of experimental studies, you may skip this step and move to Step 4.

Thought for the Day

The people who get on in this world are the people who get up and look for the circumstances they want, and, if they can't find them, make them.

—George Bernard Shaw

WHY EXPERIMENT?

The purpose of running experiments is to investigate and to learn. Either we want to know how the product or process does what it does or we want to know how certain factors influence the outcome. The types of experiments we generally run and a few experimental terms are described here.

Investigative Experiment. When we are less constrained by time than we would be if we were in a research activity, or when there is an urgent need to determine what causes the performance expected or a problem, we run experiments. In such experiments we try to identify *factors* (synonymous with *inputs, causes, variables, parameters, ingredients, constituents,* etc.) and determine their influence on the result. Here we evaluate one design and compare with another. All planned experiments carried out under some form of design of experiments (DOE) are of investigative type and are the main subject of discussion in this book.

The challenges in investigative experiments lie in how to evaluate performance, how to define evaluation criteria, and how to assess overall performance. When we want to compare one product with another, their performances can be measured in any units appropriate. But when the performance of a group of products (population of products) is compared with that of another, there is a need for an appropriate statistical index of the performances measured. How population performances are compared has been described in Step 2.

Although experiments usually are done with real products (or processes) and representative hardware, they do not have to be. When reliable analytical simulations (mathematical models) are available, they can be reliable substitutes at less cost. The experimental techniques and analysis methods discussed in this book apply equally to experiments with hardware and to analytical simulations. Most industrial products or processes that are subjects of investigation have many associated factors. As analytical simulation of performance is not always available, experimenting with physical parts is common practice. Running experiments to study the effects of all factors involved can be quite costly. The goal is to run the minimum number of experiments to get the most information.

Demonstrative Experiments. These experiments are conducted to confirm the *results* (synonymous with *outcomes, outputs, responses, objectives,* etc.) of experiments. All validation *tests* (or *experiments*) run at the end of an engineering or scientific development, but before production, are of this nature. Here we simply evaluate performance. There is nothing to compare it with because the samples tested are all from the same design. The questions we ask are like: Is it what we expect? Does it meet the specifications? How to evaluate the performance of a population of samples tested under demonstrative testing is also discussed briefly in this step.

LANGUAGE OF EXPERIMENTS

Here are a few terms that are commonly used to describe causes, effects, and measurement of effects.

Factor. This is anything that is suspected to have an influence on the performance of the product or process under study. It is an item on the input side of the system view of the product or process. It is something you can touch, feel, control, set, or adjust before carrying out the experiment. Depending on the application, many alternative terms, including *inputs, causes, variables, parameters, ingredients,* and *constituents,* are used to describe it. For simplicity, the term *factor* will be used throughout this book. Some examples of factors are time, temperature, pressure, thickness, container, vendor, and type of material.

Since the term *factor* is closely synonymous with *input,* it is important to realize that whether something is a factor or not depends strictly on the project under study. In other words, variables such as temperature, pressure, and thickness are not factors in all studies. It depends on the application. For example, the air–fuel mixture is output in a carburetor design optimization study, but the same air–fuel mixture is input in an engine combustion process study. So what is input and what is output depends on the viewpoint and can be determined clearly only by defining the project first.

Depending on how easily a factor can be adjusted for experimentation and availability, factors are grouped into two categories:

1. *Continuous factors.* Factors are considered continuous when it is possible to adjust values in a continuous manner to carry out experiments. For example, *baking temperature* (a factor) in an experiment on the baking process, *amount of carbon* (a factor) in a heat treatment process, and *time of exposure* (a factor) in a photographic film development process are all continuous factors. Typically, the value of a continuous factor is measurable in quantitative terms.

2. *Discrete factors.* A discrete (or fixed) factor is one that can only jump from one state to another. Discrete factors include *type of flour* (coarse or fine) in a cake baking process, *machine brand* (three brands of machines) used in a grinding process study, and *tool type* (steel or carbide tip) in a metal-cutting process study. Generally, discrete factors are not quantifiable, or if their status is quantified, its use is restricted to a fixed number of statuses (e.g., two knob settings for low and high *energy levels*).

Level. The value or status that a factor holds within an experiment is called its *level.* If milk is one factor in a cake baking process, *one cup* and *two cups* are two levels of the factor *milk.*

For convenience, character notations are generally used to designate factors and levels. For example, if A: *time* (meaning that A represents the factor description *time*) is a factor and 10 minutes and 20 minutes are its two levels, the notation a_1 could be used to mean 10 minutes and a_2 to represent 20 minutes. If the notation B were used to represent the factor *pressure,* b_1, b_2, b_3, and b_4 could represent the four levels of the factor. Some examples of factors and their levels are shown in Table 3.1. Factor notations such as A, B, C, and so on, are helpful and quite common when dealing with many factors. They will be used extensively in later steps in our discussion of DOE.

Factors may each have two or more levels. If a factor is present in an experiment but is unchanged (held fixed), it is not considered a factor under study. So how many levels should you use for a factor? It will depend on a number of things. A consider-

TABLE 3.1   Examples of Typical Factors and Their Levels

Factor	Level			
	1	2	3	4
A: Time (minutes)	10	20		
B: Pressure (lb/in^2)	200	225	275	350
C: Temperature (°F)	150	200	250	
D	d_1	d_2	d_3	d_4

able amount of time will be spent to address this issue in this and later steps. A number of examples of processes with many possible factors and levels are shown in Table 3.2.

Result.   A result is a measure of performance. Most results are expressed in quantitative terms in some unit of measure (e.g., hardness or a Rockwell number, strength in pounds per square inch, dimensional accuracy in meters). There are many other situations where the result will be expressed in terms of qualitative (called *attribute data*) expressions such as *pass/fail, superior/good/bad,* and *go/no-go.* Even if the results are qualitative in production practice, for experimental analysis purposes they can always be expressed in terms of numerical values that allow arithmetic manipulation. For example, if the results of test samples are evaluated in terms of *pass or fail,* the numbers 1 and 0 can be used to designate these performances. It is generally more beneficial to expand the scale to, say, 10 to 0 instead of 1 to 0, such that you have room to define performance that may be evaluated between pass and fail. But if you did so, you would need to be able to define what 5 is, what 7 is, and so on.

Result is synonymous with *response,* which is the term generally used in books on statistical science. The term *result* is the preferred term in this book. For a single test sample, there is always a single result (a quantity) that represents a measure of performance. Measuring performance, which determines how well an objective is achieved, is a straightforward matter when there is only one objective. When there are multiple objectives that are evaluated by different criteria of evaluation, measuring the overall performance requires extra planning. Whether you have single or multiple objectives, for purposes of analyzing experimental results, a single sample will always be represented by a single result. Therefore, in cases where you chose to combine multiple evaluations into a single index such as the overall evaluation criterion (OEC) described in Step 2, the result for the test sample is its OEC.

Quality Characteristic (QC).   While the *result* expresses performance in quantitative terms, it alone does not give an understanding of the level of achievement of the objective. Associated with each result is a sense of desirability, without which it will be difficult to compare one performance with another. Depending on the

TABLE 3.2   Examples of Common Products or Processes with Likely Factors and Their Levels

Product or Process Description	One of the Likely Factors	Levels of the Factor
Pound cake baking process	Amount of sugar	4 and 6 tablespoons (two levels of *sugar*)
Plastic part molding process	Injection pressure	150, 175, and 210 lb/in² (three levels of *pressure*)
Automobile generator design	Stator structure	Epoxy-coated and present design (two levels of *stator structure*)
Engine block casting process	Metal temperature	1430 and 1475 °F (two levels of *temperature*)
Engine idle stability adjustment	Spark advance	20°, 30°, and 40° (three levels of *spark advance*)
Cam lifter design study	Cam profile	Shapes 1 and 2 (two levels of *cam profile*)
Solder flux cleaning process	Rinse time	30, 50, 70, and 110 seconds (four levels of *rinse time*)
Hermetic seam sealing process	Pulse width (electric current)	2, 3, and 4 milliseconds (three levels of *pulse width*)
Cold forming design	Metal chemistry	38% and 40% carbon (two levels of *metal chemistry*)

desirability of results QC can be of type *bigger is better*, *smaller is better*, or *nominal is best*. QC is described in step 2.

Example 3.1 Engineers involved in the production of an automobile body component by SMC (sheet molding compound) processing identified edge pops (air bubbles formed after surface painting) as a major cause of rejects. Hoping to determine experimentally some way to fix the problem, they identified for study two types of resins, three amounts of prepolymer (100, 200, and 300 grams), and a few other factors.

   (a)   How would performance be measured (result)? In what units?

   (b)   What is the quality characteristic (QC) of the result?

   (c)   What are the first two factors to be identified and their levels?

Solution

   (a)   The test result can be measured by counting the number of edge pops observed. In this case the result will have no units. If, on the other hand, the sizes of edge pops are measured in terms of millimeters, then it will in units of millimeters.

(b) As the occurrence of edge pops is undesirable, QC is of the *smaller is better* type.

(c) The first factor, resin, has two levels, types 1 and 2, and the second factor, prepolymer, has three levels: 100, 200, and 300 grams.

INVESTIGATING ONE FACTOR AT A TIME

In this experiment, the goal is to learn about one factor. It is the most common form of experiment for engineers, scientists, and experimenters of all kinds. Its popularity is due to the fact that it does not require a complex formula. It is quick and cost-effective.

Here, the factor level selected is changed while all other factors known to affect performance are held fixed. At least two experiments at two different levels of the factor are necessary to draw any conclusions about its influence. Additional experiments with more than two levels of the factor, as well as multiple tests at the same level, produce better information but cost more.

Suppose that engineers involved in the project described in Example 3.1 were interested to determine the effect of resin on the result, which was measured in terms of the number of edge pops. Two samples tested at two levels of resin produced the following results:

Sample 1: resin type 1, result = 7

Sample 2: resin type 2, result = 3

Often, test results contain more information than was being sought in the tests. In the tests with resin above, it is possible to select the level of factor that is desirable just by looking at the number. Since fewer edge pops (QC = S) is desirable, type 2, which produced 3 edge pops, will be preferred. This is, of course, better portrayed graphically by drawing a graph of the effect of a factor on the two levels, as shown in Figure 3.1.

The graphs of factor influence (also called average factor effect, main effect, factorial effect, column effect, etc.) are more often plotted as some measure of results in the y-axis against factor levels along the x-axis. To plot the graph, results (or average of results) at each level are joined by straight lines (two or more lines for three or more levels). The graph so generated shows the trend of influence of the factor as it is changed from one level (resin type 1) to another (say resin type 2). The line from point 1 to 2 (Figure 3.1), which goes down, indicates graphically the trend of influence of the factor. Is going down (i.e., when the line has a negative slope with reference to the positive direction of x-axis) better or worse? It depends on the QC. Since the QC is smaller in this case, the lowest point in the graph is selected as the most desirable condition.

The straight line representing the average effect can be sloping up or sloping down. A decision about the desirable level would always be based on the corresponding QC. If the line happens to be horizontal (slope = 0), indicating that the result is unaffected

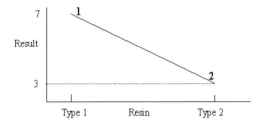

Figure 3.1   Trend of influence of factor resin.

by the change of the factor within the range bounded by the levels, either of the two levels can be chosen. Such a flat response would also indicate that this particular factor is not responsible for any variation of results and may not be the focus of future investigations. Since it can be set to any level, such factors provide opportunities to save cost or time in production.

The trend plot also allows us to make a prediction about what would happen to the result if there was a level, say resin type 3 between types 1 and 2. Based on the straight-line behavior of influence, the answer could be obtained by a point on the line. This discussion, however, is not appropriate for the factor resin type tested above, as it is a discrete factor. In other words, no third type of resin is present, and if it was present, whether it would be between the two levels tested is an impractical assumption. Use of the trend of influence line to predict the results of an arbitrary level by means of extrapolation (on the line, but beyond the endpoints) and interpolation (points between the two endpoints) may be possible when the factor is of the continuous type.

Consider now another set of tests conducted to determine the influence of the amount of prepolymer (100, 200, and 300 grams) described in Example 3.1. To save time and money, two tests at the two extreme levels were conducted first, which produced the following results:

100 grams of prepolymer: result = 2 edge pops

300 grams of prepolymer: result = 6 edge pops

In this case, the influence plot is a line (from point I to point II in Figure 3.2) that has a positive slope. When comparing the numerical values of the result or the line with the QC, a prepolymer level of 100 grams will obviously be selected. How confident can we feel about this selection? Does this mean that no other amounts of prepolymer between the two levels tested could be more desirable? If there are only two results (two data points), such as points I and II in Figure 3.2, they can be joined only by a straight line. This would force one to assume that results from any level between the two levels tested must be on the line I–II. In other words, the result expected from a test with 200 grams of prepolymer will be 4, as obtained by linear interpolation. But do we know that the influence of the factor is linear? Can we be assured that the result 200 grams of prepolymer will be 4?

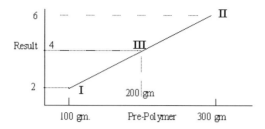

Figure 3.2   Trend of influence of prepolymer factor at two levels.

Most often, the reason you run a test is because you don't know how the result will behave when you change a factor. Without prior knowledge as to the nature of influence of a factor, it is harder to decide how many levels to test. What if the factor influence is nonlinear? How can we be sure if it is or it isn't? A factor must have at least two levels and is the least expensive way to test. But to learn about nonlinearity, at least three levels are needed. Trying to take a shortcut and doing tests at two levels instead of three could be misleading if the influence is nonlinear. You could end up selecting an incorrect level as the preferred factor level under certain conditions. The way to find out if a factor influence is nonlinear or if the correct level is chosen is to test with three or more levels.

Suppose that subsequent to the first two tests with prepolymer (Example 3.1), an additional test was conducted with 200 grams. The results of the three tests are as shown below. For a test with two levels:

100 grams of prepolymer: result = 2 edge pops

300 grams of prepolymer: result = 6 edge pops

For a test with a third level:

200 grams of prepolymer: result = 1 edge pop

Observe that levels of 100 and 300 grams were tested earlier. Now that another level, 200 grams, is considered, for plotting purposes it must be placed in its numerical order. In other words, 200 grams must be between and in the middle of 100 and 300 grams. This consideration would, of course, apply only when the factor is continuous, as it is in this case. The plot of the factor influence is shown in Figure 3.3 by the solid straight lines connecting points I, II, and III. Clearly, the plot shows that level II (i.e., 200 grams of prepolymer) produces the best results (as QC = S). This selection differs from that obtained by testing at only two levels of the factor, as seen earlier (Figure 3.2). Because the factor influence is nonlinear, the result at a point (200 grams) between the two endpoints (100 and 300 grams) (result 1) happens to be even smaller than the results at the end levels (results 2 and 6).

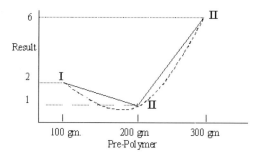

Figure 3.3 Trend of influence of factor prepolymer at three levels.

Since the results plotted in the three levels (points I, II, and III) are not on a straight line, the influence of the factor is considered nonlinear. Thus, as a general observation, in tests with three levels, if the result for the middle level is not on a straight line joining the other two, the trend of influence is not linear. But not all nonlinear situations will affect the preferred level selection as it did in this case. For instance, if the result of 200 grams were at 3 instead of 1, it would not have changed the selection of the desired level (QC = S, result $1 < 3$).

Although it is only logical to show the influence between two levels by a straight line joining two points (as in I–II and II–III in Figure 3.3), the actual influence is unlikely to follow the straight lines. The behavior of a continuous factor is more likely to be a smooth line (continuous). Thus an approximation, represented by a curve (quadratic or least-squares fit), as shown by the dashed solid line, is more likely. The exact shape of the influence is known only by conducting a larger number of tests on many levels of the factor. If the actual behavior of the factor influence is sinusoidal (highly nonlinear; Figure 3.4), four levels of the factor will have a better chance of identifying the trend of influence. Realize, of course, that even if four levels are tested, it is never a guarantee that the levels will be at points where the true maximum or minimum result is.

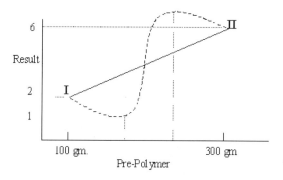

Figure 3.4 Sinusoidal trend of influence (four levels for the factors).

How are we to know how nonlinear the factor influence is? How should we decide how many levels to test? Out of the discussion of factors and their nature of influences above, we can establish these guidelines for investigative experiments:

- Experiments on at least two levels are necessary to learn about a factor's behavior.
- For continuous factors, at least three levels should be considered when nonlinear influence is suspected.
- When extreme nonlinearity is suspected, experiments at four levels would be desirable.
- When nothing is known about the factor behavior and you have limited time and money, proceed with two-level experiments.
- Consideration for nonlinear behavior will apply only when the factor is continuous.

Finding the Desirable Factor Level from Multilevel Experiments

Consider that you are still dealing with one factor at a time but you have the option to run a larger number of tests. Furthermore, assume that the factor is part of a new product under development for which you are not sure what will work well. What level (value) of the factor should you consider for the design of the product?

Example 3.2   Electrical engineers at a consumer electronic company wishing to determine a suitable input voltage for a power supply unit design conducted several experiments with various input voltages. The main output of the unit is utilized to control the picture quality of the projection tube. Assuming that there is flexibility to specify any value of the input voltage, determine the operating range from the following test data (picture quality measured in some units on a scale of -4 to 12 such that QC = B):

Input voltage	10	20	30	40	50	60	70	80	90	100	110
Picture quality	-2	-1.3	0	1.1	2.2	6.2	8.6	10.1	10.5	10.2	9.5

Solution: Data gathered from experiments at a wide range of factor levels gives a much better picture of its trend of influence. The factor level that produces the best result with the least variation is generally preferred. The factor level that is expected to produce the least variation in response is easily obtained by plotting the result against the levels and then drawing a tangent to the curve (*LMN*) as shown in Figure 3.5. In this case, as QC = B, the tangent line (*PQ*) is drawn at the highest point of the curve. Conversely, if we were after QC = S, we would then seek a lower tangent point on the curve. If we now draw a line (*RMN*) parallel to the tangent line, at about 5% below the result at the tangent point (10.5), it intersects the factor influence curve at

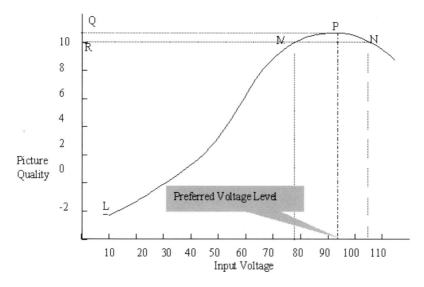

Figure 3.5 Operating factor level for multiple-level experiment of Example 3.2.

two points, M and N. The factor level corresponding to points M, P, and N (78, 93, and 105 volts) represents the operating voltage and the acceptable range.

Investigating Several Factors One at a Time

From the discussion earlier you have learned that it is always possible to find out about the influence of a factor by running two tests at two of its levels. Although never adequate for complete information about the factor influence, two tests for each factor are the minimum number of tests that would allow you to compare the results and determine the desirable level for the best (optimum) performance. This means that when you have several factors to study, you could investigate all of them by planning to run two experiments for each. But suppose that you have a time constraint and your project urgently needs design input as soon as possible. Is there a way to get something quick with fewer experiments?

Experienced experimenters will tell you that under certain conditions it is possible to study N factors by doing $N + 1$ experiments instead of $2 \times N$ experiments. To see what your layout should be, let us consider that you have three factors (A, B, and C, as shown in Table 3.3) under study. Let's see how four ($3 + 1 = 4$) experiments can produce what we want instead of six ($3 \times 2 = 6$). From simple experiments such as these, our goal still is to know for each factor the desirable level and its trend of influence.

Using the notations for factors and their levels as shown in Table 3.3, the four experiments can be described and carried out as shown in Table 3.4. Observe that in the description of experiment 1, where all factors are held at level 1 ($a_1 b_1 c_1$), may be considered as a reference level (say, current condition). In the second experiment, only

TABLE 3.3 Example Factors and Levels

Factor	Level	
	1	2
A: time (minutes)	4 (a_1)	8 (a_2)
B: temperature (°F)	176 (b_1)	225 (b_2)
C: pressure (lb/in²)	1500 (c_1)	1750 (c_2)

the level of factor A is changed; all other factors are held at level 1. In the next experiment, factor A is brought back to level 1 and factor B is changed to level 2. This process is repeated for all other factors. Note that the results (R_1, R_2, etc.) shown for each experiment are performance evaluations in units of measure such that bigger result is desirable (QC = B).

The experiments are described as follows:

Experiment 1:   4 minutes, 176 °F, 1500 lb/in²

Experiment 2:   8 minutes, 176 °F, 1500 lb/in²

and so on.

The effects of individual factors can be extracted by comparing the four test results shown in Table 3.4. Comparison of experiments 1 and 2 shows the difference that factor A makes while others are kept the same. For factor A:

1	a_1	b_1	c_1	45 (R_1)
2	a_2	b_1	c_1	60 (R_2)
Difference	($a_2 - a_1$)		$=$	($R_2 - R_1$) = 60 − 45

The effects of factors B and C are obtained similarly, by comparing (calculating the difference) experiment 3 with experiment 1 and experiment 4 with experiment 1, respectively. In numerical terms, the factor effects can be expressed as follows. (While computing the difference in results, it is customary to subtract the first-level result

TABLE 3.4 Example Experiment Description

Experiment	Factor			Result
	A	B	C	
1	a_1	b_1	c_1	45 (R_1)
2	a_2	b_1	c_1	60 (R_2)
3	a_1	b_2	c_1	20 (R_3)
4	a_1	b_1	c_2	35 (R_4)

from the second, and also, to leave the original results in the expression without showing the net difference.

Effect of factor A: result of experiment 2 – result of experiment 1 = 60 – 45 (not 15)

Effect of factor B: result of experiment 3 – result of experiment 1 = 20 – 45 (not –25)

Effect of factor C: result of experiment 4 – result of experiment 1 = 35 – 45 (not –10)

Notice that even though all factors are studied together here as part of the project, it is still looked at one factor at a time since only one factor is changed between the comparing experiments. The factor influences can also be plotted to show the trend of influence as done before (Figure 3.6). The factor influence plots show how each factor will affect the result. The trend is a straight line, as there are only two points (at levels 1 and 2 for each factor). We cannot be sure that the factor influence is truly linear unless additional tests are run. Prediction about an expected result at any other level of the factor can only be done based on the linear behavior. Interpolation for a level within the range of levels tested is quite often done satisfactorily. For example, if you were to calculate the result expected when factor A: time is held at 6 minutes, you will do so by considering that 6 minutes is halfway between the two levels tested, which allows you to calculate the expected result by interpolation as follows:

Result at 4 minutes: $R_1 = 45$

Result at 8 minutes: $R_2 = 60$

From a linear interpolation, the result at 6 minutes is calculated as

Result at 6 minutes: $R = 45 + (6 - 4) \times (60 - 45)/(8 - 4) = 52.5$

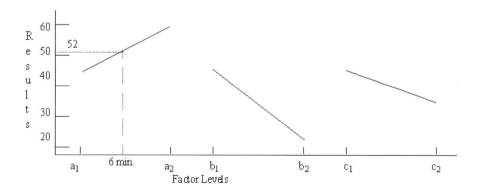

Figure 3.6   Influence plots of factors studied as a group.

This answer is also easily obtainable by drawing a vertical line from the 6-minute point between levels a_1 and a_2, as shown in Figure 3.6. From the straight-line plots of the factor influences, results at any other levels for factor A and any levels of all other factors can be determined in the same way. In addition to the trend of factor influence, the plots (Figure 3.6) help you select the levels for each factor that you prefer for determining their combination for the best (optimum) performance. Since QC = B, levels a_2, b_1, and c_1 (8 minutes, 176 °F, 1500 lb/in^2) are selected as the best factor levels.

Do you find anything inappropriate about this kind of experiment? Would you feel comfortable drawing conclusions from results from such tests? Would you be able to predict the expected result from the best level selected? Right or wrong, a vast majority of experiments in industry and research are run this way. Perhaps the simplicity of planning, the smaller number of experiments, and being able to draw conclusions based on common sense are what make the method more attractive. Unfortunately, such methods are more often wrong than not. Sure, you will learn a little but will be mislead frequently, as conclusions drawn from such tests will generally have less chance of being true (lack of reproducibility).

Lack of Reproducibility

The knowledge gathered about the factors by one-factor-at-a-time experiments such as the one under discussion is accurate when each factor behaves independently. This means that each factor influences the result the same way no matter how many factors are present. Unfortunately, this is never the case. Most factors in industrial experiments behave differently when other test factors change their level. In real experiments it is hard to find factors that are truly independent of each other. This nature of factor influence between factors is known as the *interaction effect*. This topic is discussed in detail in later steps. In addition to such mutual effects among factors (interaction), each factor may also not behave in a straight line as we are forced to assume for testing at two levels. The assumption of linearity affects reproducibility some but not to the extent that the presence of interaction does. So let's examine how the reproducibility is affected.

The lack of reproducibility comes from the way the factor influences are determined. Earlier in this example, the influence of factor A was determined by comparing the effect of the factor at level 2 with that at level 1. In both experiments 1 and 2, the levels of factors B and C were held at level 1. But we have no knowledge that factors B and C would always be at level 1, nor do we know how different factor A behaves when one or both of factors B and C are held at their second levels. Because interactions among factors are found to occur frequently in industrial experiments, what we find if we disregard it totally may not be reliable.

It is always desirable to make design decisions by findings from experiments. It is even more desirable that results of experiments are reproducible when tested under similar conditions by others at different times. In other words, experimental results are valid and useful only when they are reproducible. Although a smaller number of experiments are attractive, the lack of reproducibility is something no experimenter should settle for.

So how do we keep the experiments small, yet increase reproducibility? This question is the primary motivation for development of the design of experiments (DOE) technique. The objective has been to keep the number of experiments down, but also be able to derive conclusions that have better chance of being true. Taguchi's standardized method of DOE attempts to achieve the objectives and are topics of discussions in the later steps. Such one-factor-at-a-time studies can be carried out with any number of factors. For the case with three factors discussed so far, four ($3 + 1 = 4$) experiments were needed. In other words, the minimum number of experiments required to calculate the effects of a factor would be one plus the number of factors to be studied. Therefore, to investigate seven factors, at least eight experiments will be necessary. You will need 12 experiments to learn about 11 factors, and would have to run at least 32 experiments to study 31 factors. No matter how inappropriate, such experiments are likely to be better than guesswork or deciding things simply by judgment.

ASSESSING THE STATUS OF PERFORMANCE FROM MULTIPLE SAMPLE TESTS

Multiple sample tests are demonstrative experiments that are conducted either to verify the performance of a finalized design or to ascertain the variability present in the performance of the product. Here no changes are made to any of the influencing factors. The aim is to test a number of identical samples, record performance, and express the population performance behavior primarily in terms of numbers that include the effect of change of the mean (or average) performance as well as the variation (standard deviation) in performance around the mean. There are a number of performance indices (C_p, C_{pk}, MSD, S/N, loss function, etc.) that one way or the other combined these two items are used in the industry today to express or compare status populations of test performances. All of these indices can be derived from two basic statistics such as average (a) and standard deviation (S) calculated from the original performance test data. Conversely, when any of these performance indices is known, with some assumptions, the mean and variation in performance can be calculated analytically. Some characteristics of each of these indices are described below. *(Details of mathematical formulations are purposely avoided because it is either considered outside the scope of the book or the intent is to compute them using the software.)*

Normal Distribution. The *normal distribution* (*Gaussian distribution* or *Gaussian error law*) is a mathematical definition of a curve that simulates the shape of the bell-shaped histogram of a population of data. The normal distribution curve for a population can be completely defined and plotted (*y*-axis as frequency of occurrence, *x*-axis as data on the same scale) by knowing the average and standard deviation of a representative sample of data from the population. Although the true nature of distribution is never known (not all distributions are normal distributions) until an entire population is tested, the normal distribution allows a logical simulation of the distribution expected. The term *distribution* refers to the normal distribution curve that is defined in terms of two quantities: standard deviation (S) and average (or mean);

often, the term refers to these two quantities: the location of the average, which shows where the distribution curve is with respect to the target (when present), and the standard deviation (*S*), which indicates how narrow the distribution is. The property of the normal distribution is such that 99.7% of the area under the curve is bounded by the (average − 3*S*) and (average + 3*S*) lines.

***Capability Index* (C_p).**   The capability index is expressed in terms of a number (ratio) that indicates how narrow the population performance distribution is within the upper and lower specification limits (USL, LSL). The data set with larger C_p value is considered more desirable [QC = B (*bigger*)]. This number does not recognize the position of the mean value with respect to the target (when present) or the center of the specification limits.

***Capability Index* (C_{pk}).**   Like C_p, this is also expressed as a number (ratio) that captures the position of the mean performance as well as the variation of the data within the specification limits. A larger value of this index is preferred over a smaller one (QC = B). Between the capability indices, C_{pk} is used more than C_p, as it is capable of crediting a population whose performance distribution is more central. Both C_p and C_{pk} depend on upper and lower specification limits, as the calculations require use of these limits.

Mean-Squared Deviation (MSD).   The mean-squared deviation is a number (no units) representing the average deviation of the results from the target (or the average, in absence of a target) and is strictly a function of the average and the standard deviation. It is independent of the specification limits, as it does not make use of them. For comparison purposes, a set of data with a lower MSD is preferable [QC = S (smaller)].

***Signal/Noise Ratio* (S/N).**   There are several expressions for S/N ratios. The one we use throughout this book is a logarithmic transformation version of MSD. In other words, the S/N is the same as the MSD of the data set plotted in a log (to the base 10) scale with a −10 multiplier. The negative multiplier changes the desirability from *smaller is better* for MSD to *bigger is better* (QC = B) for the S/N ratio. Naturally, the S/N ratio also is independent of the specification limits.

Loss Function ($L/unit$).   The loss function allows you to calculate an estimated value of the loss associated with products that deviate from the target. The loss is calculated in terms of dollars of additional production cost per unit of product and bears the units $/part. The Taguchi loss function is expressed such that when all parts are made on target, the loss is zero, and when all parts are made beyond the specification limits, the loss equals the cost of production per part. Thus the loss associated with a distribution can also be used as an indicator of how good the performance is (a lower loss value is more desirable, QC = S).

NOTE: Capability statistics (C_p and C_{pk}), MSD, *S/N* ratio, and loss function will not be the subject of discussion in this step. They are introduced here briefly as they appear in the QT4 screens used for distribution statistics.

Example 3.3 The stripping torque (torque at which the threads strip) for self-tapping screws used in assembly of electronic components on the aluminum extruded base is shown below.

> *Measured torque*: 6.5, 5.2, 6.8, 7.5, 6.1, 5.5, 6.2, 7.0, 5.3, 6.4, 6.3, and 5.4 inch-pounds

Determine the range of the stripping torque and the distribution of the torque (assume normal).

Solution: Run QT4 and go to the *Experiment configuration* screen. From the *Loss/savings* menu item, select the *Population statistics* option (Figure 3.7). You need not open or work with a specific experiment file to utilize this program capability. When in the *Population statistics* screen:

1. Enter the 12 torque data one at a time in the spaces under *User data* (Figure 3.8).
2. Check the *Bigger is better* box under *Quality characteristics*. (For this problem it is immaterial whether you check the *Bigger is better* or *Smaller is better* box.)
3. Click the *Calculate* button and note the average and standard deviation values (average = 6.183 and standard deviation = 0.724).
4. Click the *Plot normal distribution* button to plot the distribution.

The distribution of the torque data is shown in Figure 3.9. The distribution is defined by average = 6.183 and standard deviation (S or σ) = 0.724. The target, for which a default value is set to zero, is not applicable in this example. The range of the torque is determined by selecting the lowest and highest value of the torque from the distribution plot. Unfortunately, the normal distribution curve has no highest or lowest points. The leading and trailing edges of the curve are asymptotic (tangential) to the

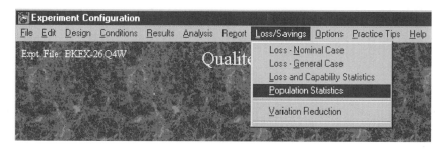

Figure 3.7 Menu option for population statistics.

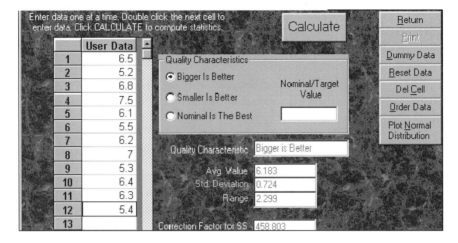

Figure 3.8 Screens for population statistics calculation for Example 3.3.

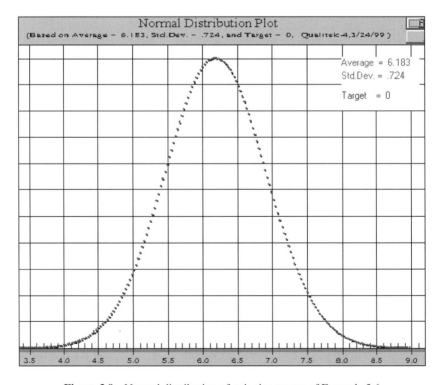

Figure 3.9 Normal distribution of stripping torque of Example 3.6.

x-axis with no definite contact points. Thus a common practice is to accept the $\pm 3S$ boundary points, which cover 99.7% of the data points. Other limits suitable for the application can be derived from the standard limits of areas under the normal distribution curve shown in Figure 3.10. Since $S = 0.724$, the probability is over 99.7% that all measured torque will fall between the lower and upper limits calculated here.

$$\text{Lower limit} = \text{average} - 3 \times S$$

$$= 6.183 - 3 \times 0.724$$

$$= 4.011 \text{ inch-pounds}$$

$$\text{Upper limit} = \text{average} + 3 \times S$$

$$= 6.183 + 3 \times 0.724$$

$$= 8.355 \text{ inch-pounds}$$

Example 3.4 The life data of a rated 125-hour tool was processed to determine the average tool life (120 hours) and the standard deviation (5 hours). If the originally recorded data were assumed to follow a normal distribution, what is the actual plot of the distribution?

Solution: Run QT4 and go to the *Experiment configuration* screen (main screen). From the *Loss/savings* menu item, select the *Loss and capability statistics* option (Figure 3.11). For this example you will only use the bottom part of the *Loss and capability statistics* screen:

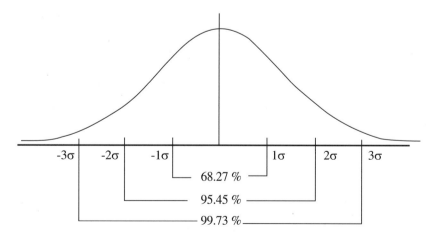

Area under Normal Distribution

Figure 3.10 Area under the normal distribution curve bounded by units of standard deviation.

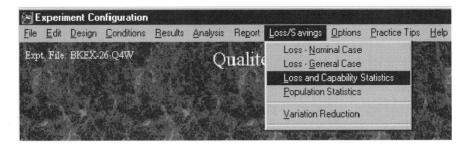

Figure 3.11   Option for plotting normal distribution from performance statistics.

1. Enter 120 for average, 125 for target, and 5 for standard deviation in the spaces as shown in Figure 3.12.
2. Click the *Plot normal distribution* button to plot the distribution.

The shape of the normal distribution curve for the tool life data that produced the average and standard deviation given is shown in Figure 3.13.

Example 3.5   In two plants producing an identical plastic part, the distance between two mounting surfaces, which has a nominal value 85 millimeters (mm), is considered critical and is used as a major quality inspection item. Parts measuring smaller than 78 mm (LSL) or larger than 92 mm (USL) are rejected. In a typical day, the following measurements for the sample parts from the two plants were obtained:

Parts from plant 1: 84, 81, 83, 82, 84.5, 77, 84.5, 85, 84.5, 84, 92, 85, and 83 mm
Parts from plant 2: 88, 89, 82, 85, 86, 87, 83, 84, 85, 86, and 84.5 mm

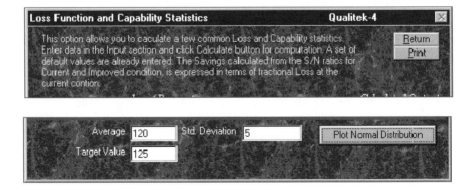

Figure 3.12   Plotting normal distribution from known average and standard deviation values for Example 3.4.

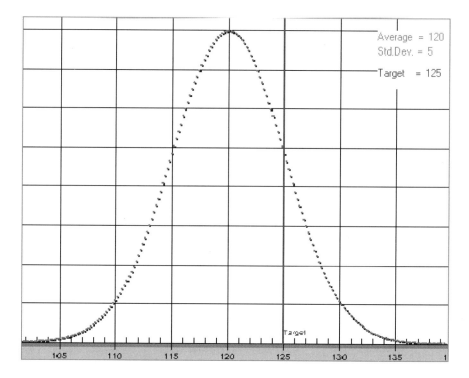

Figure 3.13 Normal distribution for data in Example 3.4.

Plant 2 recently implemented a process improvement, which is credited with reduced rejects. **(a)** Compare the parts data distribution of the two plants. **(b)** Express the improvement in quantitative terms.

Solution: Run QT4 and go to the main screen. From the *Loss/savings* menu item, select the *Population statistics* option (see Figure 3.7).

1. Enter the 13 part dimensions for plant 1 in the spaces under *User data* (Figure 3.14).

2. Check the *Nominal is best* box under *Quality characteristics* and enter 85 as its value.

3. Click the *Calculate* button and note the following for plant 1 data as shown in Figure 3.14:

$$\text{average} = 83.807 \qquad \text{standard deviation} = 3.294$$

$$S/N = -10.586$$

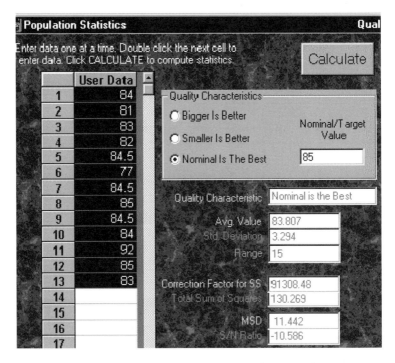

Figure 3.14 Statistics for plant 1 data of Example 3.5.

4. Click the *Reset* button and enter the 11 part dimensions for plant 2 in the spaces under *User data* (Figure 3.15).

5. Check the *Nominal is best* box under *Quality characteristics* and enter 85 as its value.

6. Click the *Calculate* button and note the following for plant 1 data as shown in Figure 3.15:

$$\text{average} = 85.409 \qquad \text{standard deviation} = 2.083$$

$$S/N = -6.143$$

7. Click *Return* to go back to the main screen.

8. While at the main screen, select the *Variation reduction* option from the *Loss/savings* menu item (Figure 3.16).

9. Click *OK* at the prompt about completing the analysis of experiments (see Figure 3.17, referring to DOE).

10. Under *Current status*, enter the values average = 83.807, nominal = 85, and *S/N* = -10.586 (Figure 3.18). The standard deviation value is not needed as input, as it is included in the *S/N* value.

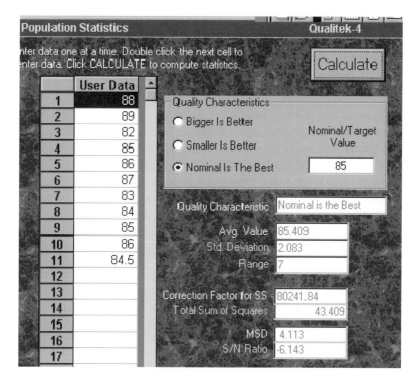

Figure 3.15 Statistics for plant 2 data of example 3.5.

Figure 3.16 Menu option selection for variation reduction plot.

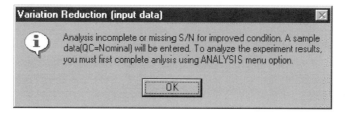

Figure 3.17 Reminder about completing analysis of results.

11. Under *Improved status*, enter the values *S/N* = –6.143 and select the *Nominal is best* quality characteristic. (For calculations of C_p and C_{pk}, the distribution is assumed to be centered.)

12. Enter LSL = 78 and USL = 92 under *Specification limits*.

13. Click the *Plot* button to display the distribution plot.

The variation reduction plot shown in Figure 3.19 makes use of the data statistics calculated earlier to generate the normal distributions and display. Using *S/N* ratios for the two distributions, QT4 also calculates the corresponding C_p, C_{pk}, and resulting savings from the loss functions. This screen shows a lot more information than we need to answer the questions in Example 3.5. In answer to the questions:

(a)   The parts data distributions for the two plants (the plant 2 distribution is assumed on target by QT4) are as shown in Figure 3.18. (Indicate the average and standard deviation associated with each plot.)

(b)   The improvement can be quantified by using C_p, C_{pk}, or *S/N* ratio. For *S/N* ratios, the improvement for plant 2 is –6.143 (*S/N* can be either positive or negative) a 42% improvement (–10.585 + 6.143)/(–10.585) over the –10.585 for plant 1. Similar improvements can be estimated using C_p and C_{pk} values for

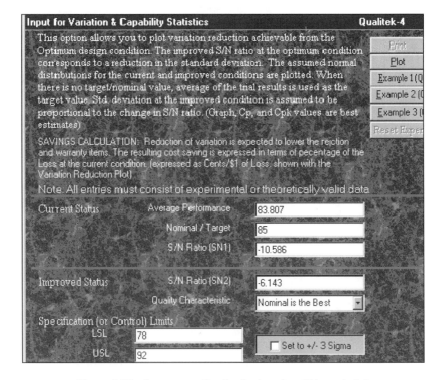

Figure 3.18   Input screen for distribution plot of Example 3.5.

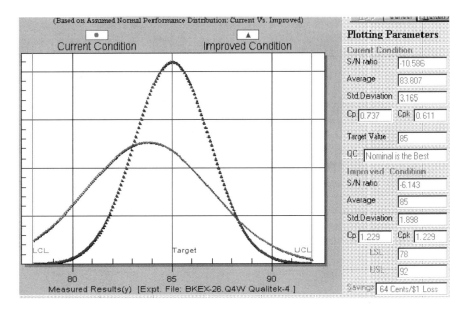

Figure 3.19   Distribution plot showing variation reduction for Example 3.5.

the two distributions. Note that the C_p and C_{pk} values shown in Figure 3.19 are calculated based on an assumed distribution on target (average = 85) and are different from the observed values shown (average = 85.409) in Figure 3.15.

Example 3.6   Mr. Eveready has installed a new on-line monitoring system that measures output voltages of the 9-volt-battery manufacturing line. By the press of a button, a computer-controlled instrument automatically monitors a sample of 100 batteries and instantly reports average and S/N ratio of the samples observed. One such sample reading is

$$\text{average voltage} = 9.162 \quad \text{and} \quad S/N \text{ ratio} = 5.174$$

If the production plan calls for all batteries to have a nominal voltage of 9 volts with upper limit = 10.5 volts and lower limit = 7.5 volts, determine the standard deviation, C_p and C_{pk} of the population of batteries from which the sample performance was obtained.

Solution: Run QT4 and go to the *Experiment configuration* screen (main screen). From the *Loss/savings* menu item, select the *Loss and capability statistics* option (see Figure 3.11). For this problem you will only use the top part of the screen. Proceed as follows:

1. Click and set *Quality characteristic* to *Nominal is Best* (Figure 3.20).
2. Enter average, nominal, and so on, as appropriate.

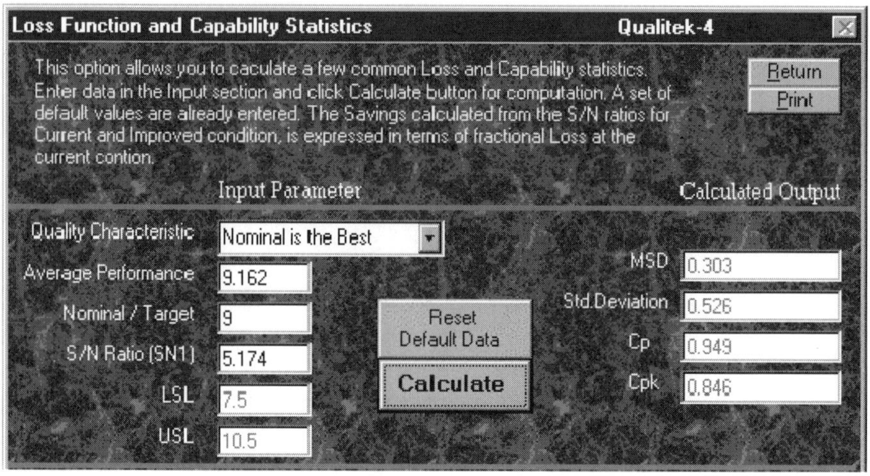

Figure 3.20 Computation of capability statistics from the S/N given in Example 3.6.

3. Click the *Calculate* button for calculation of standard deviation, C_p, and C_{pk} values.

The results are standard deviation = 0.303, C_p = 0.949, and C_{pk} = 0.846.

SUMMARY

In this step you have learned how commonly done single-factor experiments can show trend of factor influence. You are now aware of the two major categories of factors, discrete and continuous, and know how to decide the number of factor levels necessary to study factors with linear and nonlinear behavior. You have also learned the quick way to study multiple factors and risks associated with it. You now are able to use a few QT4 capabilities to calculate and compare population performance data.

REVIEW QUESTIONS

Q. What is the minimum number of levels of a factor that you should test?

A. The minimum number of levels is two. Results of a one-level test renders it to be a test for performance demonstration only. For investigative tests, two results are needed to determine the influence by comparing the two.

Q. What is the highest number of levels you can test?

A. Unlimited. Three levels are highly desirable. More than four levels is highly unusual and should be preferred only when sinusoidal behavior is suspected.

Q. What should be considered in deciding the number of levels?

A. While studying the influence of a factor, if we decide to test it at two levels, only two tests are required, whereas if three levels are included, three tests have to be performed.

Q. What is the loss function?

A. The loss function makes it possible to estimate the dollar loss associated with a performance distribution. It can also be used to compare the improvement of one set of performance data over another.

Q. What do we mean by *distribution*?

A. The term *distribution* refers to the normal distribution curve, which closely matches the data set under consideration. Since the mathematical expression for the normal distribution curve is defined in terms of two quantities: standard deviation and average (or mean), the term *distribution* also refers to these two quantities. The location of the average shows where the distribution is with respect to the target (when present), and the standard deviation indicates how narrow the distribution is.

Q. Since all the indices discussed measure improvement, is there a need to know all of them?

A. No. Depending on your project communication needs, you will need one or more of the indices. In DOE applications, a graphical representation of the distribution becomes good information for those who are interested in variation reduction.

Q. When do one-factor-at-a-time experiments work well?

A. Only when factor influences are linear, that is, when all factors act independent of each other.

Q. What are the risks of running one-factor-at-a-time experiments?

A. The risks are primarily in the following two areas: (1) conclusions may be misleading, and (2) results may not be reproducible.

EXERCISES

3.1 A team consisting of engineers from a production group and an equipment supplier identified temperature as the primary factor for the heat treatment process optimization study. At issue was selection of quenching bath temperature in the range 250 to 300°F by doing some quick and inexpensive experiments. The production group and the supplier carried out separate sets of tests, which were measured in terms of Rockwell hardness (R_c) number:

Production group test results (two tests): 85 R_c at 250°F and 90 R_c at 300°F.
Supplier test results (three tests): 83 R_c at 250°F, 92 R_c at 275°F, and 89 R_c at 300°F.

   (a) For a higher hardness number, what is the best temperature level?

   (b) Was the supplier justified doing an extra test using three temperature levels instead of the two levels used by the production group?

TABLE 3.5 Factors and Levels for Exercise 3.2

		Level
Factor	1	2
A: contact pressure	Current condition	10% higher pressure
B: cooling and ventilation	Current condition	New flow path
C: probe diameter	Current diameter	Smaller diameter

3.2 In an effort to investigate the causes of a high rate of rejects of soldered PC (printed circuit) boards, the production engineers considered three suspected factors (Table 3.5) and ran four experiments (Table 3.6).

   (a) Is the number of experiments enough to determine the trend of influence of all three factors?

   (b) Based on the available data, what is the factor-level combination for the least expected rejects?

3.3 The fuel economy data collected from the field for a newly introduced SUV (sport utility vehicle) using 87 octane gasoline are as follows (data in miles per gallon): 14.5, 12.3, 16.4, 17.2, 18.1, 14.6, 13.8 12.2, and 15.0. What is the expected gas mileage range (low and high) for over 99% of the SUVs sold?

3.4 Driving habits dictate how the same automobile component suffers different wear and tear in the hands of owners. Hoping to determine the frequency of brake applications per mile, the engineers involved in development of a new braking system instrumented a large fleet of vehicles from the preceding model year. The data collected showed that average frequency = 16.5 and standard deviation = 2.6.

   (a) Plot the normal distribution for the population of vehicle performance.

   (b) Estimate the frequency of brake application by the 99th percentile user.

TABLE 3.6 Experiments for Exercise 3.2

	Factor			
Experiment	A	B	C	Results
1	Current	Current	Current	32
2	10% higher	Current	Current	19
3	Current	New flow path	Current	42
4	Current	Current	Smaller diameter	12

EXERCISE ANSWERS

3.1 (a) The temperature level for the best performance (from among the results of the factor levels tested) is at 275°F (QC = B).

(b) It becomes obvious from the plot of the result (supplier test results at three levels) against the temperature levels that the factor influence is nonlinear. The best level would not have been identified from the tests at two levels. Thus the tests at three levels were justified.

3.2 (a) Yes.

(b) 10% Higher pressure, current condition, and smaller diameter ($a_2b_1c_2$).

3.3 Use the QT4 *Population statistics* option (*Loss/savings* menu option from the main screen) to calculate average and standard deviation (Figure 3.21). Then calculate the low and high range by subtracting and adding three times the standard deviation from the average value as calculated here.

$$Low = 14.9 - 3 \times 2.04 = 8.78 \text{ miles}$$

$$High = 14.9 - 3 \times 2.04 = 21.02 \text{ miles}$$

3.4 Use the QT4 *Loss and capability statistics* option (*Loss/savings* menu option from the main screen) to generate the normal distribution plot (Figure 3.21).

(a) Click the *plot normal distribution* button (Figure 3.22) to display the distribution.

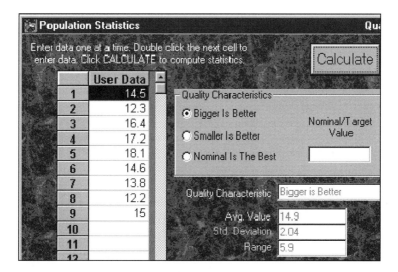

Figure 3.21 Population statistics for Exercise 3.3.

Figure 3.22 QT4 option of distribution for Exercise 3.4.

(b) Calculate the 99th percentile application by calculating the highest range (average + 3 × standard deviation, which gives 99.7% of the population, 16.5 + 3 × 2.6 = 24.3).

Experimental Design Using Orthogonal Arrays

What You Will Learn in This Step

In this step you will learn about the basic concepts in design of experiments (DOE) and how experimental possibilities expand when all factors are considered to have influence on the performance. You will also learn about several orthogonal arrays that have been created to design experiments that are smaller in number but produce more reproducible performance than the one-factor-at-a-time experiments discussed in Step 3. By the end of this lesson you will understand how easy it is to design your experiments, the simplicity of the experiment design process when orthogonal arrays are used, and how each of the arrays can be used to accomplish many experimental situations. By using QT4, you will be able to learn how to design your experiments and how to carry out analysis of the results.

Thought for the Day

As I grow older, I pay less attention to what men say. I just watch what they do.
—Andrew Carnegie

EXPERIMENTS WITH MULTIPLE FACTORS

Experiments for investigations will be much simpler if we were always to have only one factor influencing the result. If you have only one factor of concern, the question is only about how large or small you want to make your experiments. That is, the challenge is to decide whether to run two experiments (which is the minimum size) at two factor levels, or to run more than two experiments to capture its nonlinear effect on the result. For a single-factor study, there will be no need to complicate our lives and be concerned with any statistical or smarter technique. But that is not the case with most of the problems we face. In most industrial projects, more than one factor controls the outcome. When several factors are suspected to have influence on the results, it is tempting to run a one-factor-at-a-time experiment, as it may appear to be simpler to perform, and in some cases, the number of experiments may be smaller. But such experiments usually fail to test for all possible factor combinations and produce reliable results. What we need is experiments of limited size that allow us to test factors at more combinations and make better predictions about the factor behavior.

Experiments That Look At All Possible Factor Combinations

Suppose that in a soldering process, temperature (factor A) is the only factor of concern at this time. The operating range of this temperature is 600 to 680°F. In this case the effect of temperature can be found by running two experiments, one at 600°F (a_1) and another at 680°F (a_2), or by taking more levels within the range. At this point you may have questions in your mind as to how we decide which are the factors in the project, what their level ranges are, and how many levels to study. Although we have discussed briefly how to select the number of levels for a factor, we will need to discuss the selection of factors and the factor level in more detail later. For the purpose of understanding design of experiments (DOE) concepts, we will assume that the factors and their levels are already known. Further, while factors may be tested at any number of levels, discussion in this book is limited to two-, three-, and four-level factors. We also use character notation for factors (A, B, C, etc.) and their levels (a_1, a_2, b_1, etc.) to describe experimental conditions and the possibilities of many conditions. Thus, to investigate factor A at two levels, a minimum of two experiments will be needed, which we write in condensed form as A: a_1 and a_2.

Suppose that there are two factors, A (levels a_1 and a_2), and B (levels b_1 and b_2), to be studied together. Since either level of A could go with either levels of B, it poses four possibilities (experiment 1, experiment 2, etc.), which are obtained by combining the two factor levels as shown in Figure 4.1. Similarly, three factors A, B, and C tested at two-levels each requires eight experiments:

Factors: A (a_1, a_2), B (b_1, b_2), and C (c_1, c_2)

Experiments: $a_1b_1c_1$ $a_1b_1c_2$ $a_1b_2c_1$ $a_1b_2c_2$

           $a_2b_1c_1$ $a_2b_1c_2$ $a_2b_2c_1$ $a_2b_2c_2$

The eight experiments for three factors at two levels each can also be written using notations 1 for level 1 and 2 for level 2 of the factor assigned to the column, as shown in Table 4.1. This kind of notation makes description of larger experiments easier and is used throughout the book. Reading a row at a time, the experiments are described as follows:

	a_1	a_2		
b_1	Expt. 1	Expt. 3	Factor	A => a_1 and a_2
b_2	Expt. 2	Expt. 4	Factor	B => b_1 and b_2

Four experiments: a_1b_2 a_1b_2 a_1b_2 a_1b_2

Figure 4.1   All possible experiments with 2 two-level factors.

TABLE 4.1 Experiment Descriptions Using Level Notations[a]

Experiment	Factor		
	A	B	C
1	1	1	1
2	1	1	2
3	1	2	1
4	1	2	2
5	2	1	1
6	2	1	2
7	2	2	1
8	2	2	2

[a]The number in the table represents the level of the factors occupying the column.

Experiment 1: 1 1 1   as $a_1 b_1 c_1$

Experiment 2: 1 1 2   as $a_1 b_1 c_2$

and so on.

In the language of DOE, the possible combinations of factor levels are called *full factorial* combinations. If you were to be able to carry out all such possible combinations, you would be doing a full factorial experiment. If you were to test all possible combinations of factor levels, the average behavior determined would be more reliable. But conducting full factorial experiments is often costly and prohibitive. Given a number of factors and their levels, the total number of possible combination is easily determined by using the combination formula:

$$\text{total number of combinations} = (\text{number of levels})^{\text{number of factors}}$$

When all factors are at two levels, the total number of combinations will depend on the number of factors:

$$2^2 = 4 \qquad \text{for 2 two-level factors}$$

$$2^3 = 8 \qquad \text{for 3 two-level factors}$$

and so on. Other experimental conditions all with two-level factors, and the number of full factorial experiments are listed in Table 4.2.

The experimental conditions when working with three- and four-level factors can be generated similarly. Examine the following two situations.

1. Two three-level factors (A and B) will produce nine (3^2) combinations as follows:

TABLE 4.2 Full Factorial Experiments with Two-Level Factors

Experimental Condition	Full Factorial
3 factors at 2-level	$2^3 = 8$
4 factors at 2-level	$2^4 = 16$
7 factors at 2-level	$2^7 = 128$
15 factors at 2-level	$2^{15} = 32,768$

Factors: A $(a_1a_2a_3)$ and B $(b_1b_2b_3)$

Combinations: a_1b_1 a_1b_2 a_1b_3

            a_2b_1 a_2b_2 a_2b_3

            a_3b_1 a_3b_2 a_3b_3

2. Three four-level factors (C, D, and E) will produce 64 ($= 4^3$) combinations as follows:

Factors: C $(c_1c_2c_3c_4)$, D $(d_1d_2d_3d_4)$, and E $(e_1e_2e_3e_4)$

Combinations: $c_1d_1e_1$ $c_1d_1e_2$ $c_1d_1e_3$ $c_1d_1e_4$

            $c_1d_2e_1$ $c_1d_2e_2$ $c_1d_2e_3$ $c_1d_2e_4$

            $c_1d_3e_1$ $c_1d_3e_2$ $c_1d_3e_3$ $c_1d_3e_4$

       $\vdots$

            $c_4d_4e_1$ $c_4d_4e_2$ $c_4d_4e_3$ $c_4d_4e_4$

So it is quite clear that the number of experiments to cover all possible combinations becomes very large even with a limited number of factors. The total number of possibilities tends to grow at a much faster rate for three- and four-level factors. Obviously, most often it is impossible to carry out the complete investigations unless the experiments are done using analytical simulation of the process or system under study.

Sir R. A. Fisher (1890–1962) of England was first to introduce the idea of designing experiments to cover all possible combinations. He did so in his studies on agricultural production. To reduce the number of experiments and get important information about a project, Fisher attempted to utilize the work of the Swiss mathematician Leonard Euler (1707–1783) with Latin squares (also known as Greco-Latin squares, and later, orthogonal arrays). In search of smaller than full factorial experiments, Frank Yates and Oscar Kempthorne expanded Fisher's design of experiments technique and developed the fractional factorial experiment. In this technique the experimenter can use only a fraction of all possible combination and still be able to identify the major effects.

Toward the end of World War II, Dr. Genechi Taguchi of Japan carried out extensive research using the DOE technique. Along with a new approach to quality improvement as described in Step 1, his major contribution has been in developing and using a special set of orthogonal arrays for designing experiments. *Orthogonal arrays* are a set of tables of numbers each of which can be used to lay out experiments for a number of experimental situations. Use of these arrays to design experiments is the key to learning DOE through the Taguchi approach. Today when one wants to learn the Taguchi experimental technique, it is to learn how to use the orthogonal arrays in the way that Taguchi has prescribed. Much of this step is dedicated to a discussion of orthogonal arrays and their use.

Things You Already Know

- Full factorial experiments are too numerous to do. Fractional factorial experiments and orthogonal arrays were developed to make the DOE technique more applicable by reducing the size of the experiments.
- Dr. Genechi Taguchi of Japan proposed a special set of orthogonal arrays to standardize fractional factorial designs.

SHORTCUTS TO DESIGN OF EXPERIMENTS

Now that you know how to determine all possible factor-level combinations, you will appreciate the need to find a shortcut to get the job done. In other words, how to do the fewest experiments and still get the most information is of immediate concern. If every time you encountered a new experimental situation, you had to find a shortcut yourself, it would be a tedious job. Fortunately, this has already been done for you, and if you can use a set of the orthogonal arrays perfected by Taguchi, you will save a lot of time and money.

The word *design* in *design of experiments* refers to the layout that describes a combination of the factors included in the study. Like an engineering design document, experiment design will contain all the information needed to proceed with the next phases. This means that when an experiment is designed, you will know the number of experiments to carry out and how to carry out each experiment. In short, the design process answers two questions: (1) how many experiments to do, and (2) in what manner to do them.

A number of orthogonal arrays have been created to handle most common industrial experiments: for two-, three-, and four-level factors. A few arrays have special restrictions and others have built-in mixed levels. You learn the use of 10 orthogonal arrays in next few steps.

PROPERTIES OF ORTHOGONAL ARRAYS

Orthogonal arrays have been created to help accomplish experiment designs. A number of arrays are available for the purpose. Each array can be used to suit a number of experimental situations. The smallest orthogonal array is an L-4, which is used to design experiments to study three or 2 two-level factors, as shown in Table 4.3.

Orthogonal arrays are designated by the notation L (L for Latin squares) with a subscript or dash (L_4 or L-4; both are in common use). The subscript refers to the number of rows in the table, which indicates the number of combinations the design will pre-

TABLE 4.3   L-4 (2^3) Orthogonal Array

Experiment	Column		
	1	2	3
1	1	1	1
2	1	2	2
3	2	1	2
4	2	2	1

scribe. For example, an L-4 will have four rows, L-8 will have eight rows, and L-n will have n rows. In addition to the character notation, arrays contain some numerical notations, which indicate the number of factors involved and the full factorial combinations. The designation (2^3) with L-4 carries these meanings. First, $2^3 (= 8)$ indicates that the number of possible combinations is eight. Second, it indicates that the array can be used to design an experiment for up to three (exponent 3 in 2^3) two-level (base of 2 in 2^3) factors. Since the number of factors with which an experiment can be designed depends on its number of columns in the array, the exponent 3 in 2^3 also indicates the number of columns of the array. These interpretations will hold true for all orthogonal arrays discussed in this book.

Looking into the array, L-4 has four rows and three columns numbered in sequence. In the body of Table 4.3, there are three columns of numbers alternating between 1 and 2. These numbers describe the levels of factors assigned to a column. In a design, factors under study are assigned to the available columns of the array and the individual experiments are described interpreting the numerical level notations in the column (reading across, experiment 1 has levels 1, 1, and 1).

Let's see if you can put the description above to use by trying to learn about L-8 (2^7). From the notations, you should recognize that the L-8 array is a table of numbers that has eight rows and seven columns, and that it can be used to design experiments with up to 7 two-level factors. You can also say that the eight experiments that the L-8 array will prescribe are among the 128 (2^7) possible combinations of the seven factors under study.

Although you do not need to memorize any of the properties of orthogonal arrays, it is good to be aware of some of their common properties. Later in this step we learn more about these properties.

Orthogonal Properties of Arrays

The word *orthogonal* has a special meaning in the specific field of application. In coordinate geometry it has one meaning; in matrix algebra it means something else. What it means in terms of the array is that the columns of the arrays are *balanced*. The word *balance* again has two meanings. First, each column is balanced within itself. This means that within a column, there are an equal number of levels. For example, in all columns of L-4, there are two 1's and two 2's. Looking at the L-8 array in Table 4.4, there

TABLE 4.4   L-8 (2^7) Orthogonal Array

Experiment	Column						
	1	2	3	4	5	6	7
1	1	1	1	1	1	1	1
2	1	1	1	2	2	2	2
3	1	2	2	1	1	2	2
4	1	2	2	2	2	1	1
5	2	1	2	1	2	1	2
6	2	1	2	2	1	2	1
7	2	2	1	1	2	2	1
8	2	2	1	2	1	1	2

are four 1's and four 2's in each of the seven columns. In addition, the numbers in the column follow a certain order instead of being distributed randomly. You may get a sense of this order by reading the columns (starting with column 1) vertically, from top to bottom. This balanced arrangement in each assures that each level of the factor, no matter to which column it is assigned, has equal opportunity to influence the results.

The second meaning of the word *balance* is that any two columns in the arrays are also balanced. The balance in this case indicates that the combinations of the levels between the columns considered are also equal in number. Consider the property of orthogonality or balance between columns 1 and 2 of the L-8 array (Table 4.4). Since both these columns have 1 and 2, they can have four separate combinations, as shown. When two two-level columns are balanced, each of these combinations will be present in equal numbers (Table 4.5). Between columns 1 and 2 of the L-8 array, these combinations are present two times each. These combinations are also present two times between any two arbitrary columns of the array. You can easily check to see that the level combinations above are present one time each in an L-4 array. This property of

TABLE 4.5   Balanced Columns

Combination	Column	
	1	2
1	1	1
2	1	2
3	2	1
4	2	2

balance between any two columns assures that all possible factor combinations exist in equal number.

Unrelated to the orthogonal property, a couple of general observations are made by reading the array along the rows (see Tables 4.3 and 4.4). The first row of all arrays has all 1's. The second observation is that there is no row that has all 2's, 3's, or 4's. These observations hold true regardless of the array is intended for two-, three-, or four-level factors.

Common Orthogonal Arrays and Their Special Properties

In this book we make use of only 10 orthogonal arrays that will enable us to design frequently occurring experimental situations in the industry. The list includes five arrays for two-level factors, three for three-level factors, and two for four-level factors, as shown in Table 4.6.

- The 5 two-level arrays, L-4, L-8, L-12, L-16, and L-32, each has one fewer column than row. Thus L-4 has three columns, L-8 has seven columns, and so on.

- The three- and four-level arrays do not have such a rule. You have to make yourself familiar with their column sizes, such as L-9 for 4 three-level columns, L-18 for eight columns, and L-27 for 13 columns.

- The 2 four-level arrays (see Table 4.6) are modified forms of the corresponding two-level array. The first one, L-16 (4^5), has five columns and is used to design experiments with 5 four-level factors. The second one, L-32 (2^1, 4^9), has 10 columns and can accommodate 9 four-level factors and 1 two-level factor.

TABLE 4.6   Orthogonal Arrays Most Commonly Used for Experiment Design

Array	Intended Use (Experiments With:)	
L-4 (2^3)	3 two-level factors	
L-8 (2^7)	7 two-level factors	
L-12 (2^{11})	11 two-level factors	Two-level arrays
L-16 (2^{15})	15 two-level factors	
L-32 (2^{31})	31 two-level factors	
L-9 (2^4)	4 three-level factors	
L-18 (2^1, 3^7)[a]	1 two-level and 7 three-level factors	Three-level arrays
L-27 (3^{13})	13 three-level factors	
L-16 (4^5) modified	5 four-level factors	
L-32 (2^1, 4^9)[a] modified	1 two-level and 9 four-level factors	Four-level arrays

[a]Also called a mixed-level array, as it accommodates 1 two-level factor in the first column.

USING ORTHOGONAL ARRAYS TO DESIGN EXPERIMENTS

You will remember from Step 1 that there are five steps in applying DOE in a project. Logically, the first step is the task of experiment planning. It is the most important part of the entire application and should never be sidetracked. The purposes of planning are many, and how it should be carried out needs to be described in detail. A few of the things that we get after the experiment planning sessions are the factors and their levels considered for the study. It is only when the factors and their levels are known that we can consider designing experiments. But since experiment planning requires complete knowledge of the orthogonal arrays and the design process, it is necessarily kept for discussions in later steps.

Using orthogonal arrays, designing experiments is to follow three simple steps, as you will see soon. For example, in a chocolate chip cookie baking process, 3 two-level factors, A: sugar ($\frac{1}{2}$ cup, $\frac{3}{4}$ cup), B: chips (small, large), and C: butter (1 stick, $1\frac{1}{2}$ sticks) were identified for study. Experiment design is desired, as it will describe the number of times the cookies will have to be baked and the recipe for each process.

First Step in Design: Select an Orthogonal Array.   The array selected is the smallest one to do the job. In this case there are 3 two-level factors. There is no mathematical formula for selection of the array. You have to know what to look for and then do it intuitively. Later, when designing more complex experiments, there will be some rules for selection. For now, the best way to decide which array will be suitable for 3 two-level factors is by asking such questions as: Which array has three or more two-level columns? L-4 is the smallest among several arrays that meet the requirement. Thus the L-4 array is selected. Note that by selecting the L-4 array, the maximum number of experimental conditions is already known, as it equals the number of rows (the number associated with the array designation, L). Also, you never need to memorize the array—you simply need to get hold of a set of tables, such as those in the Appendix of this book, and assign the factors when you want to do it by yourself. Through the examples and exercises in this book, you will learn how to design experiments using the QT4 software.

Second Step in Design: Assign Factors to the Columns.   The three factors A, B, and C are assigned to the three columns of the array as shown in Table 4.7. At this point no particular care is needed to assignment. Any factors can be assigned to any of the three columns.

Third Step in Design: Describe the Experiments.   Once the factors are assigned to the columns, individual experiments are easily described. The individual experiments are represented by the rows, which can be read in cryptic notations first, followed by description of the four experiments in plain language, as shown here. In cryptic notation:

TABLE 4.7 L-4 (2^3) Orthogonal Array with 3 Two-level Factors Assigned

Trial	Factor		
	A	B	C
1	1	1	1
2	1	2	2
3	2	1	2
4	2	2	1

Experiment 1: 1 1 1   or a_1 b_1 c_1

Experiment 2: 1 2 2   or a_1 b_2 c_2

Experiment 3: 2 1 2   or a_2 b_1 c_2

Experiment 4: 2 2 1   or a_2 b_2 c_1

In plain language:

Experiment 1: $\frac{1}{2}$ Cup of sugar, small chips, and 1 stick of butter

Experiment 2: $\frac{1}{2}$ Cup of sugar, Large chips, and $1\frac{1}{2}$ sticks of butter

Experiment 3: $\frac{3}{4}$ Cup of sugar, small chips, and $1\frac{1}{2}$ sticks of butter

Experiment 4: $\frac{3}{4}$ Cup of sugar, large chips, and 1 stick of butter

From the three steps involved in design, it should be obvious that it is the first step, selection of the array to be used to design the experiment, is the most difficult task. Let us understand the array selection process from the following examples.

Example 4.1 Determine the number of experiments that will be necessary to study the following situations.

(a) 8 two-level factors

(b) 10 two-level factors

(c) 15 two-level factors

(d) 3 three-level factors

(e) 1 two-level factor and 5 three-level factors

(f) 4 four-level factors

(g) 1 two-level factor and 9 four-level factors

(h) 1 four-level factor and 4 two-level factors

Solution: You will need to review closely the list of arrays listed in Table 4.6 to answer these questions.

(a) If the most factors are at two levels, you should look for the array from among the group of 5 two-level arrays. For 8 two-level factors, you will need an array that has at least 8 two-level columns. The smallest array that has eight such columns is an L-12. Your design will use any eight of the 11 columns in the array. The level numbers in empty columns of the array are all changed to zero (0), which is interpreted as non-existent when describing the experiments. The number of experiments in this case will be 12.

(b) You can use the same L-12 array to design experiments with up to 11 two-level factors. The number of experiments needed is also 12 in this case.

(c) The array L-16 has 15 two-level columns. It will be your selection and the number of experiments will be 16.

(d) Among the available 3 three-level arrays, the smallest one that has at least 3 three-level columns is L-9. The three factors under study can be assigned to any three of the 4 three-level columns of the array, leaving one column empty. The number of experiments is 9.

(e) This experiment includes factors that are not all at one fixed level. Luckily, in the list of 10 arrays, there are two arrays that have a mixed-level column. Since predominant factors are at three levels (5 three-level factors), you should look for the mixed-level array among the three-level arrays. Notice that L-18 has enough three-level columns and 1 two-level column. You will be able to design this experiment with an L-18, which will require 18 experiments.

(f) The 4 four-level factors can be handled by a modified L-16 with 5 four-level columns. The number of experiments you will need to do is 16.

(g) Your selection in this case has to be a modified L-32. You will design the experiments by assigning the two-level factor to the single two-level column, and 9 four-level factors to any nine of the available 9 four-level columns. The number of experiments you would do is 32.

(h) You do not yet have the technique to design experiments in this situation. Not solved.

If all these rules and how to determine the size of arrays appears confusing to you, take heart. There is help. You will be able to rely on computer software that can easily remember the guidelines and design experiments to suit your needs. Given the number of factors you wish to study, the automatic design features of Qualitek-4 software can select the array and assign the factors to the correct columns. The correctness of the column assignment, which is not an issue yet, will play a greater role in advanced experiments discussed later in this book. Array selection for various experimental situations is demonstrated in the following example.

Example 4.2 How many separate experiments would you need to study the projects involving following factors?

(a) 6 two-level factors

(b) 1 two-level factor and 5 three-level factors

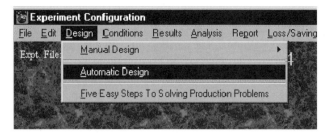

Figure 4.2 *Automatic design menu* option of the QT4 software.

(c) 12 three-level factors

(d) 5 four-level factors

Solution: Run QT4 and go to the *Experiment configuration* screen (main screen). It is not necessary for you to open a particular file. At any point in the program, you can return to the main screen by clicking *Cancel* buttons repeatedly. To proceed to the next screen, always click the *OK* button.

(a) Select the *Automatic design* option from the *Design* menu (Figure 4.2). Click *Yes* at prompt about the new experiment (Figure 4.3). To specify the number of factors and their type, click on the box for the number of two-level factors. Observe that the option box at the bottom of the box disappears. This and many other options in the screen will be of interest later. Now we only need to indicate the number of two-level factors by clicking the up button of the counter control, as shown in Figure 4.4. The number of factors at different levels is sufficient information for the software to determine the array most suitable for the design. As indicated in Figure 4.5, experiments designed using an L-8 array requiring eight separate conditions will be prescribed. (Ignore statements in Figure 4.5 about interaction and outer array and click *No* and stay in the *Automatic experiment design* screen.)

(b) From the main screen, select the *Automatic design* option from the *Design* menu and proceed to specify factors for the experiment. Specify factors by checking

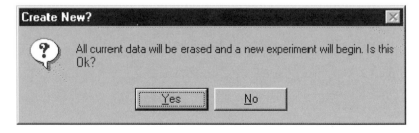

Figure 4.3 Standard prompt before a new experiment.

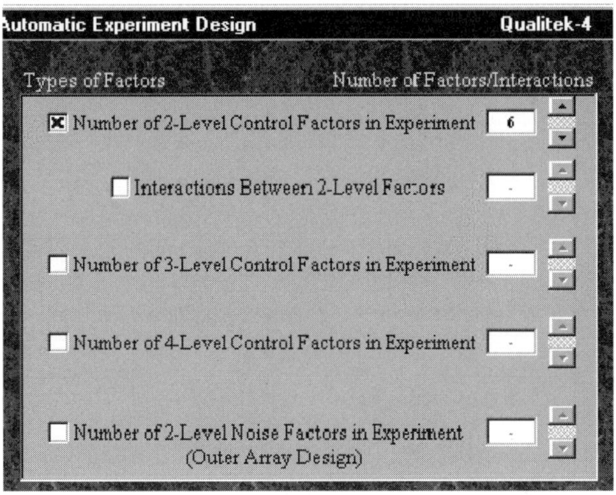

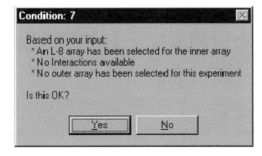

Figure 4.4 Factor specification for experimental design with 6 two-level factors.

boxes and setting numbers as shown in Figure 4.6. The array selected for the design is an L-18, indicating that there will be 18 experiments in this study (Figure 4.7).

(c) From the main screen, select the *Automatic design* option from the *Design* menu and proceed to specify factors for the experiment. Specify factors by checking boxes and setting numbers as shown in Figure 4.8. The array selected in this case is an L-27 (Figure 4.9), which will require 27 separate experiments.

(d) From the main screen, select the *Automatic design* option from the *Design* menu and proceed to specify factors for the experiment. Specify factors by checking boxes and setting numbers as shown in Figure 4.10. The necessary array in this case is a modified L-16 (labeled as M-16 in QT4), which would prescribe 16 separate experiments for the study (Figure 4.11).

Figure 4.5 Array selected for experimental design with 6 two-level factors.

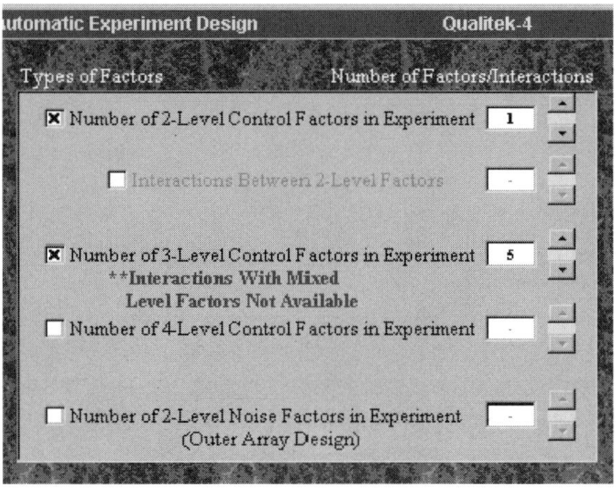

Figure 4.6   Specifications for experimental design with 1 two- and 5 three-level factors.

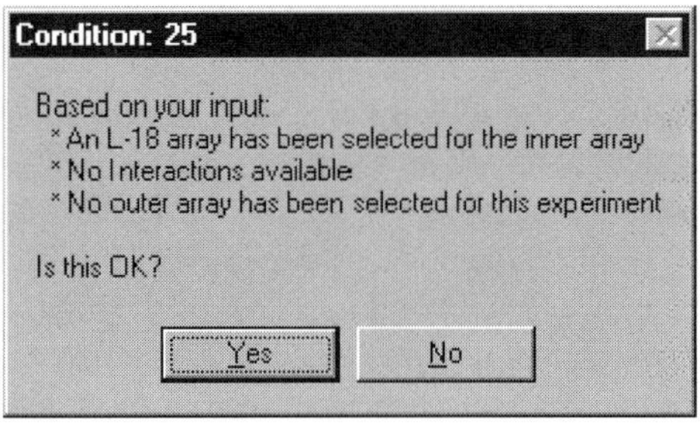

Figure 4.7   Array selected for experimental design with 1 two- and 5 three-level factors.

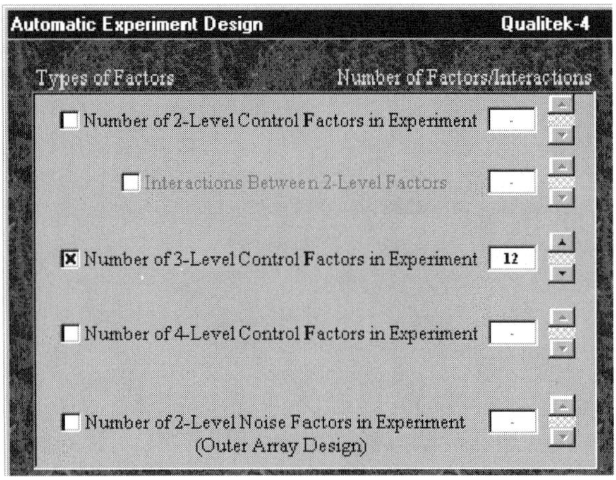

Figure 4.8   Specifications for experimental design with 12 three-level factors.

Condition: 27 ☒

Based on your input:
* An L-27 array has been selected for the inner array
* No Interactions available
* No outer array has been selected for this experiment

Is this OK?

[Yes]   No

Figure 4.9 Array selected for experimental design with 12 three-level factors.

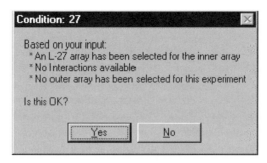

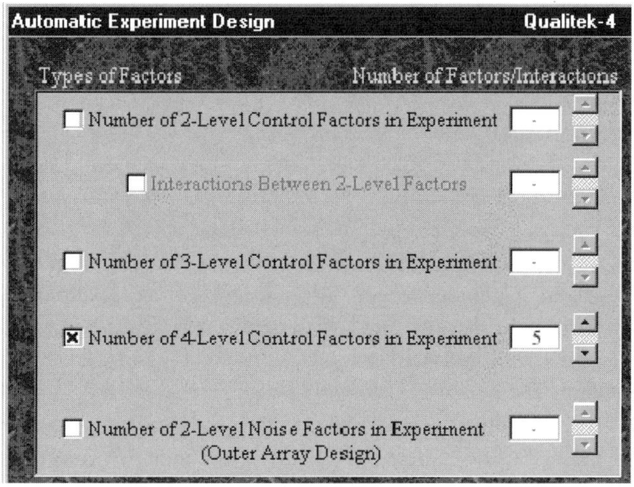

Automatic Experiment Design   **Qualitek-4**

Types of Factors   Number of Factors/Interactions

☐ Number of 2-Level Control Factors in Experiment   [-]

☐ Interactions Between 2-Level Factors   [-]

☐ Number of 3-Level Control Factors in Experiment   [-]

☒ Number of 4-Level Control Factors in Experiment   [5]

☐ Number of 2-Level Noise Factors in Experiment   [-]
            (Outer Array Design)

Figure 4.10 Specifications for experimental design with 5 four-level factors.

Condition: 28 ☒

Based on your input:
* An M-16 array has been selected for the inner array
* No Interactions available
* No outer array has been selected for this experiment

Is this OK?

[Yes]   No

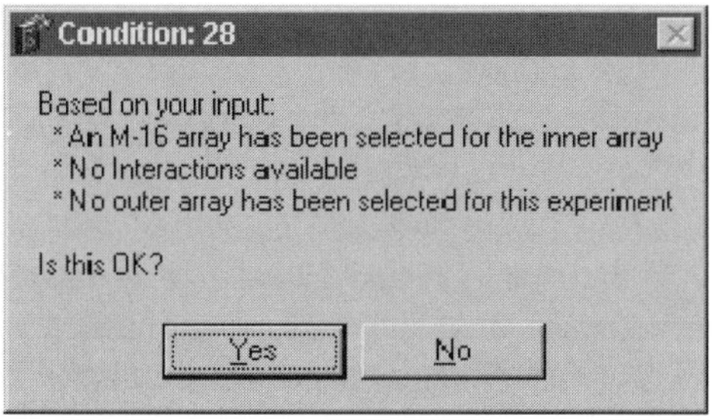

Figure 4.11 Array selected for experiment design with 5 four-level factors.

EXPERIMENT PLANNING: FIRST STEP IN DOE APPLICATION

Before an experiment can be designed, it must be planned. Careful planning is the key to success in any DOE application. This step is also known as *brainstorming*, indicating that it is a task that is better done as a group than alone. Indeed, it is strongly recommended that DOE projects be done as a group. The reasons for this are two. First, there are areas of compromises between what is desired and what is practicable. Often, project objectives involve subjective measurements that become more useful when evaluated by more than one person. The second, decisions about all aspects of the experiments, make the experiment more likely to be carried out and results implemented when they represent the team consensus. When the scope of the experiment and its purpose are agreed upon beforehand, less time is needed to gain acceptance of the recommendations derived from the experiments. So for most benefit, you must consider applying DOE in your project as a group, particularly when the beneficiary of the experiments are many people other than yourself.

A target DOE project can be for product or process optimization or problem resolution. Once the project is identified and the team is formed, you (the project leader) should invite the team for a planning session and schedule it when most members of the team can attend. Some background data gathering is a good idea as long as such facts do not inhibit the free nature of discussions and input from others. You should prefer to have someone outside the team facilitate the experiment planning session. If this is not possible and you end up facilitating your own session, try to be a little extra humble, such that all in attendance feel like wanting to offer their input to your project. The thing to remember is that you don't want to be a "fox guarding the chicken coop."

Your DOE projects would be most successful when you follow the new ways of working as a team. The new way is to work as a group (3 to 15 people) where everything is decided democratically, where each person has one vote, and all things are decided by consensus of the group. Typically, the new way would differ from conventional group tasks in two areas: work with more than a few people, and things decided by the group rather than being dictated by one person. The true nature of the experiment planning session cannot be described in this book. A few key elements of the planning session are pointed out.

- *What is the role of the project leader?* The project leader owns the project. He or she proactively initiates the project, forms a team, schedules a date of meeting, and does everything possible to make sure that everybody is a willing participant in the project. A good leader gathers facts about the project, shares them with all, but does so by being one among equals in the planning session.

- *How large should the group be, and who should be included?* All who are involved directly or have firsthand knowledge about the project should be invited to join the team. For a typical project design optimization project, potential members are the design engineers, test engineers, validation engineers, salespeople, production personnel, and internal and external customers (when possible) of the product. The idea is to involve as many people as you can, but keep it down to a few who will have information to contribute. A team size of fewer than 10 works well. If the team is larger than 15 people, problems may tend to overwhelm your meeting.

- *Who makes a good facilitator, and what is his or her role?* The facilitator must be knowledgeable in the DOE/Taguchi method and be willing to work with groups. He or she need not be familiar with the project at all and preferably not take part in the decision-making process. The facilitator should moderate discussions, devise ways to reach consensus decisions where needed, and help determine all necessary input for the experimental design. He or she must do these carefully on a predefined time schedule.

- *What is expected of the participants?* Participants need not have a background in DOE and do not need any background study on the project. They simply need to dedicate the day designated for the planning session, participate in the discussions, and offer their input based on experience. It is generally not a good idea to bring written reports or research materials and offer them to the team for their review. There is usually not enough time for all to digest reports and agree on them. Expert participants should concentrate on sharing technical knowledge verbally as much as possible. A good participant will always respect group consensus over claimed individual expertise.

- *How long should a planning session last?* Most planning sessions should be a one-day event, no more. For a simple project with a small number of team members, it may be done in half a day. Generally, it will take one entire day, and it is better to solicit and secure undivided attention of all participants for the whole day. It is never a good idea to leave the planning session unfinished, so it is very important that the discussions in the planning session are kept on schedule.

- *What are the topics and their order of discussions?* Based on many years of application experience, the author has found the following topics and their sequence helpful for planning sessions. All topics and discussions are carefully ordered and help efficiently determine all that are necessary to design, conduct, collect results, and analyze results of experiments for the study.

Topics of Discussion

- Project objectives (2–4 hours)
- Factors (same as variables, 1–2 hours)
- Levels of the factors ($\frac{1}{2}$ hour)
- Interactions (between 2 two-level factors, $\frac{1}{2}$ hour)
- Noise factors and robust design strategy ($\frac{1}{2}$–1 hour)

A significant amount of the time of the planning sessions should be devoted to a discussion of project objectives and their method of evaluations. Single objectives evaluated by a single criterion of evaluation requires the least amount of discussion. When different participants have different performance expectations, it will not be unusual to absorb most of the first part of the day in such a discussion. Models of multiple criteria of evaluations were presented in the preceding chapter. The leading questions that drive the discussion in this phase of the planning session are:

Project Objectives (2–4 hours)

- What are we after? How many objectives do we wish to satisfy?
- How do we measure the objectives?
- What are the criteria of evaluation and their quality characteristic?
- When there are more criteria than one, should we want to combine them?
- How are the various evaluation criteria weighted?
- What is the quality characteristic for the overall evaluation criterion (OEC)?

Discussions of project objectives and arriving at a group consensus are specific to the project and cannot be generalized except to "leave no stone unturned" in attempting to capture what everybody present has in mind about the project goals. Depending on whether the team agrees on one or many objectives, it may take 2 to 4 hours. It is a good idea to defer any unfinished items in this part to a future meeting and not delay the present session beyond four hours. Although consensus about the objectives is very important, as it helps team members to think along the same "wave length," it is not needed immediately for experiment design. What is essential for the experiment are the factors. Discussion of the factors must therefore follow immediately after discussion of the objectives.

The time spent on discussion of objectives serves well, as it allows team members to see the project as most others see it. This, of course, increases the chance of identifying key factors for the study. While initiating discussion of the factors, it is important to remind all participants to think about the objective agreed upon and then look for factors that influence one or more of the objectives. The key questions that drive this discussion are shown here.

Factors (Same as Variables, Parameters, or Input; 1–2 hours)

- What are all the possible factors?
- Which factors are most important?
- How many factors can we include in the study?

A common practice is first to write (on a blackboard or flipchart) all ideas (names of potential factors) suggested by all. It is important to encourage more ideas from all and seek a quantity of ideas and not worry about its quality at first. The facilitator must assure that no suggestions are subject to immediate criticism and that everybody has a chance to offer input. Once all have had their say, it will be the time to scrutinize the list, one at a time, to see that all ideas listed are indeed factors. To qualify for a factor, it must be an input, not an output or result. A factor is something that can be controlled or set to a desired value while conducting an experiment. Items that do not qualify as factors are crossed off the list. Those that are factors but cannot be controlled or which you do not want to control, are saved in a separate list of uncontrollable factors. In this way, the final list of factors should look much smaller than the original list of ideas.

The list of factors represents the variables (factors) that as a team you identified and suspect will influence the results. Whether they do influence, and whether or not such influence is significant, is to be determined from the experiments that follow. A more immediate challenge is to determine whether all the factors in the list can be included in the study, and if not, how many of them can be included. This is decided primarily by the time and money available for the project as the number of experiments directly influence both. Basically, you will be seeking answers to such questions as: How many separate experiments can we possibly run? The project leader and a few people close to him or her will have good input on this. If the answer is, say, 8 to 10, the largest array for the experiment would be an L-8. The array could be L-16 for the design if the project budget allowed doing over 16 separate experiments. Once the array is identified, the number of factors that the array can handle is already known. For instance, if L-8 were the largest experiment size, it would limit you to 7 two-level factors. Although we don't know how many levels each factor will have until we discuss it, we must still establish a limit on the number of factors and run the risk of utilizing an incorrect number until we review it later.

Once we have established a tentative number of factors for study, as dictated by the array selected, the next task is to decide which factors to include. Suppose that you qualify 13 factors and the largest array you can go for is L-8. This would mean that the team would have to determine 7 of the 13 factors used in the study. Which seven factors should you include? Which six should you omit? There is no science to this process. You probably do not have prior knowledge regarding the influence of the factors. Even if you know about some factors, they are probably going to behave differently in the presence of the others. A pareto diagram (highest to lowest influence) would be highly desirable, as you would then be able to take the first seven factors—but you will not have the time for that. Your best option could be to go around the table and ask each person for his or her opinion of the most influential factor and then take the first seven. Another approach would be to ask everyone to identify the least influential factors and slowly eliminate the factors you cannot afford to include in your study.

It is only after the factors for the study are selected that you can address the issue about their levels. All you know at this point is that you have limits on the number of factors and their level as dictated by the properties of the array. This discussion must necessarily follow factor selection; otherwise, time may be wasted talking about factors on the list that may not be included in the study. Proper selection of the number of levels of the factor is important, as it controls the kind of information obtained from the experiment. Four levels for each factor are better than three, and three are better than two, which is the minimum number of levels. As always, the larger the number of levels, the larger is the size of the experiment and the higher is the cost of project. Although it is desirable to include all the factors identified, cost considerations will force us to limit the number of factors in the study. A balance must be drawn. This is where team understanding of the details of experimental methodologies is helpful. A general rule of thumb is to keep the experiment size small: Take all factors at two levels unless prior experimental knowledge indicates nonlinear behavior, caused by continuous factors. Usually, there is not much choice about the levels of the discrete factors—it is what it is. Remember that the notion of linear or nonlinear behavior is not applicable to this kind of factor. You must keep the levels between two and four to stay within the scope of this book.

Determining the levels of the continuous factors is the next issue. Suppose that for the factor *pressure*, three levels are desired. It is known that the current working pressure is 250 lb/in^2. It is a good idea to keep one level at the current value and take the other two on either side of it. The three levels then can be low, current (250 lb/in^2), and high. But how far on either side should you go? There are no scientific recommendations. Experience must dictate the decisions here. A few basic guidelines need to be followed. The levels should be as far away from the current working level as possible, but they must be workable. All levels tested must be something that allows experiments to be performed (good or bad). They must also be something that if selected as desirable can be released for the production design.

What if you chose only two levels for the same factor (e.g., pressure)? Where should the two levels be? Should you take one at the current level and another lower or higher? Suppose that you take level 1 as the current working value and level 2 a value higher (exact value decided by the team) than the current value. This surely assumes that the most desirable condition is higher than the current working value, which may or may not be true. It will not be a good idea to preclude lower than current value as one possible desirable condition. So what is recommended is to take the two levels one at a lower and another at a higher than current value. How low and how high are still dictated by what extremities are known or expected to work. The key areas of discussion of levels will be guided by the following questions.

Factor Levels ($\frac{1}{2}$ hour)

- How is each factor suspected to behave?
- How many factor levels can the array accommodate?
- What is the trade-off between levels and factors?

Earlier in this step, you have learned how to design experiments using the 10 standard orthogonal arrays. Experiments that can be designed using these arrays as is are generally referred to as *simple experiments*. If all your experiments fell into this category, the topics covered in the planning session will be sufficient. In closing of the planning session, as with any team meeting, there should be clear understanding about who supplies the test samples, who conducts the experiments, who will analyze the results, and so on. Before the meeting is disbursed, the facilitator should be able to explain to the group how many experiments are going to be involved and how the data from the experiments should be collected.

During the planning session the team should also determine a few more items that dictate the level of complexity of the experiment design. Before adjournment, the facilitator should involve the group in a discussion that identifies potential interactions among factors. If interactions are included in the study, the size and scope of the experiment may change. The group should also engage in discussions about the levels of factors and decide whether there will be a need for three and four levels for some factors. A mixed-level factor in the study may require a different size of experiment (array). The discussion about possible uncontrollable factors (called *noise factors*) should be the last item of discussion. For a group with previous experience, identifying noise factors and including them in the experiment may indeed be quite attractive.

Who should decide whether to pursue such discussions? The facilitator should decide, which is why the facilitator for a planning session must have application knowledge of DOE. These additional topics of discussion in the planning become part of planning session discussions and are covered in later steps.

COMPLETING EXPERIMENTS AS PLANNED

Generating necessary information for the experiment design process is the main purpose of the planning session. Other information, such as objectives and how they are evaluated, can be postponed if it is necessary to make more time for discussions of factors and levels, which are necessary for experiment designs. The project leader and facilitator should take special care to ensure that factors and levels are clearly defined before the end of the planning session. It is not a good idea to hold several sessions for experiment planning. You should put all your energy into getting the most out of the day so that you are ready immediately to design the experiment. As discussed earlier, the use of DOE/Taguchi in typical projects involves five distinct phases: I, planning the experiments (brainstorming); II, designing the experiments; III, conducting the experiments; IV, analyzing the results; and V, confirming the results predicted. You can get an understanding of the tasks involved in these five phases by reviewing the following case study.

CASE STUDY 4.1: PART STRENGTH STUDY

The manufacturer of a popular brand of lawn mower has evolved a new design for a plastic engine cover. Because of concerns about durability and vibration loosening, the supplier was asked to exceed a specified part strength characteristic along with meeting other geometry requirements. The manufacturer's quality department also pointed out that the part strength is the primary cause of field failure and that the supplier is required to institute continuous long-range product design improvement strategy to reduce part variations in future years. Inspired by customer emphasis on improvement of part strength, the supplier responded quickly by undertaking a project to optimize the engine cover molding process using DOE. A team was formed consisting of personnel from design, production, sales, and customer relations and a project leader appointed. The leader followed the five application phases described here.

Phase I: Planning the Experiments

Purpose: To agree on objectives and identify factors and levels.

Part strength was considered as the single goal in everybody's mind. It was agreed that part strength would be measured in terms of force in pounds needed for a deflection $\frac{1}{8}$-inch at a predefined part location. All were also in agreement that a higher transverse load is indicative of an overall increase in the strength of the part. Brainstorming about the possible factors resulted in a list of seven factors:

1. Resin type
2. Mold temperature
3. Injection pressure
4. Set time
5. Filler level
6. Dye cleanliness (uncontrollable)
7. Time of the day (uncontrollable)

Among the factors, dye cleanliness and time of the day were recognized as uncontrollable. Frequent cleaning of the dye is a time-consuming proposition. To meet the high volume demanded, it was considered necessary to run the production line 24 hours a day. Ultimately, the process must have to be insensitive to the influence of environmental changes during the day. But for now, running a simpler set of experiments and establishing that DOE is a feasible way to optimize the process were considered the right thing to do. It was agreed that the experiments would be conducted under a real production environment, which can only be halted for half a day. No more than five separate experiments could be carried out in that period of time.

To place a limit on the number of experiments led to selection of an L-4 orthogonal array for the design. It also meant that only three factors, all of which must be at two levels, could be included in the study. After a brief discussion of the factors, *injection pressure*, *mold temperature*, and *set time* were considered for study. The factors *filler level* and *resin type* are known to have an influence, but the participant from the purchasing department indicated that contracts with suppliers would prohibit changes in these two factors in the near future.

By selecting an L-4 array for the design, the three factors included in the study have to have two levels each, as the L-4 array has 3 two-level columns. The two levels of each factor were selected at values below and above the current operating levels. For injection pressure, for example, the current production level is set at 300 lb/in^2. A production machine operator in the team recalled past experience with pressure in the range 250 to 350 lb/in^2, which produced fairly good products. So the levels of injection pressure were defined to be 250 and 350 lb/in^2, respectively. The level values of the other two factors were decided similarly (Table 4.8).

TABLE 4.8   Factors and Levels for Case Study 4.1

	Level	
Factor	1	2
A: injection pressure (lb/in^2)	250	350
B: mold temperature (°F)	150	200
C: set time (seconds)	6	9

Phase II.   Designing the Experiments

Purpose: To determine the number of experiments to run.

Selecting the array, assigning factors to columns, and then describing the experimental combination constitutes the experiment design. The array selected is an L-4 which has three columns. The 3 two-level factors can now be assigned to the three levels of the array (Table 4.9), which then allows description of the four experimental conditions. The experimental design is easily accomplished using the QT4 software. Run QT4 and go to the main screen. You need not open any particular file. Select *Inner array* in the *Manual design* option from the *Design* menu (Figure 4.12). The term *inner array* refers to the array used for the experimental design. Note that designs can be achieved using either the manual or automatic design option. The manual design option is preferred here, as it promotes clearer design methodology. Whenever a new design is started, QT4 warns you of possible loss from memory of the current experiment file (Figure 4.13). Click *Yes* at this prompt. In the *Array selection* screen, check the box next to L-4 (Figure 4.14) and click *OK* to proceed to the *Inner array design* screen. At the *Inner array design* screen, describe the three factors and their levels as shown in Figure 4.15. Click on the empty factor description box before you start typing. If you mistype any description, use the *Reset col* or *Delete cell* buttons as appropriate. Note that the term *column* in the design screen (Figure 4.15) refers to the column of the orthogonal array to which the factors (occupying rows here) are assigned. Click *OK* when done.

The next screen allows you to provide four lines of project title and description (Figure 4.16). The information in this screen is optional. Click *OK* to proceed with or

TABLE 4.9   Experiment Design with 3 Two-Level Factors

	Factor		
Experiment	A	B	C
1	1	1	1
2	1	2	2
3	2	1	2
4	2	2	1

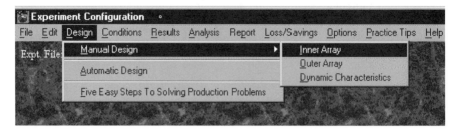

Figure 4.12 Manual design selection menu.

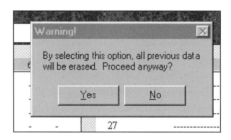

Figure 4.13 Warning at start of all new designs.

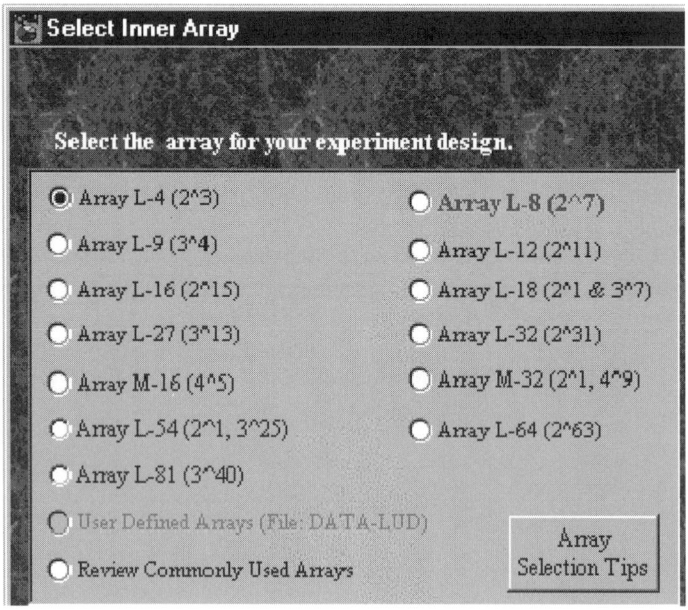

Figure 4.14 Array selection for experimental design.

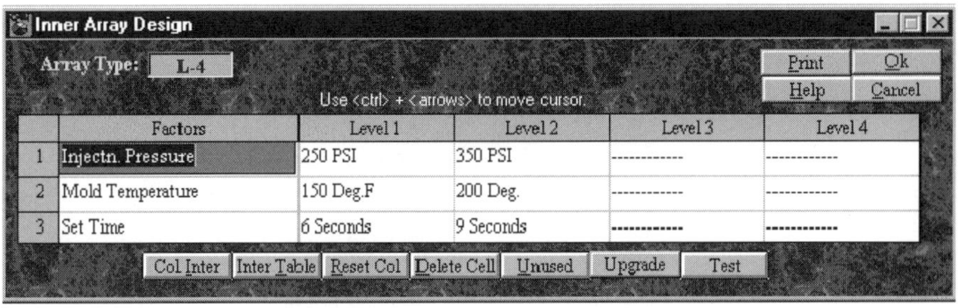

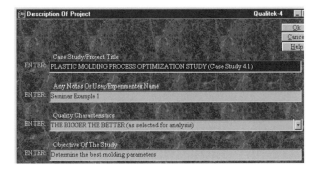

Figure 4.15 Factor assignment for experimental design.

without all the information supplied. Notice that you can click the scroll bar at the right corner of the quality characteristic arrow to set the desired quality characteristic (QC). The orthogonal array used for the experiment is displayed next for your review (Figure 4.17). The array shown should have the standard value of the table unless there are some unused columns (not present in this experiment). If, for some reason, the array does not look correct, click on the *Reset array* to revise the column to its original description. Click *OK* to move to the file-saving screen. At the *Save file as* screen shown in Figure 4.18, click *Cancel* to return to the main screen without saving the file. You will not need to save this file since it has been already created and is provided with the program. (Also, be aware that the QT4 version accompanying this book allows you to save experiment files that are created using an L-8 array only.) Click *OK* in the prompt about the file not being saved shown in Figure 4.19.

When in the main screen, select the *Open* option from the *File* menu (Figure 4.20). In the *Open experiment file* screen, make sure that you are in the *Usrfiles* subdirectory of the Q4W directory where the QT4 program is installed (see Figure 4.21). Use the scroll bar to go to the file name display list and look for *SEM-EX01.Q4W*. Select this file. Click *Open* to load this file in the program memory. This takes you back to the main screen. Back in the main screen, select the *Trial condition* option from the *Conditions* menu item as shown in Figure 4.22.

Figure 4.16 Project description for Case Study 4.1.

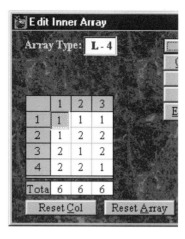

Figure 4.17 Orthogonal array used for the design.

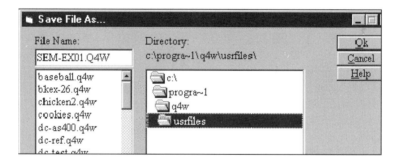

Figure 4.18 Option to save experiment file.

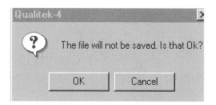

Figure 4.19 Prompt to alert users when file is not saved.

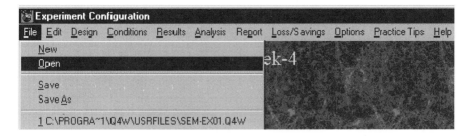

Figure 4.20 Option to open experiment file.

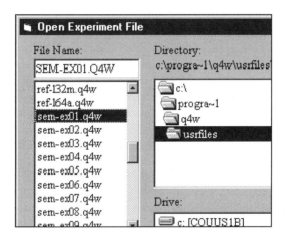

Figure 4.21 Loading experiment file for Case Study 4.1.

Figure 4.22 Option for experimental conditions.

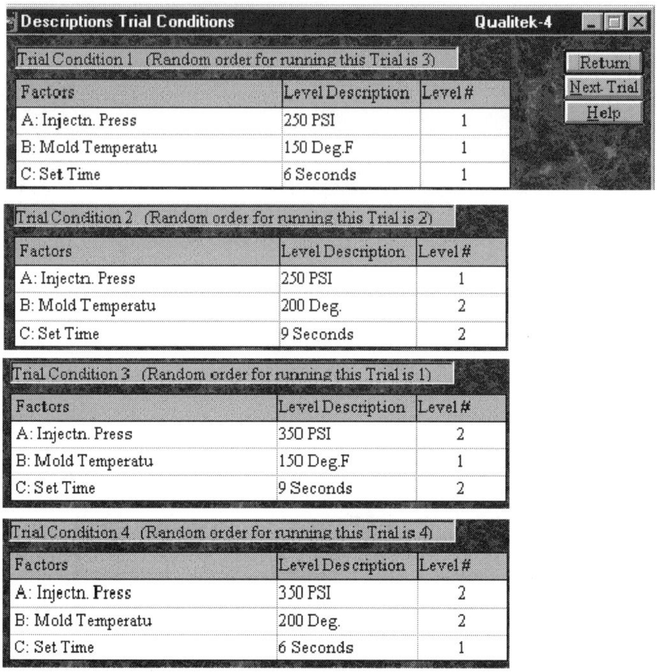

Figure 4.23 Experimental conditions for Case Study 4.1.

The four conditions of the experiments designed using an L-4 orthogonal array are shown one at a time (Figure 4.23) in the *Description of trial conditions* screen. (Consistent with traditional practice, the individual experimental conditions are usually referred to as trial conditions. *Experimental condition* and *trial condition* are used synonymously in this book.) Notice that the trial conditions are numbered in sequence, and each has a random order number indicating its sequence in the experiment (discussed next in this step). You can click *Next trial* to see other trials. You should exercise the *Print all* option to get hard-copy prints of descriptions of all the experimental conditions, which should match those shown in Figure 4.23. Click *Return* to go back to the main screen. Note that you may obtain the description of the four experiments by reading the rows of the array and interpreting the level notation for the factor levels.

Phase III: Conducting the Experiments

Purpose: To determine the sequence of running the experiments.

For the most part, tests are to be carried out in a manner that best simulates the real-life application environment. For process optimization applications, tests must be conducted in situations that closely match the production environment. If product design optimization is the goal of the project, experiments should be carried out under conditions that best simulate those during user applications. But these are not unique to

DOE. The only requirement of DOE is that experiments be carried out so as to minimize the chance for experimental or setup error.

In experimental studies, there are generally two types of experimental error: between-experiment error and within-experiment error. *Between-experiment error* is the error in the results associated with different setups. *Within-experiment error* is the error associated with repeating experiments (samples) under the same conditions. The way to keep both these types of error to a minimum is to conduct experiments in random order.

Number of Samples to Test under Each Experimental Condition. The minimum number of samples to test under each experimental condition is one. The greater the number of samples, the better the quality of information. Multiple samples are a necessity when variation from sample to sample is expected. In advanced experiment design, studied later in the book, the number of samples per experimental condition is dictated by the design itself. For simpler designs such as the one in this case study, the number of samples is decided arbitrarily by considering the cost and expected variability of results. Cheaper test samples and more variability would require more samples.

To understand the order of running experiments and the considerations recommended, suppose that in the L-4 experiment, three samples are available for test under each experimental condition. This means that there are 12 (= 4 × 3) samples for the project (Table 4.10). How should you run the tests? Which one should you run first and which second?

Naturally, there are many possible sequences for testing all 12 (S11, S12, . . . , S43) samples. The most desirable is to select the entire sample in random order (i.e., write the names of samples on 12 pieces of paper and blindly select one at a time). Conducting a test in random order is called *replication*. Replication is the most random way of testing the experimental conditions. The order in which a test is run is determined by random selection of the test specimen from among the total samples. Tests following the replication method may pose some practical difficulties. Suppose that in selecting the order, sample S33, which is part of experiment 3, is picked up first (Table 4.10). This requires that mold temperature (factor B at level 1) be at 150°F. Assume that sample S21, which is part of experiment 2, is the next in order. This condition requires that mold temperature (factor B at level 2) be set at 200°F. To set the test con-

TABLE 4.10 Planned L-4 Experiments with Three Samples in Each Condition

Experiment	Factor			Test Samples		
	A	B	C			
1	1	1	1	S11	S12	S13
2	1	2	2	S21	S22	S23
3	2	1	2	S31	S32	S33
4	2	2	1	S41	S42	S43

dition correctly (i.e., to elevate the temperature of the molding compound) may take some time. This would definitely necessitate having some extra time on hand. Who knows, the third experiment might even be randomly selected as one of the remaining samples from experiment 3, which would then call for lowering the temperature. Strict adherence to replication can become quite time consuming.

When a quick change of factor level is time consuming or otherwise difficult, a common practice is to use *repetition*, in which trial conditions are selected randomly, then all samples under the same trial condition are handled one after the other. The order of preference should be replication first, and if that is not possible, then repetition, and recognize that you may not be able to follow any such order. The best you can do is test samples as they become available. If that's how you can complete the experiments planned, so be it— *replication* or *repetition* is nice to have, not something you have to have.

Following the random order, all experiments in Case Study 4.1 were carried out and results of the test samples were measured in terms of the load measurement procedure determined in the planning session. The average of the three samples tested in each trial condition were recorded in a column to the right of the array as shown in Table 4.11. (The treatment of individual sample results is described in later steps.)

Phase IV: Analyzing the Results

Purpose: To determine the best design condition and influence of factors.

The results collected from the experiments just completed contain information that delineates the primary reasons for the experiment, and more. But the results (raw data) collected from the test samples have to be analyzed to develop key information with statistical validity. In the context of our discussion here, the term *statistical* simply signifies that conclusions and observations made from the data would be expected to be right more often than not. The experimental results 30, 25, 34, and 27 (Table 4.11) average the force measured from all samples tested in the four trial conditions. The notation Y_1, Y_2, and so on, is used to designate results in the trial conditions indicated by the corresponding subscripts and is useful when manual calculations of results are needed.

Analyses of the results produce information that can be grouped into two types. In the first is information obtained by simple arithmetic calculations. The second category requires a little more understanding of the statistical calculations.

TABLE 4.11   Experimental Design Array with Results (Case Study 4.1)

| Trial | Factor | | | Results |
	A	B	C	
1	1	1	1	30 (Y_1)
2	1	2	2	25 (Y_2)
3	2	1	2	34 (Y_3)
4	2	2	1	27 (Y_4)

Information Type 1

- Average factor effects (main effect)
- Optimum condition
- Estimated performance at the optimum condition

Information Type 2

- Relative influence of factors
- Confidence interval on optimum performance
- Test of significance of factor influence

Although the approach followed in this book is to use the software to carry out all analyses, and indeed QT4 instantly produces all of the output above as part of the standard analysis tasks, we must concentrate on understanding the simpler data manipulations first. Our focus in analysis of this experiment (Case Study 4.1) will be to obtain the first type of information. The second type is only introduced here; it is covered in later steps.

To enter and analyze the results of experiments, run QT4 and open (*Open* option from the *File* menu) experiment file *SEM-EX01.Q4W* (.QAW will be common to all files and may be dropped from the description). The results are entered using the *Enter results* option from the *Result* menu item. You will notice that the results of the experiments are already entered. You may now proceed to analyze the results. To do so, select the *Standard analysis* option from the *Analysis* menu item (Figure 4.24). Standard analysis is most common when there is just one column of results, as in this case. This can be true because there is only one test sample per trial condition or because averages of several sample results have been used for analysis purposes. Click *OK* to proceed to the next input screen for analysis. (Always click *OK* to proceed to the next step in analyses.)

There are some calculations in analyses that depend on the quality characteristic (QC) of the results, but many are unaffected by it. QT4 requires that you indicate the type of QC applicable. Of course, the results can be analyzed at different times by choosing different (bigger, smaller, or nominal) QC. Check the box next to the *Bigger is better* QC and proceed (Figure 4.25). QT4 completes a standard set of analyses without additional input from the user. As part of the standard analysis output, four

Figure 4.24   Menu option for analysis of results.

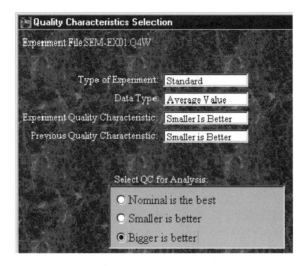

Figure 4.25 Quality characteristic selection for analysis.

screens are displayed in sequence as you click *OK* to proceed from screen to screen. The first screen shows the results of each trial condition and their averages. The second screen displays the calculated average effects of factors, which indicate their trends of factor influence. The third screen shows information obtained by statistical calculation known as analysis of variance (ANOVA). The last screen lists information about the most desirable (optimum) condition based on the QC and the estimated performance in this condition. Each screen also offers a number of options to perform additional analyses and displays.

The results of the experimental conditions and trial averages are shown in Figure 4.26. The graph of the results can also be displayed and printed by clicking on *Graph*

Experimental Results							Qualitek-4	

Expt. File: SEM-EX01.Q4W Data Type: Average Value Ok
QC Type: Bigger is Better Canc
Help
Prin

	Sample# 1	Sample# 2	Sample# 3	Sample# 4	Sample# 5	Sample# 6	Averages
Trial# 1	30						30
Trial# 2	25						25
Trial# 3	34						34
Trial# 4	27						27

Figure 4.26 Results and average of trial results (Case Study 4.1).

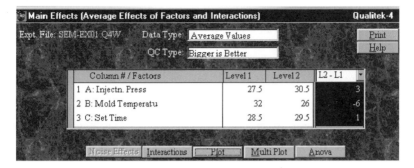

Figure 4.27   Factor average effects (main effects).

in this screen. These results, along with the experiment design information, are used to calculate the average effects of factors at each level, as shown in Figure 4.27. Click on *Multiplot* to show the average effects graphically.

Method for Calculating Average Effects of Factor Levels.   The average effect of a factor at a level is calculated by examining the orthogonal array, the factor assignment, and the experimental results shown in Table 4.11. A character with an overbar is used to express the numerical value of the average effect. For example, to calculate the average effect of factor A at level 1, all results of factor A at level 1 are averaged. The results of trial conditions 1 and 2 are those that have level 1 of factor A. Thus the results of conditions 1 and 2 are averaged. Calculations of the average effects of all three factors follow.

$$\overline{A_1} = \frac{Y_1 + Y_2}{2} = \frac{30 + 25}{2} = 27.5$$

$$\overline{A_2} = \frac{Y_3 + Y_4}{2} = \frac{34 + 27}{2} = 30.5$$

$$\overline{B_1} = \frac{Y_1 + Y_3}{2} = \frac{30 + 34}{2} = 32.0$$

$$\overline{B_2} = \frac{Y_2 + Y_4}{2} = \frac{25 + 27}{2} = 26.0$$

$$\overline{C_1} = \frac{Y_1 + Y_4}{2} = \frac{30 + 27}{2} = 28.5$$

$$\overline{C_2} = \frac{Y_2 + Y_3}{2} = \frac{25 + 34}{2} = 29.5$$

where A_1, B_2, etc. are the factor levels and Y_1, Y_2, etc. are the results at different experimental conditions.

Plot of Average Effects.   An average plot is obtained by plotting the average factor-level effect (numerical value of results along the *y*-axis) against the corresponding factor level. It is used primarily for the selection of a level (from among those levels of the factor tested) of best performance. The plot makes obvious to the experimenter two additional pieces of information about the factor. It shows the nature of the trend of influence of the factor to the result as it changes from level 1 to level 2. Second, proportional to the slope or the difference between endpoints, it indicates the variation in results for the shift in factor levels, which indicates the sensitivity of the factor's performance (results) (Figure 4.28).

Main Effect, Factorial Effect, or Column Effect.   These are synonymous terms used to refer to the trend of change of the average effect of factors. The main effect is generally expressed by the difference in the average effects at the two levels (for a two-level factor) or by plotting the average effect. It is expressed numerically and is computed by subtracting the first-level effect from that at the second level.

$$\text{Main effect of factor } A = \overline{A}_2 - \overline{A}_1 = 30.5 - 27.5.$$

Note that you should always draw a graph if the factor has three or more levels.

While in the *Main effects* screen, you can plot the main effects of several factors on one screen or of a single factor at a time by clicking the appropriate buttons from the screen. The graph of the main effects of the three factors is shown in Figure 4.29. Note that the last column in the average effects table in Figure 4.27 indicates the difference between the average effects, which give an indication of the slope of the factor influences (main effects), as shown in Figure 4.29. From a plot of main effects, the following observations can be made:

- Level 2 of injection pressure produces a higher (QC = bigger) result.
- Level 1 of mold temperature is better.
- Level 2 of set time produces a higher result.

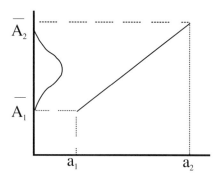

Figure 4.28   Plot of factor average effects and variability associated with slope of main effects.

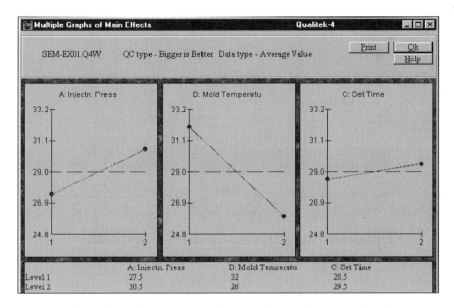

Figure 4.29   Plot main effects of the three factors for Case Study 4.1.

Click *OK* to return to the *Main effects* screen. Click *OK* (again or the *ANOVA* button) to proceed to the *ANOVA* screen. Based on the results and the experiment design (array and factor location), QT4 performs an ANOVA and displays the standard information. An ANOVA table contains information about the entire project, such as influence of individual factors and influence of factors other than those included in the study. The information contained in the ANOVA table about the project is analogous to information that an x-ray provides about a person's health. You will learn about ANOVA slowly as we cover a few more examples and read through Step 7. At this point, you should simply know that the last ANOVA column (Figure 4.30), indicates the relative percentage of influence of factors to the variation of results. You can obtain a plot of this information (Figure 4.31) by clicking either *Bar graph* (or *Pie chart*) from this screen.

Click *Return* when done reviewing the bar graph. Back in *ANOVA* screen, click *OK* (or *Optimum*) to proceed to the *Optimum* screen. The last of the four analysis screens is the screen containing information about the optimum condition (the most desirable factor-level combination) and the estimated performance in this condition (Figure 4.32). The optimum condition is described as the factor-level combination listed in the second column, which corresponds to the level numbers listed in the third column (a_2, b_1, c_2). The decision about which level of the factor is desirable is made by comparing the average factor effects and selecting the level based on the QC. This selection is consistent with the key observation made from a review of the plot of the main effects.

Col # / Factor	DOF (f)	Sum of Sqrs. (S)	Variance (V)	F - Ratio (F)	Pure Sum (S')	Percent P(%)
1 A: Injectn. Press	1	9	9	-----	9	19.565
2 B: Mold Temperatu	1	36	36	-----	36	78.26
3 C: Set Time	1	1	1	-----	1	2.173
Other/Error	0					
Total:	3	46				100.00%

ANOVA Table — Qualitek-4

Expt. File: SEM-EX01.Q4W   Data Type: Average Value   QC Type: Bigger is Better

[Main Effects] Pool Factor Auto Pool Unpool All CI Function Bar Graph Pie Chart Optimum

Figure 4.30   Analysis of variation for factors of Case Study 4.1.

Grand Average of Performance.   The average of all trial results is called the *grand average of performance.* This is a theoretical number and may or may not represent the actual average performance of the product or process under study. However, when the current performance is unavailable, it is used as a reference to compare and express improvements achieved.

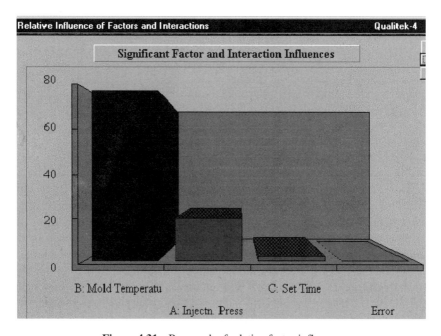

Figure 4.31   Bar graph of relative factor influence.

Optimum Conditions and Performance				Qualite

Expt. File:SEM-EX01.Q4W   Data Type: Average Value   Print

QC: Bigger is Better   Help

Column # / Factor	Level Description	Level	Contribution
1 A: Injectn. Press	350 PSI	2	1.5
2 B: Mold Temperatu	150 Deg.F	1	3
3 C: Set Time	9 Seconds	2	.5

Total Contribution From All Factors...   5
Current Grand Average Of Performance...   29
Expected Result At Optimum Condition...   34

Anova	Transform	C.I.	Report	Export Opt.	Estimate	Graph

Figure 4.32 Optimum condition and the performance estimated.

Factor Contribution. The factor contribution is the amount of improvement obtainable by setting the factor to the desired level. This improvement is measured relative to the grand average of performance (29 in this case; Figure 4.32). For example, the contribution that factor *A* (injection pressure) makes when it is set at level 2 is calculated by subtracting the grand average (29) from the average effect of factor *A* at level 2 (30.5), and is equal to 1.5 (= 30.5 − 29).

Result Expected at Optimum Condition. This is an estimate of performance at the optimum condition (Y_{opt}). In the simplest and most commonly used estimate, the performance expected is calculated by adding all improvements (contributions) from all factors to the grand average of performance. The contributions of all factors are additive only when all factor influences are linear. Since this is not always the case, the result calculated can only be expected to be *close* to the actual performance. The expected result for the problem being discussed is calculated as

$$Y_{opt} = \overline{T} + (\overline{A_2} - \overline{T}) + (\overline{B_1} - \overline{T}) + (\overline{C_2} - \overline{T})$$

$$= 29 + (30.5 - 29) + (32 - 29) + (29.5 - 29)$$

$$= 34$$

As shown in the last column, the contributions from all factors add up to make the total contribution. The grand average is increased (for QC = B) by the amount of total contribution and produces the expected result at the optimum condition (29 + 5 = 34). The expected result (34) means that when the molding process is set to the condition described by the optimum conditions (350 lb/in², 150°F, and 9 seconds), the average result will be close to 34.

Notice that the optimum condition (factor levels 2, 1, 2—third column in Figure 4.32) is the third trial (experiment 3) condition (see Figure 4.23). The chance that the optimum condition will be one of the experiments completed diminishes as the size of the experiment becomes larger. For an experiment designed with an L-4 array, four out of eight possible (full factorial) experiments are done. So in this case, there is a 50% chance that the optimum will be one of the conditions already tested. For L-8 experiments, the chance will be less than 7% (8 of 128).

Phase V: Confirming the Results Predicted

Purpose: To verify that the results predicted can be obtained.

As a rule, the optimum condition will not be one of the experiments already conducted as part of the plan. So it is a good idea to plan on running an additional few samples at the optimum condition. These confirmation tests serve two purposes. First, they establish the new performance at the new (optimum) condition, which can establish the improvement achieved. Second, they allow the experimenter to determine how close the estimate is to the results observed. The result expected (34 in this case study) is considered to be confirmed when the mean (or average) of a number of samples tested at the optimum condition falls close to it. But how close is close enough? The boundary (Figure 4.33) within which the result expected should fall is established by calculating the confidence interval (C.I.) (covered in later steps).

Case Study 4.1 Summary

- The trends of influence of injection pressure, mold temperature, and set time are known and used to determine the most desirable condition (the optimum condition).

- The optimum condition is determined to be at 350 lb/in², 150°F, and 9 seconds.

- When the process is set at the optimum condition, it is expected to improve performance by 17% (29 to 34).

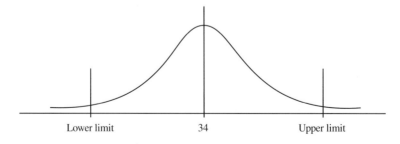

Figure 4.33   Boundary of confirmation of results expected.

SUMMARY

In this step you have learned how to design experiments using a set of orthogonal arrays. You have been aware that each of these orthogonal arrays can be used to design experiments for several different situations. By using a case study, you have seen all five phases of the application process and how the QT4 software can help you accomplish the design and analysis tasks of the experiments.

REVIEW QUESTIONS

Q. What are the five major phases in DOE/Taguchi?

A. Planning, design, conducting, analysis, and confirmation.

Q. What are the items discussed in a planning session?

A. The objectives of the study, factors and their levels, and other information that is necessary for experimental design.

Q. How are factors determined, and when are they discussed in the planning session?

A. Factors are determined by first listing all possible ideas suggested by the team members, then by scrutinizing each candidate to see if it qualifies to be a factor. Finally, the number of factors that can be included in the study is determined based on the scope of the experiment. In the planning session, discussion of the factor must always follow that for the objectives and how they are evaluated.

Q. What considerations dictate how many factors can be included in an experiment?

A. The cost of running the experiments and time available for the project usually dictate the size of the array that can be used to design experiments. The size of the array then limits the number of factors that can be included in the study.

EXERCISES

4.1 To design an experiment (i.e., to lay out the conditions of an individual trial) Taguchi uses a set of specially constructed orthogonal arrays, such as L-4, L-8, L-9, and so on.

    (a) If you used an L-8 compared to a full factorial, what will be your savings in numbers of experiments?

    (b) If you decide to perform a one-factor-at-a-time experiment, what would be the fewest number of experiments that you would need to study 7 two-level factors?

    (c) Experiments using orthogonal arrays (OAs) require about the same number of trials as in one-factor-at-a-time experiments. For which of the following reasons is OA preferred? Select all appropriate answers.

(1) Columns of OAs are balanced (i.e., there are an equal number of levels in a column).

(2) Factorial effects determined by OA design are more reproducible.

(3) Experiments using OAs produce the best possible factor combination.

4.2 An L-8 array is used to study five factors, each of which has two levels.

(a) How many trial conditions will you have to run (at a minimum)?

(b) Where would you place these factors?

(c) How do you treat the unused columns?

4.3 In an experiment involving an L-8 array, each trial condition is repeated five times.

(a) What is the total number of test runs involved?

(b) What is the most desirable way to run these tests?

4.4 In which of the following ways is the average effect of a factor (say, A) at a level A_1 determined? Select all appropriate answers.

(a) Averaging all results containing A_1.

(b) All results containing A_1 and only level 2 of all other factors.

(c) All results containing A_1 and only level 1 of all other factors.

4.5 Based on your understanding of orthogonal arrays and their properties, answer the following questions.

(a) Is the array shown in Table 4.12 fully orthogonal? If not, explain why.

(b) What are the numbers in the first row of an L-9 array?

(c) What number is in the ninth trial and first column of an L-16 array?

(d) How many columns are there in an L-12 array?

(e) What is the sum of all levels in the fifteenth column of an L-16 array?

(f) What is the sum of all levels in the first row of an L-32 array?

EXERCISE ANSWERS

Use QT4 capability to solve exercises when appropriate.

4.1 (a) You will be carrying out eight experiments instead of the 128 required for the full factorial experiment. The resulting savings will be 120 experiments.

(b) One-factor-at-a-time experiments would also need eight experiments $(7 + 1)$.

(c) OA is preferred for reasons (1) and (2).

4.2 **(a)** Eight trial conditions, since the experiment will be designed using an L-8 array.

(b) Factors may be assigned arbitrarily: say, columns 1 to 5.

(c) Unused columns are eliminated (or level numbers are changed to zero for all unused columns).

4.3 **(a)** The total number of samples is 40 (8×5).

(b) The most desirable way to run these samples is by replication.

4.4 Answer (a) is correct.

4.5 **(a)** Yes; **(b)** four 1's; **(c)** 2; **(d)** 11 columns; **(e)** 24; **(f)** 31.

Experimental Design with Two-Level Factors Only

What You Will Learn in This Step

A large number of experiments are done naturally with all factors at two levels. You will learn how the set of two-level orthogonal arrays introduced earlier is used to accomplish the design. You will be familiar with how each array is used to design experiments with a different number of factors. You will also practice how to accomplish experimental designs when you have a group of factors that are all studied at two levels.

Thought for the Day

> It is the quality of our work which will please God and not the quantity.
>
> —Mahatma Gandhi

In Steps 3 and 4 you have learned how experiments can be designed using orthogonal arrays and how to analyze the results of the experiments. The analysis technique discussed so far (standard analysis) is the same for experimental designs of all kinds. It is the design that will vary depending on the number of factors involved. In this step and Steps 6 and 7, you will learn how each array is used for a wide range of experimental situations. Complete analysis in case studies and examples is introduced only to reinforce your analysis capability.

TWO WAYS TO USE AN L-4 ORTHOGONAL ARRAY

Use an L-4 array to design experiments with:
1. 2 two-level factors
2. 3 two-level factors

Note: Factors are assigned to the three columns arbitrarily.

The L-4 orthogonal array has 3 two-level columns and is used to design experiments with either 2- or 3 two-level factors. If there are three factors (say A, B, and C), all columns of the array will be occupied (Table 5.1). The order in which the factors are as-

136

TABLE 5.1 Column Assignment for Experiment with 3 Two-Level Factors

	Factor (Column)		
Experiment	A (1)	B (2)	C (3)
1	1	1	1
2	1	2	2
3	2	1	2
4	2	2	1

signed to the columns is immaterial at this point. If, on the other hand, you have only two factors (A and B), you should place them in columns 1 and 2, leaving column 3 empty (Table 5.2). Keep in mind that when you are testing two factors at two levels each, the maximum possible number of ways they can be combined is four, which is what an L-4 experiment will do in this case. When a column is not used, as is the case for the design with 2 two-level factors shown in Table 5.2, the unused column is shown with all levels set to zero. If you are doing this by hand on a piece of paper, you may simply erase this column from the array.

Let's now see how these experiments are designed using QT4. You already know how QT4 selects the array and designs experiments in the *Automatic design* option. In this step you will also learn how to design experiments using the *Manual design* option.

Example 5.1: Brake Rotor Experiment Material properties play an important role in an automobile disc brake operation. Material engineers involved in a brake noise reduction study wish to run a few short experiments to investigate the effects on brake noise of rotor surface finish and its hardness. A third factor, pad thickness, is also identified for follow-up study, if needed. The factor levels tested are as follows:

TABLE 5.2 Column Assignment for Experiment with 2 Two-Level Factors

	Factor (Column)		
Experiment	A (1)	B (2)	(3)
1	1	1	0
2	1	2	0
3	2	1	0
4	2	2	0

 A: rotor surface finish [50 and 100 microinches]

 B: rotor hardness [205 and 225 BHN (Brinell hardness number)]

 C: pad thickness [Current thickness and thicker (optional)]

The objective of the study is to see which of the two material properties has the greater influence on operating noise.

Solution: All experiments that you will need to conduct can be designed using the *Manual design* option of QT4.

Automatic design. Run QT4 and go to the main screen (*Experiment configuration* screen). From this screen, select the *Automatic design* option from the *Design* menu item. When at the *Automatic experiment design* screen, check the box at the bottom for all two-level factors (Figure 5.1). When you have them all, you can either check the top box (as you did before) or the bottom one. The bottom option is especially designed for experiments with all two-level factors and interactions among them. In later steps you will learn about experiment design to include interaction among factors. But it would be a good idea to get used to this option now. Click *OK* to proceed to the next screen.

At the *Input of two-level factors* screen (Figure 5.2), enter descriptions of the three factors and their levels. The order in which you enter the factors is arbitrary. However, if you know or suspect that some factors are more important than others, enter the important factors first. The next screen (Figure 5.3) is for interaction selection and is not needed for this example. Click *OK* to proceed to the next screen. The factors described and their assignment to the columns of the array selected is shown on the screen in Figure 5.4. The column number just left of each factor shows the column to which the

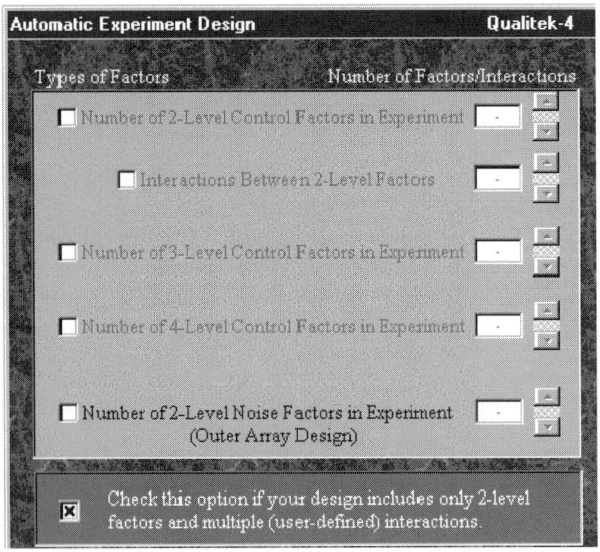

Figure 5.1 Automatic design input screen.

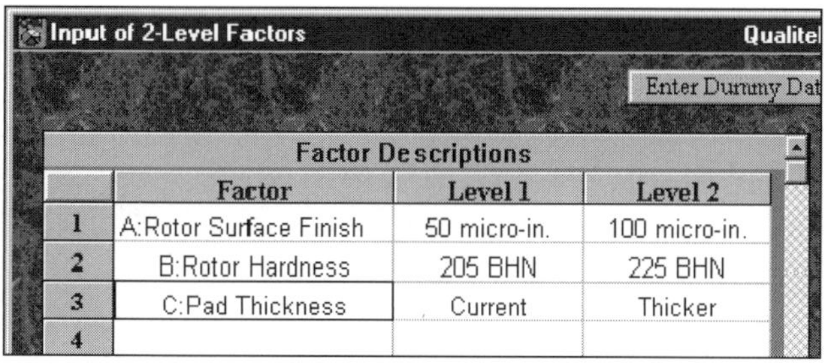

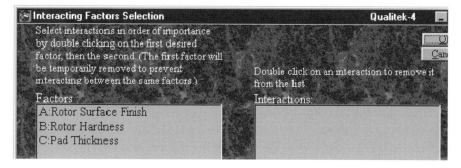

Figure 5.2   Factor description for automatic design.

Figure 5.3   Interaction selection screen.

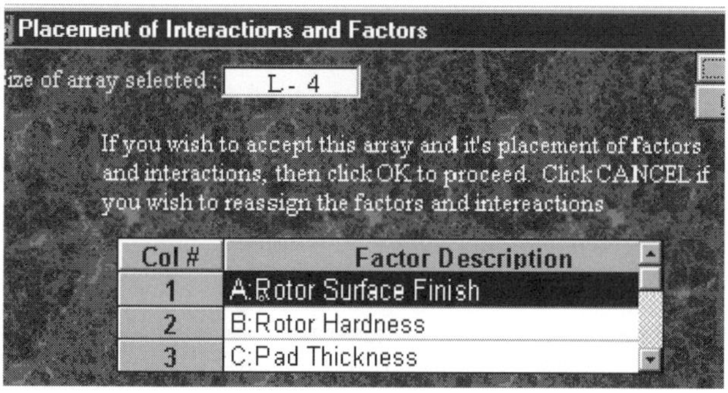

Figure 5.4   Factor assignment for the experiment of Example 5.1.

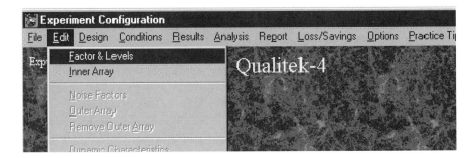

Figure 5.5   *Factor & level* editing option selection.

factor is assigned. The array selected for the experiment is also indicated in the box in the top part of the screen. To prepare for designing this experiment using the *Manual design* option, click *Cancel* several times to return to the main screen. There may not be any data shown on the main screen, and it does not matter which file is loaded into the memory of QT4 at this time.

Manual design.   The experiment for this example (with two factors) is already created and saved as *BKEX-51.Q4W*. Open this experiment file from the *File* menu. To review factor and level descriptions, select *Factor & levels* from the *Edit* menu (Figure 5.5). For the initial study, the experiment was designed to investigate only the first two factors (A and B). You can practice adding or modifying any description in this screen (Figure 5.6). To add the third factor, click on the third line (indicates column 3 of the array), on *Column unused*, then click on the *Reset col* button at the bottom of the screen. This action clears the column assignment and lets you type the factor description as shown in Figure 5.7. Try editing temporarily (for practice only) any factor or level descriptions by first clicking on

Inner Array Design

Array Type: L-4

Use <ctrl> + <arrows> to move cursor.

	Factors	Level 1	Level 2	Level 3	
1	A:Rotor Surface F	50 micro-	100 micro-in.	------------	--
2	B:Rotor Hardness	205 BHN	225 BHN	------------	--
3	COLUMN UNUSED	*UNUSED*	------------	------------	--

Col Inter | Inter Table | Reset Col | Delete Cell | Unused | Upgrade | Test

Figure 5.6   *Factor & level* description screen.

Inner Array Design			
Array Type:	L-4		

Use <ctrl> + <arrows> to move cursor.

	Factors	Level 1	Level 2	Level 3
1	A:Rotor Surface F	50 micro-	100 micro-in.	------------
2	B:Rotor Hardness	205 BHN	225 BHN	------------
3	C:Pad Thickness	Current	Thicker	------------

Col Inter	Inter Table	Reset Col	Delete Cell	Unused	Upgrade	Test

Figure 5.7 Factor and level descriptions.

the description and then modifying it as needed. If convenient, you can use the *Delete cell* button to erase a single cell completely and retype its description.

When you are done reviewing this screen, click the *Cancel* button to return to the main screen. Check out the four trial conditions in this experiment by selecting *Trial conditions* from the *Condition* menu item. These four conditions were tested and the results recorded in a subjective scale between 0 and 100 (Figure 5.8). The results were such that smaller numbers represented lower noise (QC = S).

To determine the effects of the factors studied, carry out the analysis using *Standard analysis* from the *Analysis* menu item. Be sure to select the *Smaller is better* characteristic for analysis. As you proceed by always clicking *OK* buttons, you will find

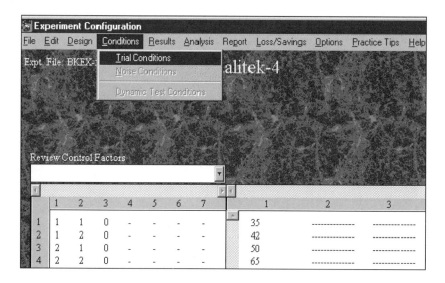

Figure 5.8 Main screen and option for trial condition.

Main Effects (Average Effects of Factors and Interactions)   Qu

xpt. File: BKEX-51.Q4W   Data Type: Average Values

QC Type: Smaller is Better

Column # / Factors	Level 1	Level 2	L2 - L1
1 A:Rotor Surface F	38.5	57.5	19
2 B:Rotor Hardness	42.5	53.5	11

Noise Effects | Interactions | Plot | Multi Plot | Anova

Figure 5.9   Main effects of factors.

the *Main effects* screen (Figure 5.9). Click on the *Multiplot* button in this screen to see the plots of the factor influences as shown in Figure 5.10. The plot of the factor effects (average effect or main effect; see Figure 5.10) shows the trend of influence of the factor as well as their relative influence to the variation of the result. By comparing the slope alone, it is evident that rotor surface finish, which has the steeper slope, is the most influential factor.

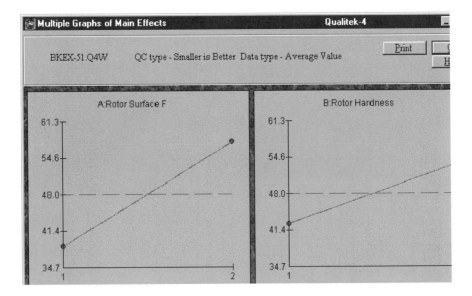

Figure 5.10   Plot of factor effects.

FOUR WAYS TO USE AN L-8 ORTHOGONAL ARRAY

Use an L-8 array to design experiments with:

1. 4 two-level factors
2. 5 two-level factors
3. 6 two-level factors
4. 7 two-level factors

Note: Factors are assigned to the seven columns arbitrarily.

The L-8 orthogonal array has 7 two-level columns and is the most widely used among all the arrays. It can be used for experimental situations over 15 different experimental situations, as you will learn before you finish this book. When all your factors have the same number of levels, L-8 array can be used to design experiments with 4, 5, 6, or 7 two-level factors. Using QT4's automatic design capability, you can easily check the design possibilities with an L-8 array. To see different experiment configurations, run QT4 and go to the main screen. Select the *Automatic design* option from the *Design* menu item. When you click *OK* at prompt, you will be in the *Automatic experiment design* screen (Figure 5.11). In this screen, check the box for two-level factors (top of the list of boxes) and click on the counter arrows to set the number of factors to be included in the experiment. The input screen for experiment design with 4 two-level factors is shown here.

Based on the number of factors, QT4 selects the array to be used for the design. As expected, the array selected for 4 two-level factors is an L-8, as indicated in Figure 5.12. At this point you should click *No* (Figure 5.12) so that QT4 may return you to the *Automatic experiment design* screen. You may now repeat designing experiments by 5, 6, and 7 two-level factors. Notice that in each of these cases, the array selected is an L-8.

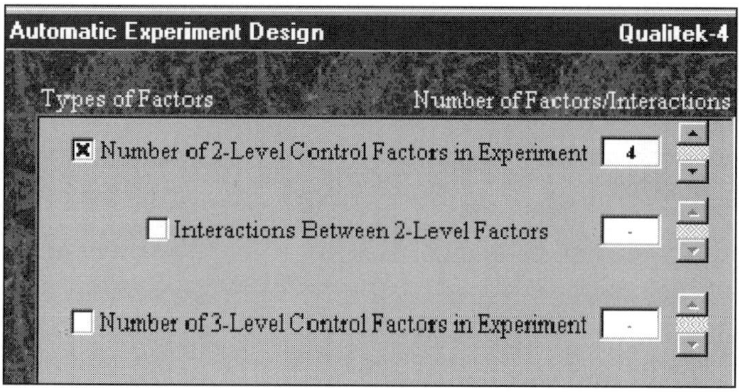

Figure 5.11   Automatic experiment design input screen.

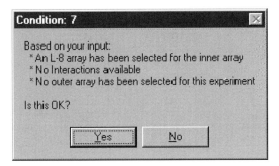

Figure 5.12 Array selected for 4 two-level factors.

Example 5.2: Joint Sealing System Experiment In an experiment with the sealing effectiveness of a movable joint, the following 6 two-level factors were studied:

A: seal material (rubber and paper)

B: articulation angle (5 and 7.5°)

C: surface finish (30 and 100 units)

D: grease type (polyurea and lithium)

E: bolted preload (20 and 30 newton·meters)

F: speed (2000 and 2500 rev/min)

The results of experiments obtained by assigning factors to columns 1 through 6 measured in terms of loss of lubricant are 40, 50, 62, 70, 60, 90, 45, and 65 milligrams. Assuming that the current average loss is near the first trail result (40), what condition is likely to produce the least amount of lubricant and the percentage of expected improvement?

Solution: Run QT4 and go to the main screen. To design the experiment, select the *Manual design–Inner array* option from the *Design* menu item (Figure 5.13). At the prompt (Figure 5.14) click *OK* to proceed to the array selection screen shown in Figure 5.15. Check the box next to the L-8 array, as this is the array needed for the design.

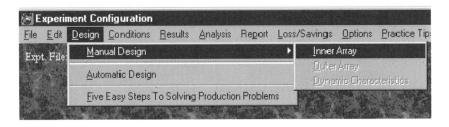

Figure 5.13 Manual design selection option.

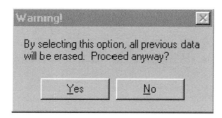

Figure 5.14 Prompt for new experiment design.

Note that when you are using the *Manual design* option, you will have to select the array and decide where the factors will be assigned. Click *OK* to proceed to the *Inner array design* screen (Figure 5.16). In this screen you will enter a description of the factors and levels in the columns of your choice. As the problem stated in this case, factors *A* to *F* will be assigned to columns 1 to 6 and column 7 will be left empty. Describe all factors one at a time as shown in Figure 5.16. After you have entered descriptions of all factors and levels, identify the seventh item, which is column 7 of the array, as unused. To do so, click anywhere in the seventh line, then click on the *Unused* button. You should never attempt to key in the words. Click *OK* to move to the screen to input experiment titles (Figure 5.17).

In this screen you can enter four lines of information related to your experiment (Figure 5.17). It is unimportant what you write in these four spaces, because they do not enter into any calculation. They are strictly for your own reference. Be sure to recognize from the problem statement that the objective is to identify the condition for the least amount of lubrication loss which will make you select the *smaller is better*

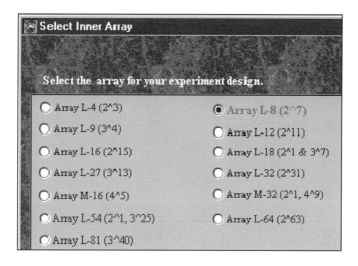

Figure 5.15 Array selection screen.

Inner Array Design

Array Type: L-8

Use <ctrl> + <arrows> to move cursor.

	Factors	Level 1	Level 2	Level 3	
1	A:Seal Material	Rubber	Paper	-----------	--
2	B:Articulation Angle	5 degrees	7.5 degrees	-----------	--
3	C:Surface Finish	30 units	100 units	-----------	--
4	D:Grease Type	Polyurea	Lithium	-----------	--
5	E:Bolted Preload	20 Nm	30 Nm	-----------	--
6	F:Rotational Speed	2000 rpm	2500 rpm	-----------	--
7	COLUMN UNUSED	*UNUSED*	-----------	-----------	--

| Col Inter | Inter Table | Reset Col | Delete Cell | Unused | Upgrade | Test |

Figure 5.16   Factor and level description screen.

quality characteristic. You could save typing by electing to click on the option icon at the right end of the third line and selecting the QC desired.

The next screen will show you the array used for experiment design (Figure 5.18). QT4 always turns the unused column to zeros (column 7), even when experiments are designed using the *Manual design* option. Notice that the bottom of each column shows the numerical totals of the levels in the column. These numbers are only for your visual awareness. In case of L-8, each column has eight numbers, four 1's and four 2's. The columns must all then have "12" as totals. Should you observe any dis-

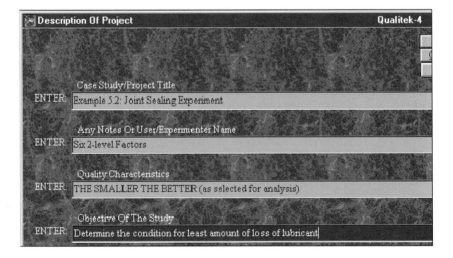

Figure 5.17   Experiment title input screen.

Figure 5.18   Orthogonal array for the experiment of Example 5.2.

crepancies, you should click on the *Reset Array* button to refresh the screen with correct numbers in the columns.

When you are done reviewing the array and its columns, proceed to save the experiment file under a name of your choice. Name the file by using eight or fewer characters and click *OK*. You do not need to enter .Q4W at the end of the file name (see Figure 5.19). The experiment file will be saved in the *Usrfiles* subdirectory under the QT4 directory in your computer. You will return to the main screen after the file is saved. You should now be able to see the conditions of the eight experiments that are prescribed by the experiment designed. To review the experimental condition, select the *Trial conditions* option from the *Condition* menu item in the main screen (Figure 5.20).

Once in the *Description of trial conditions* screen (Figure 5.21) you can review all eight trial conditions one at a time by clicking on the *Next trial* button. Note that QT4 indicates the randomly selected order of running each trial condition. For example, the suggested order of running trial condition 1 is 6. This means that trial condition 1 should be tested as the sixth experiment. Of course, you can also determine this random order by randomly pulling eight pieces of numbered paper one at a time from an enclosed box. Click the *Return* button when done reviewing the trial conditions.

When carried out, these eight trial conditions produced the results given in the problem description. Before QT4 can analyze the results, you must enter them and save the experiment file. To enter the results of experiments, be in the main screen and select the *Enter results* option from the *Result* menu item. Begin entering results one result at a time in the appropriate row and column. If there is only one sample result, which it is in this case, the results should be entered in the first column (Figure 5.22). When you are done entering results, click *OK* to move on to the file update screen.

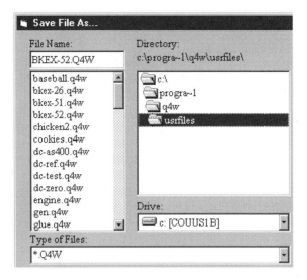

Figure 5.19 Experiment file saving option.

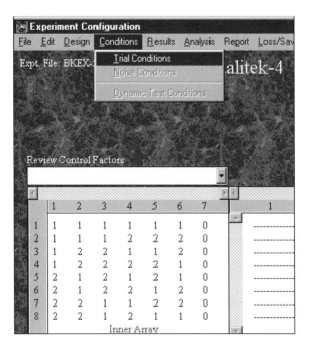

Figure 5.20 Option for trial conditions.

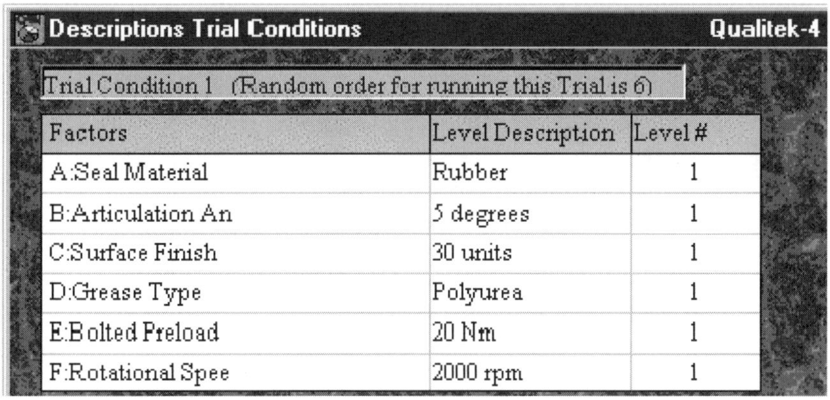

Figure 5.21   Description of trial condition.

Click *OK* here again to update the experiment file and return to the main screen. You are now ready to analyze the results.

To analyze results, select the *Standard analysis* option from the *Analysis* menu item and click *OK* to proceed to the following screens:

- *Quality characteristic* screen after checking *Smaller is better*
- *Experimental results* screen
- *Main effects* screen (click *OK* or *ANOVA* button)
- *ANOVA* screen (click *OK* or *Optimum* button)

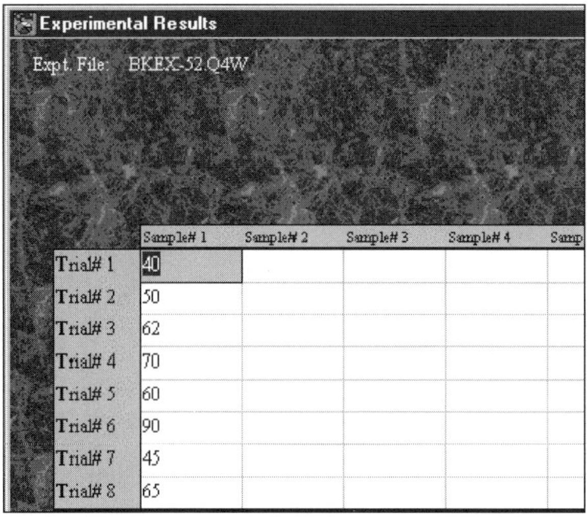

Figure 5.22   Results input screen.

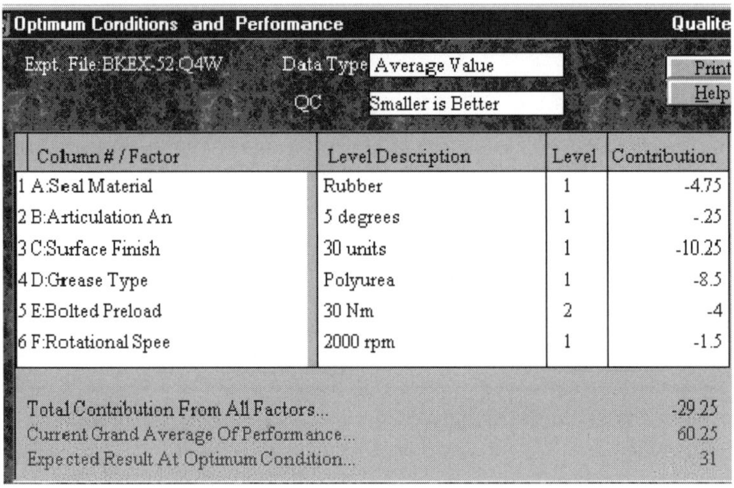

Optimum Conditions and Performance				Qualite
Expt. File:BKEX-52.Q4W	Data Type	Average Value		Print
	QC	Smaller is Better		Help
Column # / Factor	Level Description		Level	Contribution
1 A:Seal Material	Rubber		1	-4.75
2 B:Articulation An	5 degrees		1	-.25
3 C:Surface Finish	30 units		1	-10.25
4 D:Grease Type	Polyurea		1	-8.5
5 E:Bolted Preload	30 Nm		2	-4
6 F:Rotational Spee	2000 rpm		1	-1.5
Total Contribution From All Factors...				-29.25
Current Grand Average Of Performance...				60.25
Expected Result At Optimum Condition...				31

Figure 5.23   Screens for optimum condition and performance.

The four analysis steps above bring you to the *Optimum condition* screen, which contains information with respect to the best condition and improvement (Figure 5.23). The factor level description is listed in the second column (Figure 5.23). The expected performance at this condition (optimum) condition is 31, which represents $(40 - 31)/40 =$ a 22.5% improvement.

Example 5.3:   Adhesive-Bonded Joint Strength Study   The manufacturer of a popular brand of audio speakers was experiencing higher-than-normal field failure with one of their newer products. The source of the problem was identified as a bonded piece of magnetic component. The factors and their levels (Table 5.3) were studied using an L-8 array. The objective of the test was to increase the bonding strength under direct tensile (pull) loading. Three test samples in each of the eight trial conditions were built according to the design shown in Figure 5.24. The samples were subjected to destructive tests in which the tensile loads were measured using load cells. The loads (in pounds) recorded are listed in Table 5.4. Based on the experimental results,

TABLE 5.3   Factors and Levels for Example 5.3

	Level	
Factors	I	II
A: contact plate	Galvanized	Brass-plated
B: adhesive thickness	Current specifications	Thicker
C: base surface	Machined	Rough
D: curing temperature	Ambient	Air convection

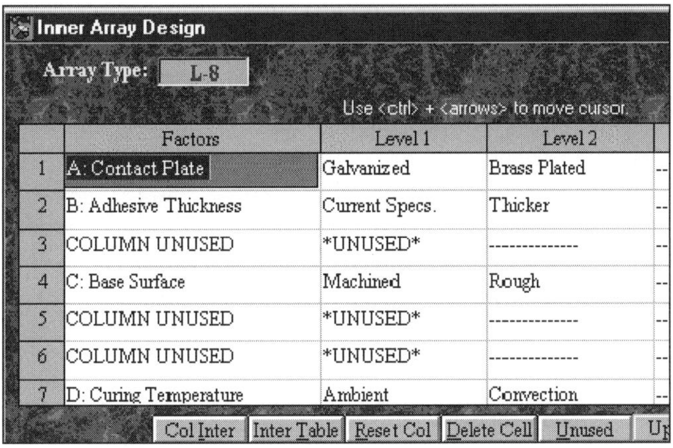

Figure 5.24   Experimental design for Example 5.3.

(a) Determine the factor that has the most influence on the strength.

(b) Determine the joint bonding condition that will produce the most strength.

(c) Determine the expected range of performance at the 90% confidence level.

(d) Identify the factor that need not be subjected to statistical process control (SPC).

(e) Determine the loss of strength when the condition of the base plate is changed from brass-plated to galvanized.

Solution: You can design this experiment by the automatic experiment design feature of QT4. Run QT4 and from the *Design* menu item in the main screen, select the *Automatic design* option. At the input for automatic design, check the box two-level factor and the number of factors as 4 (Figure 5.25). As you click *OK*, QT4 will show you the array selected for the design. Click *Yes* to proceed to describe factors and lev-

TABLE 5.4   Loads Recorded for Example 5.3

Trial	Sample		
	1	2	3
1	2025	2115	2270
2	1690	1692	1710
3	1145	1230	1168
4	1495	1280	1350
5	2550	2680	2585
6	1870	1815	1830
7	1645	2035	1830
8	1670	1100	1380

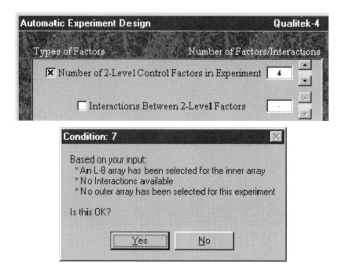

Figure 5.25   Experimental design input for Example 5.3.

els. Describe each factor and their levels in the space provided (Figure 5.26). Click *OK* when finished entering factors and levels.

QT4 uses the factor and level description to design the experiment by assigning the factors to the column as shown in Figure 5.24. Notice that unlike assigning the factor in the first four columns of the array or in random order, QT4 assigns them in an order than minimizes the influence of interaction effects between factors. The interaction effects and how to assign factors to the column are the subject of discussion in later steps. The *Inner array* screen, which shows the experiment design, is mainly for your review. Should it be necessary, you can edit all descriptions entered earlier in this screen. Otherwise, click *OK* to proceed to enter four lines of descriptions (Figure

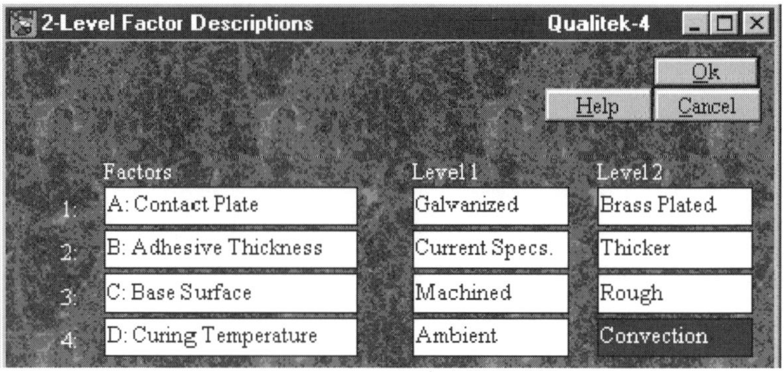

Figure 5.26   Factor and level description for Example 5.3.

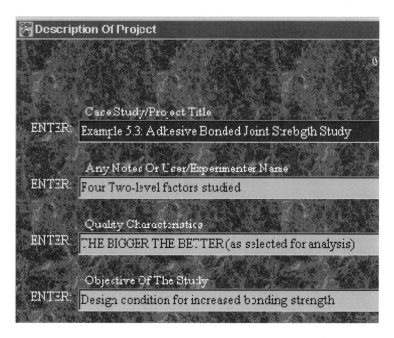

Figure 5.27   Experiment titles for Example 5.3.

5.27). Click *OK* to proceed to review the array and save the experiment file (*BKEX-53.Q4W*).

When you are done saving the experiment file, you will return to the main screen. You may now review the eight trial conditions that were carried out by selecting the *Trial condition* option from the *Condition* menu item. Recall that three samples were tested in each of these eight trial conditions; their results are as given in the example. You must now enter these results in QT4 so that you can analyze them and obtain the information desired. Select the *Enter results* option from the *Result* menu item in the main screen. Enter results in the first three columns in the *Experimental results* screen as shown in Figure 5.28. Click *OK* to update the experiment file and return to main screen.

To perform an analysis, select the *Standard analysis* option from the *Analysis* menu item. Later in this book you will learn that a better way to analyze the results of multiple sample tests is to perform analysis using the signal-to-noise (*S/N*) ratios of the results. QT4 warns you about this by the *Notice* screen (Figure 5.29). Click the *Yes* button at this prompt and check *Bigger is better* in the next screen. As you are familiar by now, QT4 requires you to indicate your selection of the quality characteristic for the analysis first. It then shows you the results for analysis. This screen is only for your review. The results should be the same as you entered earlier. Should you find any discrepancies in the results, you will need to cancel the analysis and modify the results using the *Edit* menu from the main screen. Upon performing all calculations, QT4 displays results of analysis in three screens: (1) *Main effects*, (2) *ANOVA table*, and (3)

	Sample# 1	Sample# 2	Sample# 3
Trial# 1	2025	2115	2270
Trial# 2	1690	1692	1710
Trial# 3	1145	1230	1168
Trial# 4	1495	1230	1350
Trial# 5	2550	2680	2585
Trial# 6	1870	1815	1830
Trial# 7	1645	2035	1830
Trial# 8	1670	1130	1330

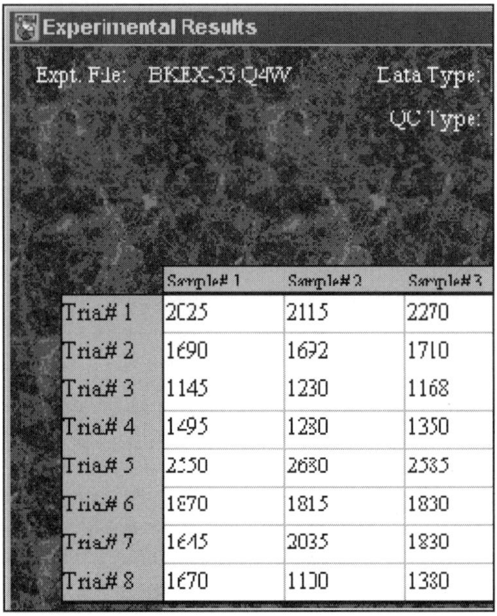

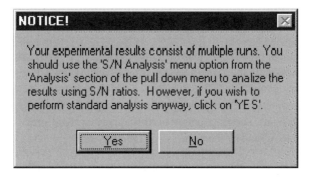

Figure 5.28 Experimental results of Example 5.3.

Optimum conditions and performance. These three screens contain answers to most questions you may have about the results of the experiment. You will also find answers to questions (a) to (e) in this example.

Proceed with the analysis and step through the analysis screens as shown in Figures 5.30 to 5.32. When at the *Optimum conditions* screen, click on the *C.I.* button to calculate the limits of the expected performance at the optimum condition. Enter 90 for the 90% confidence level and click *OK*. The boundary of performance at the 90% con-

Figure 5.29 Reminder about situations for *S/N* and standard analysis.

Main Effects (Average Effects of Factors and Interactions)				Qua

Expt. File: BKEX-53.Q4W   Data Type: Average Values

QC Type: Bigger is Better

Column # / Factors	Level 1	Level 2	L2 - L1
1 A: Contact Plate	1597.5	1915.833	318.333
2 B: Adhesive Thick	2069.333	1444	-625.334
4 C: Base Surface	1939.833	1573.5	-366.334
7 D: Curing Tempera	1796.666	1716.666	-80

Figure 5.30 Main effects of factors in Example 5.3.

fidence level is shown in Figure 5.33. The answers to the specific questions are as follows.

(a) You can identify the factor that has the most influence on the strength by carefully examining the last column of the ANOVA table. You may recall from earlier discussions that this column represents the relative influence of the factors to the variation of results. In other words, the percentages reflect how influential the factors are to the results. The most influential factor in this study, is B: adhesive thickness (see Figure 5.31), which has a 47.7% influence.

(b) The description of factor levels for the optimum condition for the joint is shown in the second column (Figure 5.31).

ANOVA Table					Qualitek-4		

Expt. File: BKEX-53.Q4W   Data Type: Average Value

QC Type: Bigger is Better

Col # / Factor	DOF (f)	Sum of Sqrs. (S)	Variance (V)	F - Ratio (F)	Pure Sum (S')	Percent P(%)
1 A: Contact Plate	1	608016.822	608016.822	11.534	555303.53	11.57
2 B: Adhesive Thick	1	2346244.909	2346244.909	44.509	2293531.617	47.787
4 C: Base Surface	1	805205.455	805205.455	15.275	752492.163	15.678
7 D: Curing Tempera	1	38399.882	38399.882	.728	0	0
Other/Error	19	1001552.547	52713.291			24.965
Total:	23	4799419.617				100.00%

Figure 5.31 ANOVA for Example 5.3.

Optimum Conditions and Performance				Qualitel
Expt. File:BKEX-53.Q4W	Data Type	Average Value		Print
	QC	Bigger is Better		Help

Column # / Factor	Level Description	Level	Contribution
1 A: Contact Plate	Brass Plated	2	159.166
2 B: Adhesive Thick	Current S	1	312.666
4 C: Base Surface	Machined	1	183.166
7 D: Curing Tempera	Ambient	1	39.999

Total Contribution From All Factors...	694.997
Current Grand Average Of Performance...	1756.666
Expected Result At Optimum Condition...	2451.663

Figure 5.32 Optimum condition and performance for Example 5.3.

(c) The range of performance expected at the optimum condition is calculated in the option available under C.I in the *Optimum condition* screen. The range obtained at the 90% confidence level is between 2282 and 2620 pounds.

(d) When a factor is found to have little or no influence, its tolerance can be removed and may also be excluded from any statistical control. Such is the case with factor *D*: curing temperature (see Figure 5.31, 0%).

(e) There are times when it is not possible to set levels of all factors as identified by the optimum condition. This could be due primarily to unavailability of the part, or the condition may be too costly to maintain. In these situations it is better to know

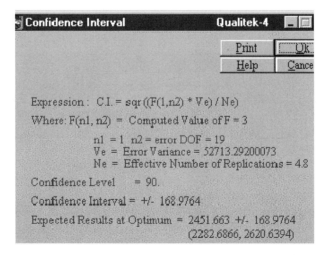

Figure 5.33 Confidence interval on performance at optimum condition for Example 5.3.

Estimate of Performance At Any Arbitrary Condition				Qualitek

Expt. File: Data Type: Average Value

Transform   QC Type: Bigger is Better   **Help**

Column # / Factor	Level Description	Level	Contribution
1 A: Contact Plate	Galvanize	1	-159.167
2 B: Adhesive Thick	Current S	1	312.666
4 C: Base Surface	Machined	1	183.166
7 D: Curing Tempera	Ambient	1	39.999

Total Contribution From All Factors...	376.664
Current Grand Average Of Performance...	1756.666
Expected Result At Optimum Condition...	2133.33

Figure 5.34   Estimated performance with other factor levels for Example 5.3.

about the lost opportunity by setting the factor at a level other than that prescribed as the optimum condition. Notice (from Figure 5.32) that factor *A*: contact plate at the brass-plated condition is identified as part of the optimum condition. What if this condition cannot be incorporated into the design at this time. For reasons of cost, *galvanized* (level 1) will have to be maintained instead. What will be the loss of strength?

The *Estimate* option from the *Optimum condition* screen allows you to estimate the expected performance with any of the possible factor conditions. Indeed, you can calculate the performance expected for the full factorial combinations ($2^4 = 16$ in this example) for the factors included in the study. To estimate the performance with level 1 of *A*: contact plate, click on the *Estimate* button in the *Optimum condition* screen. In the *Estimate of performance* screen (Figure 5.34), click on level 2 shown in the line for factor *A*: contact plate. Type number 1 in place of number 2. Notice how the level description (second column, galvanized) and the expected performance numbers are also changed instantly. Because of this level change from brass-plated to galvanized, the expected strength is reduced from 2451.66 pounds (see Figure 5.32) to 2133.33 pounds (see Figure 5.34 bottom line). The loss of strength is about $2451 - 2133 = 318$ pounds.

FOUR WAYS TO USE AN L-12 ORTHOGONAL ARRAY

Use an L-12 array to design experiments with:

1. 8 two-level factors
2. 9 two-level factors
3. 10 two-level factors
4. 11 two-level factors

Note: Factors are assigned to the 11 columns arbitrarily.

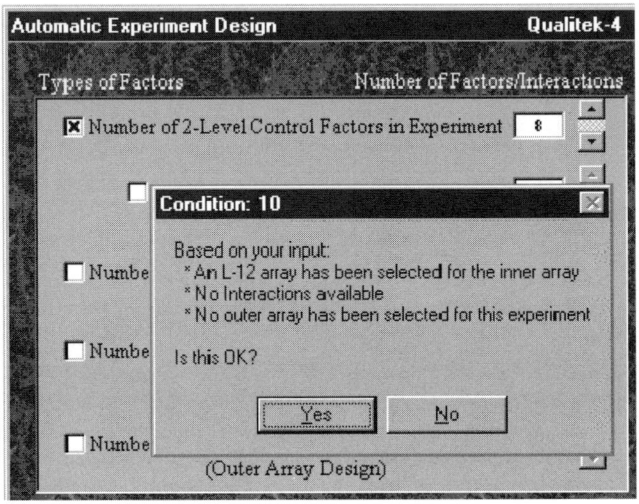

Figure 5.35 Experimental design array for 8 two-level factors.

An L-12 orthogonal array has 11 two-level columns and is commonly used to design experiments with 8 to 11 two-level factors. The methods of experiment designs and the analysis of results are the same as those you have seen in the examples so far. To confirm that QT4 utilizes the array for experimental situations including 8 to 11 two-level factors, practice designing experiments with the *Auto-*

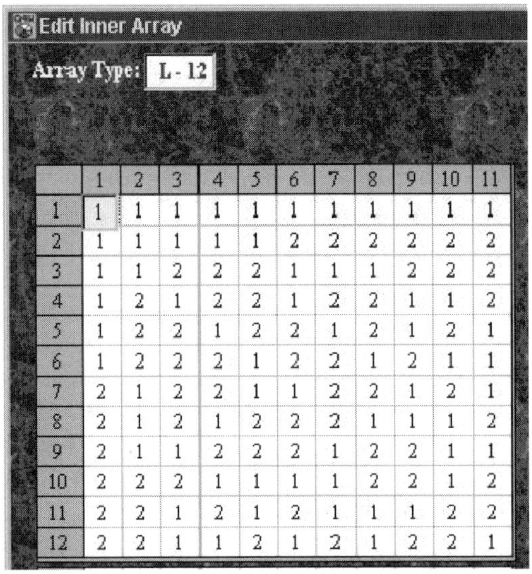

	1	2	3	4	5	6	7	8	9	10	11
1	1	1	1	1	1	1	1	1	1	1	1
2	1	1	1	1	1	2	2	2	2	2	2
3	1	1	2	2	2	1	1	1	2	2	2
4	1	2	1	2	2	1	2	2	1	1	2
5	1	2	2	1	2	2	1	2	1	2	1
6	1	2	2	2	1	2	2	1	2	1	1
7	2	1	2	2	1	1	2	2	1	2	1
8	2	1	2	1	2	2	2	1	1	1	2
9	2	1	1	2	2	2	1	2	2	1	1
10	2	2	2	1	1	1	1	2	2	1	2
11	2	2	1	2	1	2	1	1	1	2	2
12	2	2	1	1	2	1	2	1	2	2	1

Figure 5.36 L-12 orthogonal array.

matic experiment design option. Run QT4 and be in the main screen. It does not matter which file you currently have in the memory. QT4 always starts assuming that you are designing a new experiment when you select the *Design* menu item. From this menu item, select the *Automatic design* option and proceed to input the design information. In the *Automatic experiment design* screen, check the two-level factor box, set the number of factors to 8, and click *OK*. QT4 responds by showing you the array selected (Figure 5.35).

By clicking *No* at the prompt, repeat designing experiments with the number of factor settings to 9, 10, and 11. In each of these cases you will see that QT4 selects the L-12 array for the designs. Click *No* at the array selection prompt and *Cancel* at the *Automatic experiment design* screen to return to the main screen. You can check the content of the L-12 orthogonal array by referring to the Appendix or by reviewing the example *REF-L12A.Q4W* experiment file in QT4. To review this file, you must be in the main screen and open this file from the *Open* option in the *File* menu item. Once the file is loaded, select *Inner array* from the *Edit* menu item. When you are done reviewing the array (Figure 5.36), click *Cancel* to return to the main screen.

FOUR WAYS TO USE AN L-16 ORTHOGONAL ARRAY

Use an L-16 array to design experiments with:
1. 12 two-level factors
2. 13 two-level factors
3. 14 two-level factors
4. 15 two-level factors

Note: Factors are assigned to the 15 columns arbitrarily.

An L-16 orthogonal array has 15 two-level columns and is commonly used to design experiments with 12 to 15 two-level factors. The methods of experimental design and analysis of results are the same as those that you have seen in the examples so far.

To confirm that QT4 utilizes the array for experimental situations including 12 to 15 two-level factors, practice designing experiments with the automatic experiment design option. Run QT4 and go to the main screen. From the *Design* menu item in the main screen, select the *Automatic design* option and proceed to input design information. In the *Automatic experiment design* screen, check the two-level factor box, set the number of factors to 12, and click *OK*. QT4 responds by showing you the array selected (Figure 5.37).

As you proceed, click *No* at the next prompt and repeat the experiments with the number of factor settings at 13, 14, and 15. In each of these cases you will see that QT4 selects the L-16 array for the designs. By clicking *No* at the array selection prompt and *Cancel* at the *Automatic experiment design* screen, you will return to the main screen.

You can check the content of the L-16 orthogonal array by referring to the Appendix or by reviewing the example *REF-L16A.Q4W* experiment file in QT4. To review this file,

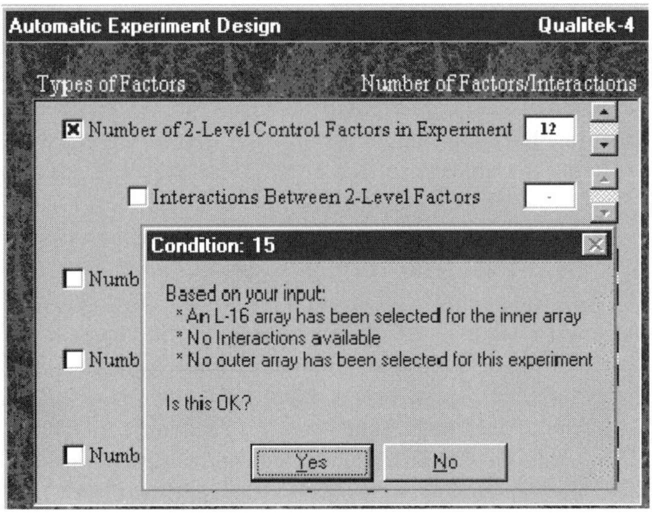

Figure 5.37 Experimental design array for 12 two-level factors.

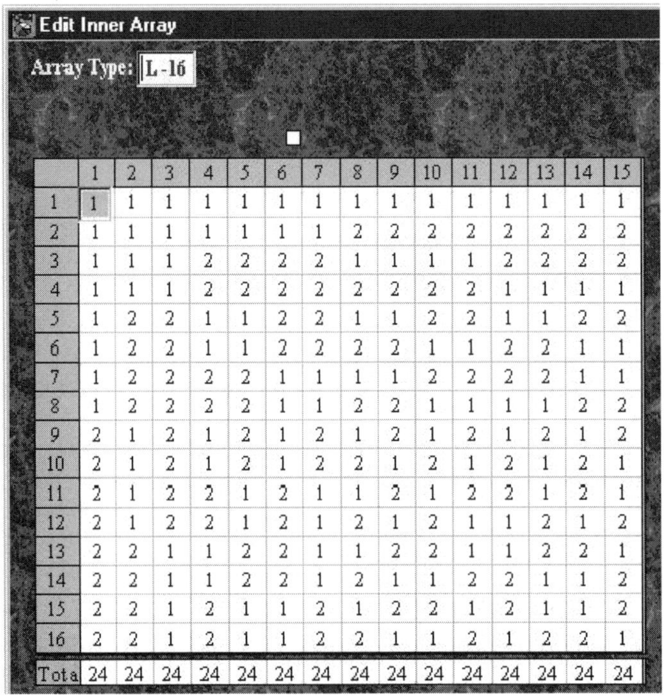

Edit Inner Array

Array Type: L-16

	1	2	3	4	5	6	7	8	9	10	11	12	13	14	15
1	1	1	1	1	1	1	1	1	1	1	1	1	1	1	1
2	1	1	1	1	1	1	1	2	2	2	2	2	2	2	2
3	1	1	1	2	2	2	2	1	1	1	1	2	2	2	2
4	1	1	1	2	2	2	2	2	2	2	2	1	1	1	1
5	1	2	2	1	1	2	2	1	1	2	2	1	1	2	2
6	1	2	2	1	1	2	2	2	2	1	1	2	2	1	1
7	1	2	2	2	2	1	1	1	1	2	2	2	2	1	1
8	1	2	2	2	2	1	1	2	2	1	1	1	1	2	2
9	2	1	2	1	2	1	2	1	2	1	2	1	2	1	2
10	2	1	2	1	2	1	2	2	1	2	1	2	1	2	1
11	2	1	2	2	1	2	1	1	2	1	2	2	1	2	1
12	2	1	2	2	1	2	1	2	1	2	1	1	2	1	2
13	2	2	1	1	2	2	1	1	2	2	1	1	2	2	1
14	2	2	1	1	2	2	1	2	1	1	2	2	1	1	2
15	2	2	1	2	1	1	2	1	2	2	1	2	1	1	2
16	2	2	1	2	1	1	2	2	1	1	2	1	2	2	1
Tota	24	24	24	24	24	24	24	24	24	24	24	24	24	24	24

Figure 5.38 L-16 orthogonal array.

be in the main screen and open this file from the *Open* option in the *File* menu item. Once the file is loaded, select *Inner array* from the *Edit* menu item. When you are done reviewing the array (see Figure 5.38), click *Cancel* to return to the main screen.

SIXTEEN WAYS TO USE AN L-32 ORTHOGONAL ARRAY

Use an L-32 array to design experiments with:
1. 16 two-level factors
2. 17 two-level factors
    ⋮
16. 31 two-level factors

Note: Factors are assigned to the 31 columns arbitrarily.

An L-32 orthogonal array has 31 two-level columns and is commonly used to design experiments with 16 to 31 two-level factors. The methods of experimental design and analysis of results are same as before.

To confirm that QT4 utilizes the array for experimental situations including 16 to 31 two-level factors, you may want to practice designing experiments as you did with other arrays. Run QT4 and go to the main screen. From the *Design* menu item in the main screen, select the *Automatic design* option and proceed to input design information. In the *Automatic experiment design* screen, check the two-level factor box, set number of factors to 16, and click *OK*. QT4 responds by showing you the array selected (Figure 5.39).

By clicking *No* at the prompt, repeat the experiments with the number of factor settings at 17 through 31. In each of these cases you will see that QT4 selects the L-32

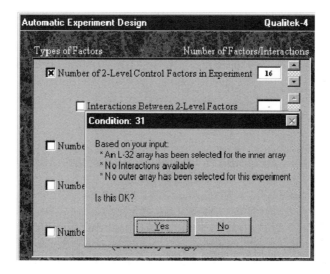

Figure 5.39   Experiment design array for 16 two-level factors.

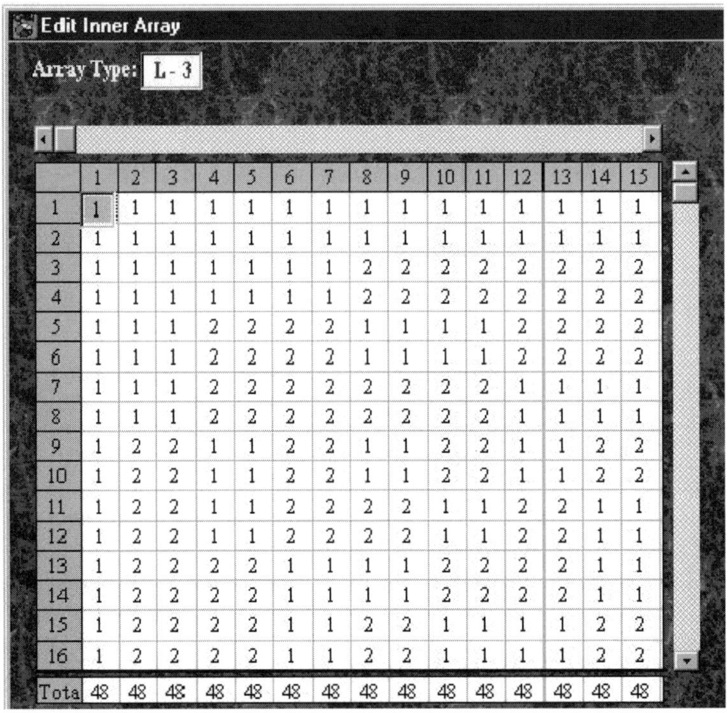

	1	2	3	4	5	6	7	8	9	10	11	12	13	14	15
1	1	1	1	1	1	1	1	1	1	1	1	1	1	1	1
2	1	1	1	1	1	1	1	1	1	1	1	1	1	1	1
3	1	1	1	1	1	1	1	2	2	2	2	2	2	2	2
4	1	1	1	1	1	1	1	2	2	2	2	2	2	2	2
5	1	1	1	2	2	2	2	1	1	1	1	2	2	2	2
6	1	1	1	2	2	2	2	1	1	1	1	2	2	2	2
7	1	1	1	2	2	2	2	2	2	2	2	1	1	1	1
8	1	1	1	2	2	2	2	2	2	2	2	1	1	1	1
9	1	2	2	1	1	2	2	1	1	2	2	1	1	2	2
10	1	2	2	1	1	2	2	1	1	2	2	1	1	2	2
11	1	2	2	1	1	2	2	2	2	1	1	2	2	1	1
12	1	2	2	1	1	2	2	2	2	1	1	2	2	1	1
13	1	2	2	2	2	1	1	1	1	2	2	2	2	1	1
14	1	2	2	2	2	1	1	1	1	2	2	2	2	1	1
15	1	2	2	2	2	1	1	2	2	1	1	1	1	2	2
16	1	2	2	2	2	1	1	2	2	1	1	1	1	2	2
Tota	48	48	48	48	48	48	48	48	48	48	48	48	48	48	48

Figure 5.40 L-32 orthogonal array.

array for the designs. As you proceed, click *No* at the array selection prompt and *Cancel* at the *Automatic experiment design* screen to return to the main screen.

Like other arrays, you can check the content of the L-32 orthogonal array by referring to the Appendix or by reviewing the example *REF-L32A.Q4W* experiment file in the QT4 program. To review this file, be in the main screen and open this file from the *Open* option in the *File* menu item. Once the file is loaded, select the *Inner array* from the *Edit* menu item. Be sure to use the scroll bars to display level numbers beyond row 16 and columns 15. When done reviewing the array (Figure 5.40), click *Cancel* to return to the main screen.

IMPROVED REPRODUCIBILITY WITH ORTHOGONAL ARRAY EXPERIMENTS

As you have seen earlier in this and previous steps, an L-4 array is necessary to study 3 two-level factors. The L-4 experiment design will call for four separate experimental conditions. If you studied the same factors using the one-factor-at-a-time approach (discussed in Step 3), you would still need the same number of experiments (four). Why, then, should you prefer to use an orthogonal array? Orthogonal arrays are preferred for experimental designs because they are more likely to produce reproducible results. In a general sense the results refer to the conclusions derived from an analysis of the measured performances of the experiments, such as conclusions, observations, and recommendations. When such results are reproducible, it is possible for another person to repeat an experiment at other times or locations and get the same outcome. To understand how orthogonal experiments are potentially superior to one-factor-at-a-time experiments, look into the ways that factor influences are calculated in each method.

Suppose that in a study of 3 two-level factors (A, B, and C) using the one-factor-at-a-time approach, the four experiments and their results are as shown in Table 5.5. From the experimental results, the influence of factor A is calculated by comparing the results of experiment 2 with that of experiment 1, as these two experiments are the same in all respects. The levels of factor A as seen in the description are:

Experiment 1: a_1 b_1 c_1,   result = 30 (has level 1 of factor A)

Experiment 2: a_2 b_1 c_1,   result = 60 (has level 2 of factor A)

The influence (also called the *main effect* or *average effect*) of factor A is expressed by subtracting the results of experiment 1 from experiment 2:

$$\text{influence of } A = 60 - 30$$

This conclusion and any recommendation that follows from the experiments would be widely accepted unless there were questions such as:

- Can we trust that the effect of factor A will always be to increase the results (60 − 30 =) 30 points when its condition is changed from level 1 to level 2?
- What were the conditions of the other factors (B and C) in the two experiments used in drawing conclusions?

Observe that in the two experiments used to draw conclusions about factor A, only level 1 of factors B and C are present. This means that we have no knowledge of how changes in factor A would influence the results when two levels of factors B and C are

TABLE 5.5 One-Factor-at-a-Time Experiments with 3 Two-Level Factors

	Factor			
Experiment	A	B	C	Result
1	1	1	1	30
2	2	1	1	60
3	1	2	1	33
4	1	1	2	45

present. Thus the second question above clearly points out the weakness of this approach.

By doing the same number of experiments designed using an orthogonal array (Table 5.6), the way that conclusions are drawn about the factor influence have a higher potential of becoming reproducible. The difference lies in the manner by which factor influences are calculated, as shown here. From the four experimental results, factor influence is calculated as follows:

Experiment 1: a_1 b_1 c_1,   result = 30
Experiment 2: a_1 b_2 c_2,   result = 48

$$\text{Average effect of factor } A \text{ at level } 1 = \frac{30 + 48}{2}$$

$$= 39$$

Experiment 3: a_2 b_1 c_2,   result = 75
Experiment 4: a_2 b_2 c_1,   result = 63

$$\text{Average effect of factor } A \text{ at level } 2 = \frac{75 + 63}{2}$$

$$= 69$$

Thus,

$$\text{influence of factor } A = \text{average effect at level } 2 - \text{average effect at level } 1$$

$$= 69 - 39$$

Notice that in calculating the average effects of factor A at both levels, both levels of factors B and C were present in the results used to compute the averages. In other

TABLE 5.6   Orthogonal Array Experiment Design with 3 Two-Level Factors

Test	Factor			Result
	A	B	C	
1	1	1	1	30
2	1	2	2	48
3	2	1	2	75
4	2	2	1	63

words, the effect of factor A is obtained in the presence of any possible influence of factors B and C (interaction effects). Because the average effect (of factor A) here is calculated from the average of results, which included both levels of the other factors (B and C), it is more likely to be close to the value expected.

ANALYTICAL VERIFICATION OF ORTHOGONAL ARRAY EXPERIMENTS

By now you are familiar with a set of orthogonal arrays that you will be using for many of your experimental needs. You also realize that an orthogonal array experiment represents the smallest possible fraction of all the possible combinations. In most cases, the number of experiments that you will need to study a set of factors will be the same if you were to do one-factor-at-a-time experiments. The number of experiments you need to do is, of course, dictated by the size of the array, which depends on the number of factors included in the study. If you consider the experimental efficiency described by a ratio of the number of full factorial experiments to the same number of orthogonal array experiments, the larger arrays usually result in higher efficiency. For example, in an L-4 experiment, four of eight possible (full factorial) experiments are done, which results in a 50% reduction in number of experiments. The reduction number for an experiment designed with an L-16 array will be over 99% (do only 16 of the possible $2^{15} = 32{,}768$).

The small size of the experiments and the fact that they seem to give satisfactory (not necessarily accurate) results are the two reasons that orthogonal arrays are preferred for experimental designs. But how accurate are the findings, and is there a way to verify them? The best verification comes from running experiments with actual parts and by carrying out confirmation tests at the predicted optimum (most desirable) condition. Short of that, you can use an analytical model, which may save some material costs. Here is a small example to demonstrate how orthogonal experiments are able to predict findings with 100% accuracy when all factors behave linearly.

Example 5.4: Occupant Stopping Distance Simulation Study During automobile barrier tests, the stopping distance (SD) traversed by test dummies after the restraint system is activated is a critical factor contributing to occupant injury. A simple analytical model in which the SD was related directly to three factors was available for study. Using two levels for each factor, all possible combinations were calculated and the conditions for maximum distance and the value were determined as described in Table 5.7. The analytical expression for the stopping distance,

$$SD = 3A - B + \tfrac{1}{2} C$$

Full factorial combinations are obtained by substituting the two values of each of factors A, B, and C in the equation above, which results in the following: $SD(a_1, b_1, c_1)$ or $SD(1, 1, 1)$ is calculated using level 1 values of factors A, B, and C. Thus

$$SD(1, 1, 1) = 3 \times 10 - 5 + 0.5 \times 10 = 30$$

TABLE 5.7   Factors and Levels for Example 5.4

	Level	
Factor	1	2
A: belt stiffness (units)	10 (a_1)	20 (a_2)
B: anchor position (inches forward)	5 (b_1)	2 (b_2)
C: locking time (milliseconds)	10 (c_1)	40 (c_2)

Similarly,

$$SD(1, 1, 2) = 3 \times 10 - 5 + 0.5 \times 40 = 45$$

$$SD(1, 2, 1) = 3 \times 10 - 2 + 0.5 \times 10 = 33$$

$$SD(1, 2, 2) = 3 \times 10 - 2 + 0.5 \times 40 = 48$$

$$SD(2, 1, 1) = 3 \times 20 - 5 + 0.5 \times 10 = 60$$

$$SD(2, 1, 2) = 3 \times 20 - 5 + 0.5 \times 40 = 75$$

$$SD(2, 2, 1) = 3 \times 20 - 2 + 0.5 \times 10 = 63$$

$$SD(2, 2, 2) = 3 \times 20 - 2 + 0.5 \times 40 = 78$$

From the calculations above of all possible combinations of the factor level, the maximum value of SD is found to be 78 at a belt stiffness of 20 units, an anchor position of 2 inches, and a locking time of 40 milliseconds (a_2, b_2, c_2; last calculation).

Solution (Using L-4 Array):   Experiments to study the same three factors (A, B, and C) using the L-4 array (Table 5.8) will require following four of the eight combinations calculated earlier:

$$SD(1, 1, 1) = 3 \times 10 - 5 + 0.5 \times 10 = 30$$

$$SD(1, 2, 2) = 3 \times 10 - 2 + 0.5 \times 40 = 48$$

$$SD(2, 1, 2) = 3 \times 20 - 5 + 0.5 \times 40 = 75$$

$$SD(2, 2, 1) = 3 \times 20 - 2 + 0.5 \times 10 = 63$$

This experiment has already been designed for you and a QT4 program for analysis is available. To review and analyze this experiment, run QT4 and load file *BKEX-54.Q4W* by selecting the *Open* option from the *File* menu item. Notice that the results

TABLE 5.8 L-4 Array Design for 3 Two-Level Factors

Trial	Factor			Result
	A	B	C	
1	1	1	1	Y_1
2	1	2	2	Y_2
3	2	1	2	Y_3
4	2	2	1	Y_4

of the four trial conditions as calculated above are already entered. You can now proceed to analyze the results and determine the optimum condition. Execute the following steps and be in the *Optimum condition* screen (Figure 5.41).

- Select *Standard analysis* from the *Analysis* menu item in the main screen.
- Check the box for the *Bigger is better* QC and click *OK*.
- Click *OK* in the *Experimental results* screen.
- Click *OK* in the *Main effects* screen.
- Click *OK* in the *ANOVA* screen.

From the optimum condition, note that the maximum distance (78) and the condition (belt stiffness of 20 units, anchor position of 2 inches, and locking time of 40 milliseconds) is the same as obtained by the full factorial experiment done before. The accuracy in this case is 100%, which is expected when the factor influences are all linear, as it is forced to be by the analytical expression.

This small example is included in this step only as a demonstration for the possibility of more advanced verification studies. If you want, you can create similar analytical

Optimum Conditions and Performance				Qualitek-4
Expt. File:BKEX-54.Q4W	Data Type	Average Value		Print
	QC	Bigger is Better		Help

Column # / Factor	Level Description	Level	Contribution
1 A:Belt Stiffness	20 units	2	15
2 B:Anchor Position	2 in.	2	1.5
3 C:Locking Time	40 mil.sec.	2	7.5

Total Contribution From All Factors...	24
Current Grand Average Of Performance...	54
Expected Result At Optimum Condition...	78

Figure 5.41 Optimum condition from orthogonal array experiments.

runs with larger array. You can also handle simulations where factor influences are nonlinear and then compare the results of the full factorial with the orthogonal array experiments to determine inaccuracies encountered. It will be unwise to assume that your finding from orthogonal arrays will be inaccurate just because factors may not behave linearly. The degree of inaccuracies depends on a number of things. The most critical among them is the nonlinearity of results when the factors vary within the levels selected for the study. Most factors in industrial projects have a nonlinear influence. Only limited cases could potentially influence your study outcome because they did not behave linearly.

SUMMARY

In this step you have become familiar with the 5 two-level orthogonal arrays. You have learned how to use these arrays through several examples. You have also learned why orthogonal experiments are preferred over one-factor-at-a-time experiments and how to verify the accuracy of such experiments using analytical simulations.

REVIEW QUESTIONS

Q. For a factor such as temperature with two levels, does the smaller value always have to be level 1?

A. No. Any level of a factor can be assigned any value.

Q. In a study with all two-level factors, is it possible to have the optimum condition somewhere between?

A. By assigning two levels to a factor, its influence is assumed to be linear, whether it is or not. DOE analysis will always identify optimum condition with one or the other level. In reality, however, the optimum condition may indeed be composed of an infinite number of combinations of level values between the two studied. You will not know whether or not such things exist until you run a larger set of experiments with more factor levels.

Q. How do you select factor levels if nothing is known about the factor before the experiments?

A. You may have to run a separate set of experiments with the factor alone. Usually, you will try a large number of levels of a factor within the working range, then select the range of values where the result is least sensitive to the factor change.

EXERCISES

5.1 Indicate the orthogonal array you will use to design experiments to study each of the following situations:

(a) 2 two-level factors

(b) 4 two-level factors

(c) 7 two-level factors

(d) 10 two-level factors

(e) 15 two-level factors

(f) 27 two-level factors

5.2 An experiment with 7 two-level factors (A, B, C, etc. assigned to column 1, 2, 3, etc.) produced the following trial (L-8 array) results: 40 (trial 1), 44, 32, 38, 50, 56, 60, and 62.

(a) Determine the grand average of performance.

(b) Determine the average effect of factor A, which is assigned to column 1, at the second level.

5.3 Find the optimum combination of levels of factors A, B, and C (Table 5.9) and the highest magnitude of performance.

$$\text{Grand average} = \frac{30 + 40 + 20 + 60}{4}$$

$$= 37.5$$

Factor averages:

$$A_1 = \frac{30 + 40}{2} = 35 \qquad A_2 = \frac{20 + 60}{2} = 40$$

$$B_1 = \frac{30 + 20}{2} = 25 \qquad B_2 = \frac{40 + 60}{2} =$$

$$C_1 = \frac{30 + 60}{2} = 45 \qquad C_2 =$$

TABLE 5.9 Array for Exercise 5.3

		Factors		
Trial	A	B	C	Result
1	1	1	1	30
2	1	2	2	40
3	2	1	2	20
4	2	2	1	60

TABLE 5.10   Factors and Levels for Exercise 5.5

	Level	
Factors	I	II
A: tool type	High carbon	Carbide tip
B: cutting speed (rev/min)	1500	2000
C: feed rate (mm/s)	2	5

Optimum condition: _____

$Y_{opt} =$

5.4 Three two-level factors, A, B, and C, were studied using an L-4 orthogonal array and by assigning them to columns 1, 2, and 3, respectively. Each trial condition was tested once and the following results were obtained: 40, 65, 55, and 70 (for trials 1, 2, 3, and 4). Assume that a higher value is desired.

(a) Determine the optimum condition.

(b) Determine the performance at optimum.

(c) Determine the main effect of factor B.

5.5 In an effort to study the production problem (high reject rate) experienced in a machining process, three factors among several possible causes were selected for a quick study. An L-4 experiment was designed to study 3 two-level factors (Table 5.10). A large number of samples were tested in each trial condition, and average performances were noted for two major objectives (Table 5.11; evaluation criteria: surface finish and capability).

(a) Describe (recipe) the factor levels used to conduct the third experiment.

(b) Determine the optimum condition for the best surface finish (QC = S).

(c) Determine the optimum condition for the best capability (QC = B).

TABLE 5.11   Array for Exercise 5.5

Trial	Factor			Results (micrometers)	
	A	B	C	Surface finish (QC = S)	Capability (QC = B)
1	1	1	1	17	1.26
2	1	2	2	12	1.32
3	2	1	2	16	0.28
4	2	2	1	20	1.16

(d)   If experiment 1 is considered the current performance, estimate the percent improvement of surface finish expected from the new process settings.

EXERCISE ANSWERS

5.1   **(a)** L-4; **(b)** L-8; **(c)** L-8; **(d)** L-12; **(e)** L-16; **(f)** L-32.

5.2   **(a)** 47.75; **(b)** 57. (You may use QT4 to solve this problem.)

5.3   Follow hints and hand calculate.

5.4   **(a)** $A_2B_2C_2$; **(b)** 75; **(c)** (67.50 – 47.50).

5.5   Calculate by hand or use QT4.
  (a)   Tool; carbide tip; speed; 1500 rev/min; feed rate; 5 mm/s (2 1 2, the third row of the array).
  (b)   Use only the surface finish results. (Run and review the experiment *BKEX-55A.Q4W* in QT4.) Tool; carbide; speed; 1500 rev/min; feed rate; 2 mm/s (2 1 1).
  (c)   Use only the capability results. (Run and review the experiment *BKEX-55B.Q4W* in QT4.) Tool; high carbon; speed; 2000 rev/min; feed rate; 2 mm/s (1 2 1).
  (d)   Improved from 17 to 12. (Run and review the experiment *BKEX-55A.Q4W* in QT4.) Improvement = (17 – 12)/17 = 29.4%.

Experimental Design With Three- and Four-Level Factors

What You Will Learn in This Step

In this step you will learn about a set of standard orthogonal arrays created to handle factors at three and four levels. When factors are suspected to influence the results in a nonlinear manner, you should consider studying designing experiments with three- or four-level arrays such as those discussed in this step. Here you will also learn about situations where studying a factor at four levels is more desirable than at three and how to make convenient use of the two special arrays that also have one two-level column each (L-18 and L-32).

Thought for the Day

> Quality is never an accident; it is always the result of high intention, sincere effort, intelligent direction and skillful execution; it represents the wise choice of many alternatives.
> —William A. Foster

For many experimental situations you will encounter in the industry, you will generally have a larger number of factors whose behavior is known to you. What you would prefer to have is an experiment that can include more factors and quickly produce some information that helps you sort out factors that have more influence on the results from those that do not. The two-level orthogonal arrays you learned about in Step 5 most often will fulfill your needs. But as you know more about the factors, or you are interested in the influence of individual factors, you may want to design your experiments with three- or four-level orthogonal arrays. Depending on the number of factors you want to study and the depth of information about their behavior you want, you will select the array from the list of five arrays; L-9, L-18, L-27, L-16 modified, and L-32 modified. As a general guideline, for higher levels of the same number of factors, you will end up running more experiments. Thus, when economy of cost and time are of primary considerations, you have to draw a careful balance between how many factors you can study and the quality of information you want.

THREE WAYS TO USE AN L-9 ORTHOGONAL ARRAY

Use an L-9 array to design experiments with:

1. 2 three-level factors
2. 3 three-level factors
4. 4 three-level factors

Note: Factors are assigned to the columns arbitrarily.

The L-9 orthogonal array has 4 three-level columns and is the most frequently used three-level array. This would become your choice when the number of factors is small and you have the budget to conduct nine experiments. For one extra experiment over an L-8 design, you can gather more valuable information for up to 4 three-level factors. You can review the content of an L-9 array by one of the example experiments provided with QT4. To review, run QT4 and go to the main screen. Select the *Open* option from the *File* menu item and load experimental file *REF-L09A.Q4W*. From the *Edit* menu item, select *Inner array*, which should bring up the L-9 array (Figure 6.1) used for this example experiment. Observe how level numbers 1, 2, and 3 are distributed along the column. You can be certain that each column has these levels in equal numbers by taking a quick look at the column totals placed at the bottom. Although you should keep in mind that just because the total is equal does not guarantee that the columns are balanced. It simply makes the likelihood of it being balanced much higher. It is always a good idea to check to make sure that the array is error free before you proceed with analysis of your experimental results. Click *Cancel* to return to the main screen when you are done reviewing the array. To design the experiment, the 4 three-level factors (*A*, *B*, *C*, and *D*) can be assigned to the columns of an L-9 array arbitrarily. If only two factors are involved, columns 1 and 2 should be preferred.

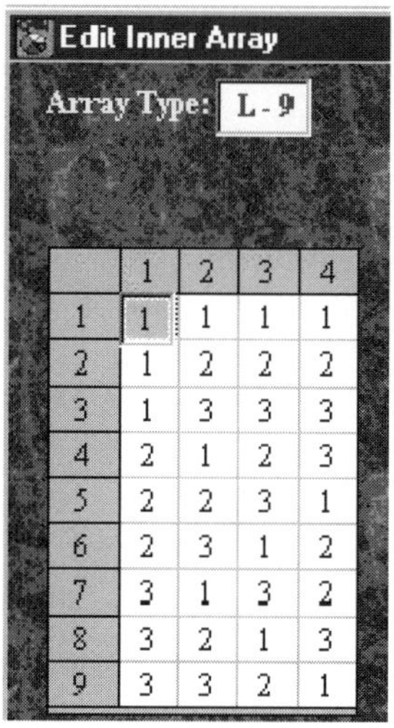

	1	2	3	4
1	1	1	1	1
2	1	2	2	2
3	1	3	3	3
4	2	1	2	3
5	2	2	3	1
6	2	3	1	2
7	3	1	3	2
8	3	2	1	3
9	3	3	2	1

Figure 6.1 L-9 orthogonal array.

TABLE 6.1   Factors and Levels for Example 6.1

	Level		
Factor	1	2	3
A: lube levels (pints)	3.5	6.5	9
B: engine speed (rev/min)	1500	2000	3000
C: torque (ft-lb)	100	600	1100
D: surface finish (μm)	10	40	60

Example 6.1:   Transmission Axle Noise Study   The suspect causes for audible noise from an automobile transmission axle design were identified (Table 6.1). The three levels for each of the factors were identified as the lower, middle, and upper ranges of application conditions. Experiments were laid out using an L-9 array, and four samples were tested in each of the nine trial conditions. The results of the tests, measured in decibels (dB), are shown in Figure 6.2. Determine the range of noise levels expected by comparing the same at the worst and best combination of factor levels.

Solution: This experiment is already predesigned and included in the program. You can follow these steps and complete the process of creating the design and entering the experimental results.

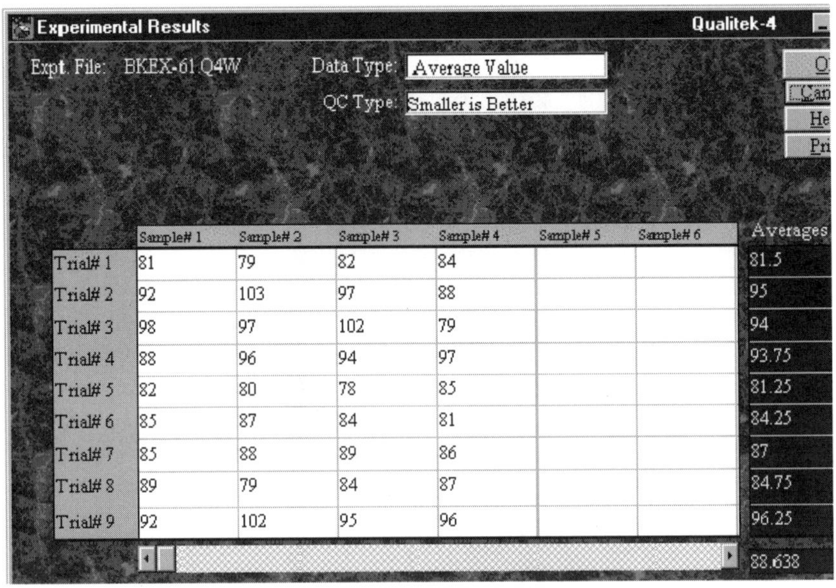

Experimental Results   **Qualitek-4**

Expt. File: BKEX-61 Q4W   Data Type: Average Value
                               QC Type: Smaller is Better

	Sample# 1	Sample# 2	Sample# 3	Sample# 4	Sample# 5	Sample# 6	Averages
Trial# 1	81	79	82	84			81.5
Trial# 2	92	103	97	88			95
Trial# 3	98	97	102	79			94
Trial# 4	88	96	94	97			93.75
Trial# 5	82	80	78	85			81.25
Trial# 6	85	87	84	81			84.25
Trial# 7	85	88	89	86			87
Trial# 8	89	79	84	87			84.75
Trial# 9	92	102	95	96			96.25
							88.638

Figure 6.2   Results of experiments in Example 6.1.

QT4 Design Steps

1. Run QT4 and go to the main screen.

2. Select *Automatic design* from the *Design* menu item.

3. Click *Yes* at the prompt for *New experiment design*.

4. Check the box for three-level factors in the *Automatic experiment design* screen (Figure 6.3) and set the number 4 in the box for number of factors. Click *OK* to proceed (also in other screens, as appropriate).

5. Click *Yes* when QT4 shows you that it selected an L-9 array for the design. Click *Yes* to proceed.

6. Type in descriptions of the four factors and the three levels for each. Click *OK* to proceed.

7. Review the descriptions of factors and levels you just entered. Your screen should look like the one shown in Figure 6.4. You may edit any description, if necessary, here in this screen.

8. Enter the project title, objectives, and the quality characteristic (QC).

9. Review the array used for the design. The array should look like the one shown in Figure 6.1.

10. When at the *Save file as* screen, click *Cancel* and then *OK* to return to the main screen without saving the file. (If this experiment were not already predesigned for you, you will need to name this experiment file.)

11. When in the main screen, load the experiment file *BKEX-61.Q4W* by selecting the *Open* option from the *File* menu item.

12. Compare the experimental results displayed next to the array in the main screen with that shown in Figure 6.2. You may have to scroll down to view trial

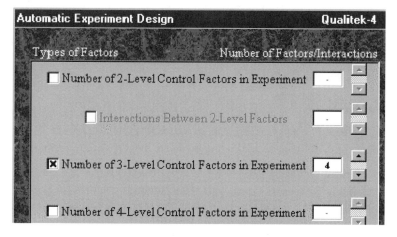

Figure 6.3 Automatic experiment design input screen.

Inner Array Design

Array Type: L-9

Use <ctrl> + <arrows> to move cursor.

	Factors	Level 1	Level 2	Level 3
1	A: Lube levels	3.5 pints	6.5 pints	9 pints
2	B: Engine Speed	1500 rpm	2000 rpm	3000 rpm
3	C:Torque Applied	100 ft lb	600 ft.lb.	1100 ft.lb.
4	D: Surface Finish	10 Micro	40 Micro	60 Micro

Figure 6.4 Descriptions of factors and levels.

numbers beyond trial 8. You can also review factors, levels, and results by selecting the appropriate options from the *Edit* menu item. You may also edit any description or results, but QT4 will not allow you to update this example experiment.

You should be familiar with the steps you need to follow for analysis using the QT4 program. As a standard format, QT4 displays analysis output in three screens: *Main effects*, *ANOVA*, and *Optimum*. Once you have selected the quality characteristic (QC) and proceeded with analysis, you will come across these screens by clicking *OK* in each displayed screen. Notice that the object of the experiment is to find the range of operating noise, which is measured in terms of decibels (dB). The level of noise is measured by recording the air pressure which the sound wave creates on a pressure-measuring transducer (e.g., a microphone). The decibel is the logarithm (to the base 10) of the sound pressure measured in terms of pounds per square inch. By definition, smaller values of noise readings in decibels represents lower noise levels.

Unlike common design optimization studies, the factors studied in this example are not design factors. Instead, these are operational factors which can assume any value within the range specified by the three levels. The two extreme values of the noise can be found by determining the expected results at the most and least desirable conditions. The most desirable condition is found by analyzing the results by selecting QC = S. To do this, you need to follow these steps:

QT4 Analysis Steps

1. Run QT4 and load the experiment file *BKEX-61.Q4W* from the *File* menu in the main screen. (You need not do this if you are already working with the file.)

2. Select *Standard analysis* from the *Analysis* menu item.

3. Click *Yes* at the *Notice* screen. This screen reminds you of the fact that you are to prefer *S/N* analysis when there are multiple sample results (four results per trial condition in this case) in each trial condition. In Step 12 you will learn about *S/N* analysis and be able to perform it with the results of this experiment.

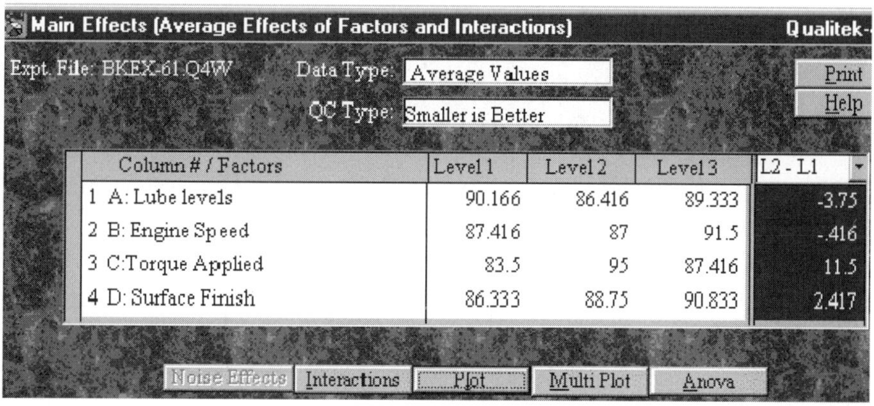

| Main Effects (Average Effects of Factors and Interactions) | | | | Qualitek- |

Expt. File: BKEX-61.Q4W Data Type: Average Values   Print

QC Type: Smaller is Better   Help

Column # / Factors	Level 1	Level 2	Level 3	L2 - L1
1 A: Lube levels	90.166	86.416	89.333	-3.75
2 B: Engine Speed	87.416	87	91.5	-.416
3 C: Torque Applied	83.5	95	87.416	11.5
4 D: Surface Finish	86.333	88.75	90.833	2.417

Noise Effects | Interactions | Plot | Multi Plot | Anova

Figure 6.5 Main effects of factors.

4. Check the box for *Smaller is better* in the *Quality characteristic selection* screen. Click *OK* to proceed.

5. Review the results of the experiments and check to see if they are the same as shown in Figure 6.2. Click *OK* to proceed.

6. In the *Main effects* screen, study the factor influences by selecting the *Plot* or *Multi plot* buttons. Your screen should match that shown in Figure 6.5. Click *OK* to proceed.

7. Examine the rightmost column of the table in the *ANOVA* screen and see how it is graphically displayed using the *Bar graph* or *Pie chart* option buttons. This screen should look like the one in Figure 6.6. Click *OK* to proceed.

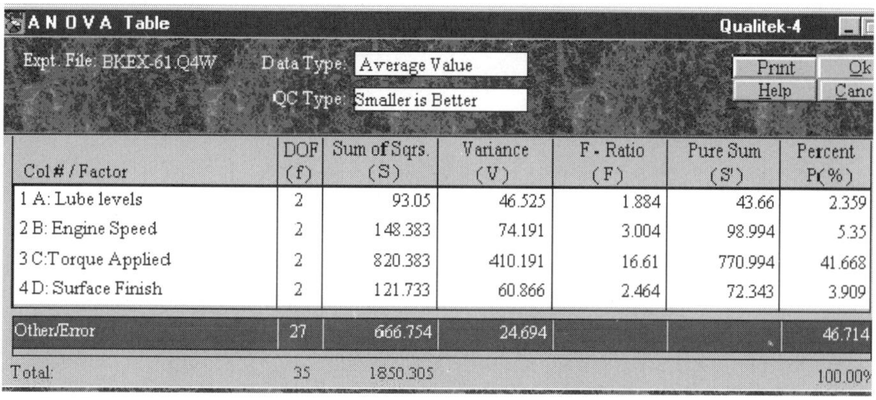

| ANOVA Table | | | | | | Qualitek-4 |

Expt. File: BKEX-61.Q4W Data Type: Average Value   Print | Ok

QC Type: Smaller is Better   Help | Canc

Col # / Factor	DOF (f)	Sum of Sqrs. (S)	Variance (V)	F - Ratio (F)	Pure Sum (S')	Percent P(%)
1 A: Lube levels	2	93.05	46.525	1.884	43.66	2.359
2 B: Engine Speed	2	148.383	74.191	3.004	98.994	5.35
3 C: Torque Applied	2	820.383	410.191	16.61	770.994	41.668
4 D: Surface Finish	2	121.733	60.866	2.464	72.343	3.909
Other/Error	27	666.754	24.694			46.714
Total:	35	1850.305				100.00%

Figure 6.6 ANOVA showing relative factor influences.

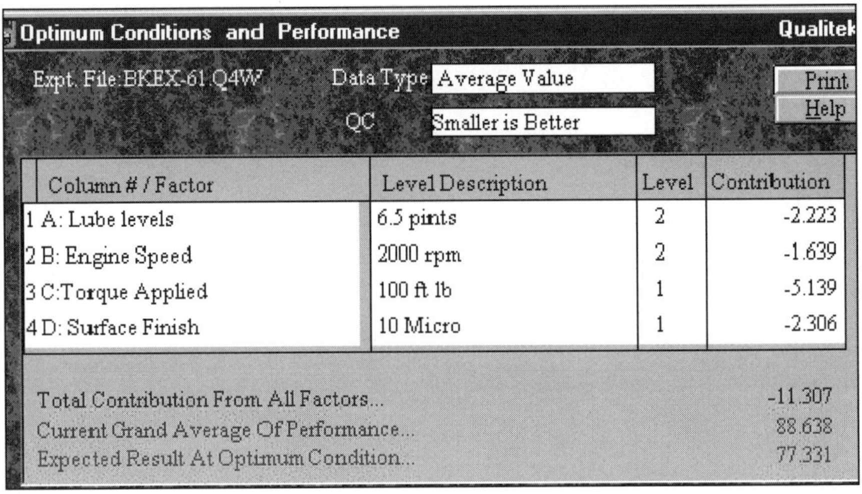

Optimum Conditions and Performance			Qualitek

Expt. File:BKEX-61.Q4W   Data Type: Average Value   Print / Help

QC: Smaller is Better

Column # / Factor	Level Description	Level	Contribution
1 A: Lube levels	6.5 pints	2	-2.223
2 B: Engine Speed	2000 rpm	2	-1.639
3 C:Torque Applied	100 ft lb	1	-5.139
4 D: Surface Finish	10 Micro	1	-2.306

Total Contribution From All Factors...   -11.307
Current Grand Average Of Performance...   88.638
Expected Result At Optimum Condition...   77.331

Figure 6.7 Optimum factor levels and performance (QC = S).

8. Your *Optimum* screen should be as shown in Figure 6.7. This screen indicates the operating condition that will produce the least audible noise (measured in decibels). Click *OK* to return to the main screen.

To analyze the results for a worst noise condition, repeat steps 1 to 3 above. Check the box for *Bigger is better* in the *Quality characteristic selection* screen. Click *OK* to proceed. Follow steps 5 to 8 above and obtain the *Optimum* screen as shown in Figure 6.8. The expected noise level and the operating conditions are determined and dis-

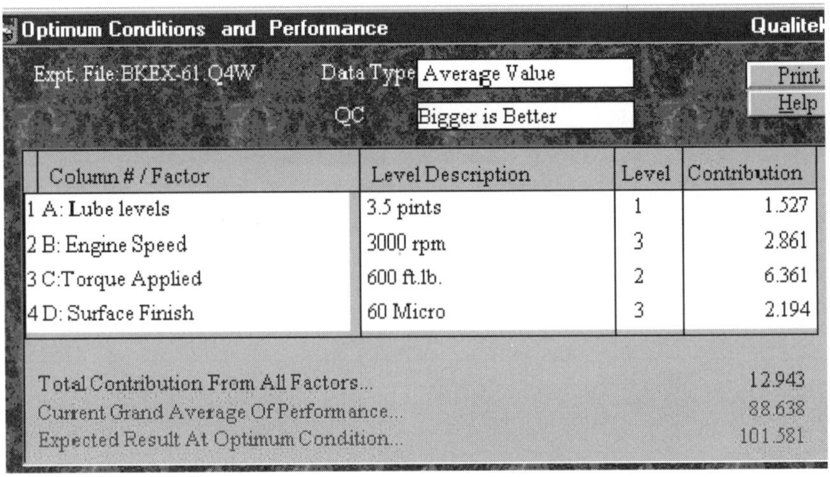

Optimum Conditions and Performance			Qualitek

Expt. File:BKEX-61.Q4W   Data Type: Average Value   Print / Help

QC: Bigger is Better

Column # / Factor	Level Description	Level	Contribution
1 A: Lube levels	3.5 pints	1	1.527
2 B: Engine Speed	3000 rpm	3	2.861
3 C:Torque Applied	600 ft.lb.	2	6.361
4 D: Surface Finish	60 Micro	3	2.194

Total Contribution From All Factors...   12.943
Current Grand Average Of Performance...   88.638
Expected Result At Optimum Condition...   101.581

Figure 6.8 Optimum factor condition (QC=B).

played in the two *Optimum* screens. Thus the minimum level of noise expected is 77.33 dB (see Figure 6.7) and the highest level of noise is 101.58 dB (see Figure 6.8). This means that at any condition of operation, the noise level will be between 77.33 and 101.58 dB.

OVER SIX WAYS TO USE AN L-18 ORTHOGONAL ARRAY

Use an L-18 array to design experiments with:

- 1 two-level factor and 4, 5, 6, or 7 three-level factors

Note: Assign the two-level factor to column 1 and the three-level factors to the other columns arbitrarily.

The L-18 orthogonal array has 1 two- and 7 three-level columns. It is an array that allows the mixing of 1 two-level factor with many three-level factors. The array is constructed such that the first column has nine 1's and nine 2's. The remaining columns all have three levels spread with equal numbers along the depth of the column. This array will be useful to you when you have more than 4 three-level factors or you exceed the capacity of an L-9 array, particularly when one among the many factors you have needs to be a two-level factor. Remember, if the factor is a continuous factor, you are better off making it a three-level factor for the experiment when columns are available. But you may not have the option available when the factor is of discrete (fixed-level) type. For example, if the factor is machine type and you have only two brands of machine, it is not possible for you to consider a third level. In a situation like this, where you have to have a two-level factor, a standard L-18 array will help you design your experiments.

It is not that you have to include a two-level factor to use the L-18 array. Its 7 three-level columns can be used to design experiments to study 5, 6, or 7 three-level factors along with or without a two-level factor. In absence of the two-level factor, the first column must always be left empty. You can review the content of the L-18 array in one of the example experiments provided with QT4. To review, run QT4 and go to the main screen. Select the *Open* option from the *File* menu item and load experiment file *REF-L18A.Q4W*. From the *Edit* menu item, select *Inner array*, which should bring up the L-18 array. Notice that the first column has only levels 1 and 2; all others have three levels. As mentioned before, the column totals shown at the bottom of each column are only for a quick review of column accuracy.

For experimental situations involving 1 two-level factor and 5, 6, or 7 three-level factors, QT4 selects the L-18 array. You can quickly try some of these situations and check the selection. To test the QT4 array selection, try designing an experiment with 1 two-level factor and 5 three-level factors.

1. Run QT4 and go to the main screen.
2. Select the *Automatic design* option from the *Design* menu item.
3. Click *Yes* to the prompt at the *Design new experiment* screen.

Automatic Experiment Design   **Qualitek-**

Types of Factors   Number of Factors/Interaction

[X] Number of 2-Level Control Factors in Experiment [1]

[ ] Interactions Between 2-Level Factors   [-]

[X] Number of 3-Level Control Factors in Experiment [5]
        Interactions With Mixed

Figure 6.9 Input for automatic experimental design.

4. Check the box for the two-level factor and set the number of factors to 1. Also, check the box for three-level factors and set the number of factors to 5 as shown in Figure 6.9. Click *OK* to proceed.

5. QT4 shows that it has selected an L-18 array for the design, as shown in Figure 6.10. Click *OK* to proceed to describe factors and levels for QT4 to complete the design.

You can repeat steps 1 to 3 and enter other factor combinations in step 4 to check that QT4 also selects an L-18 array for the design. In the *Automatic experiment design* screen you may input any of the following conditions and find that in each, QT4 displays the same array selection (see Figure 6.10):

- 1 two-level factor and 6 three-level factors
- 1 two-level factor and 7 three-level factors
- 5 three-level factors only
- 6 three-level factors only
- 7 three-level factors only

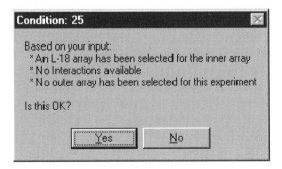

Condition: 25

Based on your input:
  * An L-18 array has been selected for the inner array
  * No Interactions available
  * No outer array has been selected for this experiment

Is this OK?

[Yes]   [No]

Figure 6.10 QT4-selected array for experimental design.

Example 6.2:   Automobile Hood Vibration Study The hood design team of a luxury automobile chose to design an experiment to study the vibration response of the hood under normal operating excitation. The objective of the study was to determine the combination of design parameters that produce the highest fundamental frequency of vibration. Being a continuous system, a structure such as a hood has an infinite number of natural frequencies. It is the lowest natural frequency (also called the *fundamental frequency*) that is of most concern, as it is closest to the excitation due to the road disturbances, suspension characteristics, and engine rotation. The further the hood natural frequency is from these disturbances, the less likely it is to resonate. Hood vibration is quite obvious to the customer and is considered a critical quality item.

Background.   The hood for this automobile, as with most, is constructed using an inner panel and an outer panel made of steel. The inner panel is shaped to provide structural strength. It is welded to the outer panel with insulating and sound-deadening materials. It is further reinforced by additional plate structures at the hinges, lock, and other support locations. The design also incorporates rubber bumpers in front and back to absorb the shock from impact while closing the hood.

In general, the hood frequency of vibration increases with stiffness and decreases with weight. Incorporating heavier gauge sheet metal to increase stiffness does not necessarily increase the frequency of vibration, as it also adds weight. The frequency of vibration is also affected by the location and structural properties of the support members. To avoid vehicle vibration excited by the low-frequency road excitations (potholes, bumps, etc.); a higher natural frequency of hood is desirable. Finding the design with the highest frequency with the least mass and cost is truly a design optimization task.

Experiment design considerations.   The team identified seven factors that are considered to influence the vibration that they have the option to control (Figure 6.11). Most of the factors, except the rear bumper mounting locations, were considered at three levels. The geometrical constraints were such that the Rear Hood had to be mounted at two distinct locations. The experiment design utilized an L-18 array, which called for 18 separate trial conditions. Eighteen separate hoods were fabricated and tested for vibration response.

Test results.   The tests were performed under real-life support structure and constraints. A vibration shaker placed at locations identified as the main transmission points excited the hood. The first natural frequency of the hood was recorded for several excitations. The average of the recorded frequency [hertz (Hz; cycles per second)] was considered as the result for the sample (Figure 6.12).

Objectives.   Laboratory and road tests showed that the 90th percentile excitation from engine vibration and road surface conditions was around 19 Hz. To stay away from this frequency, the design team was looking for the hood frequency to be over 25 Hz. Staying within the design conditions considered in the study, would they be able to achieve their objective?

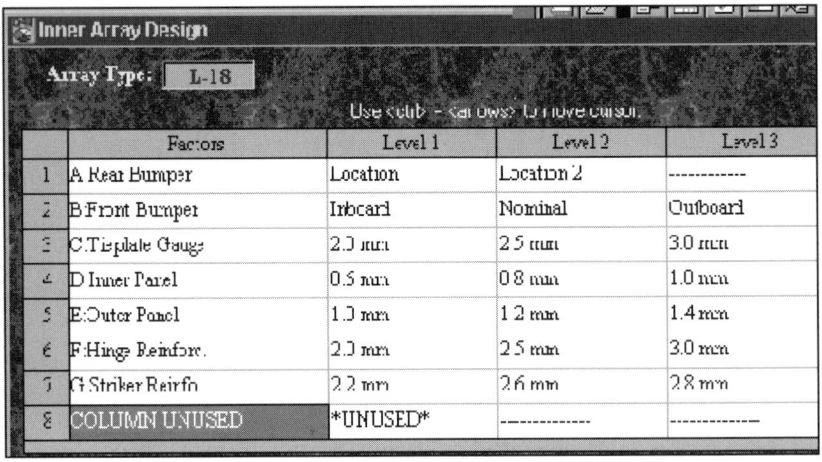

	Factors	Level 1	Level 2	Level 3
1	A:Rear Bumper	Location	Location 2	------------
2	B:Front Bumper	Inboard	Nominal	Outboard
3	C:Tieplate Gauge	2.0 mm	2.5 mm	3.0 mm
4	D:Inner Panel	0.5 mm	0.8 mm	1.0 mm
5	E:Outer Panel	1.0 mm	1.2 mm	1.4 mm
6	F:Hinge Reinforc.	2.0 mm	2.5 mm	3.0 mm
7	G:Striker Reinfo	2.2 mm	2.6 mm	2.8 mm
8	COLUMN UNUSED	*UNUSED*	-------------	-------------

Figure 6.11 Factors and levels for experiments in Example 6.2.

Solution: This example experiment has been saved as *BKEX-62.Q4W*. You may review the design and the results using the *Edit* menu option or practice the design process by executing the following steps.

QT4 Design Steps

1. Run QT4 and go to the main screen.
2. Select *Automatic design* from the *Design* menu item.

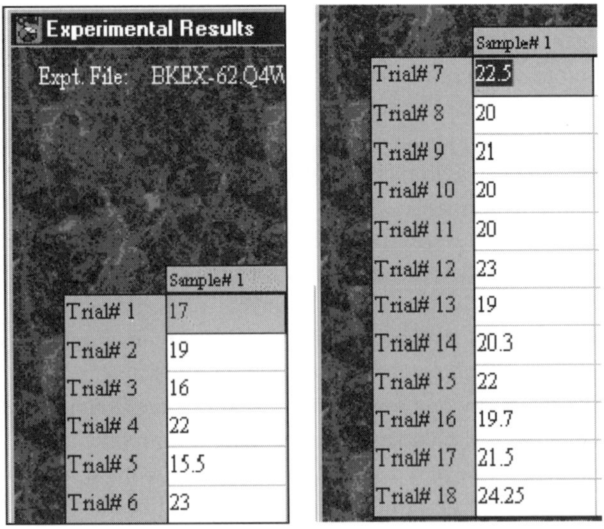

	Sample# 1
Trial# 1	17
Trial# 2	19
Trial# 3	16
Trial# 4	22
Trial# 5	15.5
Trial# 6	23
Trial# 7	22.5
Trial# 8	20
Trial# 9	21
Trial# 10	20
Trial# 11	20
Trial# 12	23
Trial# 13	19
Trial# 14	20.3
Trial# 15	22
Trial# 16	19.7
Trial# 17	21.5
Trial# 18	24.25

Figure 6.12 Experimental results for Example 6.2.

3. Click *Yes* at the prompt for *New experiment design*.

4. In the *Automatic design* screen, check the box for two-level factors and set the number of factor to 1. Also check the box for three-level factors and set the number of factors to 6. Click *OK* to proceed.

5. Click *Yes* when QT4 displays that an L-18 array has been selected for the experiment design.

6. Type in the description of 1 two-level factor and its levels. Click *OK* to proceed.

7. Type in the descriptions of 6 three-level factors and their levels. Click *OK* to proceed.

8. Review descriptions of factors and levels you just entered and the columns QT4 has assigned them. Your screen should look as shown in Figure 6.11. You may edit any descriptions, if necessary, on this screen. Click *OK* when done.

9. Enter the project title, objectives, and the quality characteristic (QC).

10. Review the array used for the design. Click *OK* to save the experiment file.

11. When prompted to save the experiment file, click *Cancel* here and return to the main screen, as this experiment has already been saved as *BKEX-62.Q4W*. (If this experiment were not predesigned, you would need to name this experiment file.)

12. When in main screen, load the experiment file *BKEX-62.Q4W* by selecting the *Open* option from the *File* menu item.

13. Compare the experimental results displayed next to the array in the main screen with that shown in Figure 6.12. You may have to scroll down to view trial numbers beyond trial 8. You can also review factors, levels, and results by selecting the appropriate options from the *Edit* menu item. You may practice editing any description or results, but QT4 will not allow you to update this example experiment.

To find out if the hood can be constructed such that it will produce the desirable frequency, you need to analyze the results of the 18 trial conditions and determine the optimum condition and performance. Run QT4 and execute the following steps to determine the performance expected.

QT4 Analysis Steps

1. Run QT4 and load the experiment file *BKEX-62.Q4W* from the *File* menu in the main screen (you need not do this if you are already working with the file).

2. Select *Standard analysis* from the *Analysis* menu item.

3. Check the box for *Bigger is better* in the *Quality characteristic selection* screen. Click *OK* to proceed.

4. Review the results of the experiments and check to see that they are the same as shown in Figure 6.12. Click *OK* to proceed.

5. In the *Main effects* screen, study the factor influences by selecting the *Plot* or *Multiplot* buttons. Your screen should match that shown in Figure 6.13. The

Main Effects (Average Effects of Factors and Interactions)				Qualitek-4
Expt. File EKEX-62.Q4W	Date Type: Average Values			Print
	QC Type: Bigger is Better			Help

Column # / Factors	Level 1	Level 2	Level 3	L2 - L1
1 A:Rear Bumper	19.555	21.083		1.527
2 B:Front Bumper	19.166	20.3	21.491	1.134
3 C:Tieplate Gauge	20.033	19.383	21.541	.651
4 D:Inner Panel	20.583	20.541	19.832	-.042
5 E:Outer Panel	21.175	20.949	18.833	-.226
6 F:Hinge Reinforc.	20.583	21.091	19.283	.508
7 G:Striker Reinfo	19.416	21.333	20.208	1.916

Figure 6.13 Main effects of factors in Example 6.2.

column labeled "L2–L1" is an additional piece of information derived from the factor average effects listed in the level 1 and level 2 columns. As a default setting, QT4 displays the difference of the average effects between levels 1 and 2. For factors with three or four levels, other level differences can be displayed by clicking the arrow just right of "L2–L1" and selecting the desired level differences. The differences in average effects can help you create a mental picture of the factor influence when graphs are not readily available. You can examine the nature of factor influences by reviewing the factor effects. To do so, click the *Plot* button. This option shows the plot of the factor average effects (also called main effects). Notice that for a factor with two levels (such as rear bumper), a single (blue color) line joins the average effects. But for factor levels over two levels, as for factor C (tieplate), as shown in Figure 6.14, a smooth curve attempts to join the points (green line). This curve is the least-squares curve fit for the points and represents the actual behavior. If the factor is of *continuous* type, the response is also likely to be continuous when it is plotted against the levels. Click *OK* or *ANOVA* to proceed.

6. Examine the rightmost column of the table in the *ANOVA* screen and see how it is graphically displayed using the *Bar graph* or *Pie chart* option button. This screen should look like the one in Figure 6.15. Notice that the inner panel has no influence on the variation in the hood frequency. This means that there is no benefit to be derived by changing the gauge of this part. You should prefer a lighter gauge or stay with the current design. Further, there is no need to hold the thickness to its precise dimension. Any tolerance on the thickness of this part can also be removed. Also observe that the total influence of all the factors amounts to about 45% (100 − 55.06). This means that 55% of the variation in the hood frequency is expected from factors other than those included in the study. Other sources of influence could be from many uncontrollable factors such as manufacturing, experimental setup, and environmental variations. How

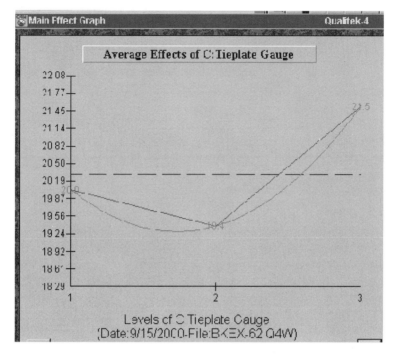

Figure 6.14 Main effect of factor tieplate.

Col# / Factor	DOF (f)	Sum of Sqrs. (S)	Variance (V)	F - Ratio (F)	Pure Sum (S')	Percent P(%)
1 A:Rear Bumper	1	10.303	10.303	3.003	7.006	7.063
2 B:Front Bumper	2	16.22	8.11	2.313	9.225	9.304
3 C:Tieplate Gauge	2	14.711	7.355	2.103	7.716	7.782
4 D:Inner Panel	2	2.132	1.066	.304	0	0
5 E:Outer Panel	2	20.028	10.014	2.863	13.033	13.144
6 F:Hinge Reinforc.	2	10.437	5.218	1.492	3.442	3.471
7 G:Striker Reinfc.	2	11.132	5.566	1.591	4.137	4.172
Other/Error	4	13.989	3.497			55.062
Total	17	99.155				100.00%

Expl. File: BKEX-62.Q4W Data Type: Average Value QC Type: Bigger is Better

Figure 6.15 ANOVA for experiment of Example 6.2.

Optimum Conditions and Performance			Qualitek-4

Expt. File: BEEX-62.Q4W	Data Type Average Value		Print
	QC Bigger is Better		Help

Column # / Factor	Level Description	Level	Contribution
1 A:Rear Bumper	Location 2	2	.763
2 B:Front Bumper	Outboard	3	1.172
3 C:Tieplate Gauge	3.0 mm	3	1.222
4 D:Inner Panel	0.6 mm	1	.263
5 E:Outer Panel	1.0 mm	1	.855
6 F:Hinge Reinforc.	2.5 mm	2	.772
7 G:Striker Reinfo.	2.6 mm	2	1.013

Total Contribution From All Factors...	6.06
Current Grand Average Of Performance...	20.319
Expected Result At Optimum Condition...	26.379

Figure 6.16   Optimum condition and performance for hood design of Example 6.2.

do these factors influence your decision about future designs? What can you do about it? These and many similar questions will be addressed later in the book. At this point, just be aware that regardless of the value of the influence of all other error sources, the relative influence of each factor is a useful number. Click *OK* or *Optimum* to proceed.

7. Your *Optimum* screen should be as shown in Figure 6.16. This screen indicates the operating condition that will produce the least audible noise (measured in decibels). Click *OK* to return to the main screen.

The expected performance at the optimum condition (see Figure 6.16) indicates that the highest frequency of the hood (26.379) exceeds the level required. The design specification for the hood is also indicated by the second column in the *Optimum condition* screen (Figure 6.16).

SIX WAYS TO USE AN L-27 ORTHOGONAL ARRAY

Use an L-27 array to design experiments with:

- 8, 9, 10, 11, 12, or 13 three-level factors

Note: Factors are assigned to the columns arbitrarily.

The L-27 orthogonal array has 13 three-level columns. You will use this array to design experiments when all your factors are at three levels and the number of such factors is between 8 and 13. This is the biggest three-level array that we discuss in this book. Naturally, you will be limited to designing experiments that can accommodate up to 13 factors. You can review the content of the L-27 array in one of the example experiments provided within QT4. To review, run QT4 and go to the main screen. Select the *Open* option from the *File* menu item and load experiment file *REF-L27A.Q4W*. From the *Edit* menu item, select *Inner array*, which should bring up the L-27 array used for this example experiment. Be sure to click the scroll bar to review all 27 rows of the array. Observe that each of the 13 columns has an equal number of 1's, 2's, and 3's distributed along the columns and that each column totals to 54.

Example 6.3:   Passenger Car Steering Wheel Crush Simulation Study The use of mathematical models to optimize designs of machines and structures is quite common. Because desktop computers are increasingly powerful, use of an analytical model to experiment with different design possibilities is much more cost-effective. The design and analysis team of a newly redesigned popular passenger car used an analytical simulation (using finite element and lumped-mass models) to optimize the design of the steering column. The model allowed analysts to input a large number of structural parameters and study, among many things, the collapse of the steering wheel/column assembly under the equivalent impact of a driver when vehicles collide. Federal Motor Vehicle Safety Standard 204 (FMVSS 204) requires that vehicles be designed such that the steering column deformation does not exceed a fixed distance (say 125 mm at 30 mph). For certification purposes, automobile manufacturers are required to demonstrate the performance by running instrumented prototype vehicles with anthromorphic dummies to a rigid wall at a speed of 30 mph. The force a driver dummy exerts on the steering wheel is obtained from the test, which usually has over 150 transducers (instruments to measure displacements, acceleration, force, etc.) placed at numerous locations on the vehicle and the dummy. This force serves as the input to the simulation model for evaluating the structural behavior of the steering column under study.

For design optimization purposes, 11 factors were identified for the study. Each factor was assigned three levels (Figure 6.17). For all factors except *K* (pencil bracket), material thickness was varied by the steps specified by their level description. Using the 11 factors and levels, QT4 designed an L-27 experiment. The 27 separate trial conditions (see trial 1 in Figure 6.18) described by QT4 provide input to set the computer runs (program executions) for the simulation model.

Objective.   The crush distance from the simulation model for the 27 separate runs were recorded and entered into QT4 experiment file *BKEX-63.Q4W* (Figure 6.19). The results were expressed in millimeters of crush distance of the steering wheel face with respect to an undisturbed reference point. The objective of the experimental study has been to reduce the crush distance at a point below the legal limit of 125 mm. The current performance has been observed to be about 135 mm.

(a) From the analytical model results, identify the three factors that have the most influence.

Factors	Level 1	Level 2	Level 3
1: A:Capsule Bracket	5.0 mm	5.5 mm	6.0 mm
2: B:Upper Jacket	1.5 mm	1.88 mm	2.00 mm
3: C:Toe Plate	3.5 mm	4.00 mm	4.56 mm
4: D:Lower Jacket	1.50 mm	1.88 mm	2.12 mm
5: E:Exten. Bracket	1.5	1.88 mm	2.12 mm
6: F:Qtr. Beam-Upper	2.00 mm	2.50 mm	3.00 mm
7: G:Qtr. Beam-Lower	2.00 mm	2.50 mm	3.00 mm
8: H:Retainer Sleeve	1.50 mm	1.88 mm	2.26 mm
9: I:Rim Wire Dia.	10 mm	11 mm	12 mm
10: J:Spoke Wire Dia	6 mm	8 mm	10 m
11: K:Pencil Bracket	Absent	Flat	Narrow

Figure 6.17 Factors and level descriptions (Example 6.3).

(b) Is it possible to achieve the desired result (crush distance below 125 mm), if only these three factors are adjusted?

Solution: This example experiment has been designed using QT4 and is saved as *BKEX-63.Q4W*. You may review the design and results using the *Edit* menu option or practice the design process by executing the following steps.

Descriptions Trial Conditions		Qual
Trial Condition 1 (Random order for running this Trial is 3)		
Factors	Level Description	Level #
A:Capsule Bracket	5.0 mm	1
B:Upper Jacket	1.5 mm	1
C:Toe Plate	3.5 mm	1
D:Lower Jacket	1.50 mm	1
E:Exten. Bracket	1.5	1
F:Qtr. Beam-Upper	2.00 mm	1
G:Qtr. Beam-Lower	2.00 mm	1
H:Retainer Sleeve	1.50 mm	1
I:Rim Wire Dia.	10 mm	1
J:Spoke Wire Dia	6 mm	1
K:Pencil Bracket	Absent	1

Figure 6.18 Sample input for computer simulation run.

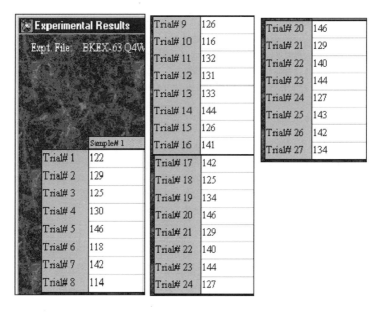

Figure 6.19   Results of simulation runs.

QT4 Design Steps

1. Run QT4 and go to the main screen.

2. Select *Automatic design* from the *Design* menu item.

3. Click *Yes* at the prompt for *New experiment design.*

4. In the *Automatic design* screen, check the box for three-level factors and set the number of factors to 11. Click *OK* to proceed.

5. Click *Yes* when QT4 displays that an L-27 array is selected for the experimental design.

6. Type in the descriptions of 11 three-level factors and their levels. When you are done typing all the descriptions, your screen will contain all that is shown in Figure 6.17. Click *OK* to proceed.

7. Review the descriptions of factors and levels you just entered and the columns QT4 has assigned them. Notice that QT4 has assigned all the factors in the first 11 columns and left the last two columns unused. You may edit any description, if necessary, here in this screen. Click *OK* when done.

8. Enter the project title, objectives, and the quality characteristic (QC = S).

9. Review the array used for the design. Note that the QT4 has modified the levels of the last two columns to zeros, as there are no factors assigned to them. Click *OK* to proceed to save the experiment file.

10. When prompted to save the experiment file, click *Cancel* and return to the main screen, as this experiment has already been saved as *BKEX-63.Q4W*. (If this experiment was not predesigned, you would need to name this experiment file.) Once you are in the main screen, load the experiment file *BKEX-63.Q4W* by selecting the *Open* option from the *File* menu item.

Experiment Review

1. Upon experiment design completion, you can review the 27 trial conditions that served as the input to set up the simulation runs. To review the trial condition, select *Trial conditions* from the *Conditions* menu item. QT4 always starts with the first trial condition. You should see a screen like Figure 6.18. You may now click on the *Next trial* button to view all other conditions, one at a time. Click *Return* when you are done.

2. Compare the experimental results displayed next to the array in the main screen with that shown in Figure 6.19. You may have to scroll down to view trial numbers beyond trial 8. You can also review factors, levels, and results by selecting the appropriate options from the *Edit* menu item. You may edit any descriptions or results in these screens, but QT4 will not allow you to update this example experiment.

The answers to the questions posed in this example can be obtained by analyzing the simulation model results of the 27 trial conditions. Run QT4 and execute the following steps to determine the factor influences and performance expected at the optimum condition.

QT4 Analysis Steps

1. Run QT4 and load the experiment file *BKEX-62.Q4W* from the *File* menu in the main screen (you need not do this if you are already working with this file).

2. Select *Standard analysis* from the *Analysis* menu item.

3. Check the box for *Bigger is better* in the *Quality characteristic selection* screen. Click *OK* to proceed.

4. Review the results of the experiments and check to see that they are the same as shown in Figure 6.19. Click *OK* to proceed.

5. In the *Main effects* screen, study the factor influences by selecting the *Plot* or *Multi plot* buttons. You can examine the nature of factor influences by reviewing the factor effects. To do so, click the *Plot* button. Notice that factors *A*, *C*, and *H* behave almost linearly. In their cases, the *least-squared curves* are almost like the straight lines. Do nothing at this point. You have data that establish that the crush distance is dependent linearly on these factors. For future studies you will be able to treat these factors confidently as two-level factors and perhaps reduce the size of your experiment. You will know how nonlinear all other factors in the experiment are by examining each carefully. You can also click on the *Multi*

plot button to look at several factor influence plots at a time. Click *OK* or *ANOVA* to proceed.

6. The ANOVA table shown in Figure 6.20 shows the effects of all factors assigned to columns of the array zero and nonzero influence. The numbers in the right-hand column (you will learn about other columns later in the book) shows how sensitive the result (crush distance in this case) is to the change of levels of the factors. As you notice in this and other problems, not all factors will ever have equal or nonzero influence. Rather, it is more natural that approximately half the factors will have an important influence; the other half will not. Whether a factor is important or not is found by a test of significance, which you will learn later. In this example, you will simply practice ignoring (pooling) factors when they are deemed unimportant. The most obvious place to start revising (cleaning up) ANOVA calculation is by pooling factors, one at a time, starting with the one with the least influence. The relative influence can be determined by comparing the percent column or, when necessary, by comparing the numbers under the sums of squares (S). To pool factors, start with factor *C*: toe plate (S = 0.51), as it has the least influence. Click on the description that is on *C*: toe plate in column 1. Click *Yes* when prompted about pooling. Pool factor *K*: pencil bracket next. Continue to pool in this way until you have only three factors left. You can expedite this process by pooling factors automatically below a certain

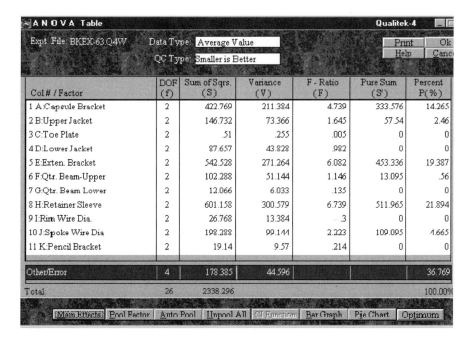

Col# / Factor	DOF (f)	Sum of Sqrs. (S)	Variance (V)	F - Ratio (F)	Pure Sum (S')	Percent P(%)
1 A:Capsule Bracket	2	422.769	211.384	4.739	333.576	14.265
2 B:Upper Jacket	2	146.732	73.366	1.645	57.54	2.46
3 C:Toe Plate	2	.51	.255	.005	0	0
4 D:Lower Jacket	2	87.657	43.828	.982	0	0
5 E:Exten. Bracket	2	542.528	271.264	6.082	453.336	19.387
6 F:Qtr. Beam-Upper	2	102.288	51.144	1.146	13.095	.56
7 G:Qtr. Beam Lower	3	12.066	6.033	.135	0	0
8 H:Retainer Sleeve	2	601.158	300.579	6.739	511.965	21.894
9 I:Rim Wire Dia.	2	26.768	13.384	.3	0	0
10 J:Spoke Wire Dia	3	198.288	99.144	2.323	109.095	4.665
11 K:Pencil Bracket	2	19.14	9.57	.214	0	0
Other/Error	4	178.385	44.596			36.769
Total:	26	2338.296				100.00%

Figure 6.20 Unrolled ANOVA table for Example 6.3.

confidence level. To pool automatically, click on the *Auto pool* button. QT4 prompts you for a confidence level number. Key in 90 (for a 90% confidence level) or click *OK* if it is already there. QT4 automatically pools all factors with a confidence level below 90% and displays the revised ANOVA as shown in Figure 6.21. You may try this again by clicking on the *Unpool all* button and pool again. As you may have noticed already, the numbers in the percent column will be slightly different depending on whether you *pooled* automatically or *pooled* one factor at a time. Do not be alarmed about this discrepancy. When you are done with the automatic *pooling*, click *OK* to proceed to the *Optimum* screen.

7. QT4 is programmed to calculate performance at the optimum condition by considering only the factors that are significant. Understandably, factors that are determined to be insignificant are pooled out. The optimum condition and the expected performance with the three factors adjusted to the condition desired are as shown in Figure 6.22.

In answer to the questions:

(a) The three factors that have the most influence are:

H: retainer sleeve   21.98%

E: extension bracket   19.38%

A: capsule bracket   14.26%

(b) By adjusting these three factors alone, the crush distance can be reduced to 116.29 mm (see Figure 6.22), which is below the value desired.

Col # / Factor	DOF (f)	Sum of Sqrs. (S)	Variance (V)	F - Ratio (F)	Pure Sum (S')	Percent P(%)
1 A:Capsule Bracket	2	422.769	211.384	4.739	333.576	14.265
2 B:Upper Jacket	(2)	(146.732)		POOLED	(CL=70.31%)	
3 C:Toe Plate	(2)	(.51)		POOLED	(CL= *NC*)	
4 D:Lower Jacket	(2)	(87.657)		POOLED	(CL= *NC*)	
5 E:Exten. Bracket	2	542.528	271.264	6.082	453.336	19.387
6 F:Qtr. Beam-Upper	(2)	(102.288)		POOLED	(CL=59.96%)	
7 G:Qtr. Beam-Lower	(2)	(12.066)		POOLED	(CL= *NC*)	
8 H:Retainer Sleeve	2	601.158	300.579	6.739	511.965	21.894
9 I:Rim Wire Dia.	(2)	(26.768)		POOLED	(CL= *NC*)	
10 J:Spoke Wire Dia	(2)	(198.288)		POOLED	(CL=77.99%)	
11 K:Pencil Bracket	(2)	(19.14)		POOLED	(CL= *NC*)	
Other/Error	20	771.834	341.32			44.454
Total:	26	2338.296				100.00%

Figure 6.21 Pooled ANOVA (Example 6.3).

Column # / Factor	Level Description	Level	Contribution
1 A:Capsule Bracket	5.0 mm	1	-4.63
5 E:Exten. Bracket	2.12 m	3	-5.852
8 H:Retainer Sleeve	1.50 mm	1	-5.852

Optimum Conditions and Performance   **Qualitek**

Expt. File:BKEX-63.Q4W   Data Type Average Value   Print

QC   Smaller is Better   Help

Total Contribution From All Factors...		-16.334
Current Grand Average Of Performance...		132.629
Expected Result At Optimum Condition...		116.295

Figure 6.22   Optimum condition and performance with three factors.

FOUR WAYS TO USE A MODIFIED L-16 ORTHOGONAL ARRAY

Use a modified L-16 array to design experiments with:

- 2, 3, 4, or 5 four-level factors

Note: Factors are assigned to the columns arbitrarily.

The L-16 modified orthogonal array has 5 four-level columns. This array is said to be *modified*, as it is prepared by upgrading columns of the standard L-16 array with 15 two-level columns. (*Note*: QT4 uses the M-16 designation for this array.) Later you will learn how to modify standard L-8, L-16, and L-32 arrays to upgrade their two-level columns into three- or four-level columns. When you have established the need to study factors at four levels, you will find this array to be quite convenient to design your experiments with 2 to 5 four-level factors.

Higher factor levels will usually be expensive. There are very few occasions where you will need to run factors at four levels. Usually, results of three-level studies will clearly establish whether or not the factor behaves nonlinearly. It is only when your prior studies or strong conviction tells you that the factor may have a sinusoidal behavior within the range of levels selected for the study that you should insist on having four levels for the factor. The other occasions where you would like to study factors at four levels is when time or the number of experiments is not an issue. Studies with four-level factors, however, will provide you with most information about factor behavior. You may prefer to adopt this strategy when you have the option to experiment with analytical simulations where the cost of computer runs is not costly and you have a larger number of factors.

Although you may use this array for 2 four-level factors, the QT4 *Automatic design* option does not offer this design. Should you have such a situation, you should use the *Manual design* option and assign the factors to the first two columns of the array. You can review the content of the modified L-16 array by one of the example experiments provided with QT4. To review, run QT4 and go to the main screen. Select the *Open* option from the *File* menu item and load the experiment file *REF-L16M.Q4W*. From the *Edit* menu item, select *Inner array*, which should bring up the modified L-16 array (Figure 6.23). Observe that each of the five columns has an equal number of 1's, 2's, 3's, and 4's distributed along the columns and that each column has a total of 40.

Example 6.4: Chemoprevention of Colon Cancer—Laboratory Experiment Research associates at a local medical school routinely carry out studies to determine the effects of various chemicals and nutrients on colon cancer. A major part of their experiments is carried out using laboratory-bred rats with a life span of two to four years. The test rats are first injected with tumor-causing drugs that produce tumors of different numbers and sizes. Each tumor is characterized by its percentage of malignant cells (malignant tumor cells lose their normal appearance; benign cells do not). During the study, tumor-carrying rats are fed and injected with a variety of chemicals and nutrients. Changes in tumors (number, size, and degree of malignancy) are monitored over a period of time.

In one such study, the effects of five factors (chemicals and nutrients) were studied using a modified L-16 array. The factors included in the study and their levels are as

Edit Inner Array

Array Type: **L-1**

	1	2	3	4	5
1	1	1	1	1	1
2	1	2	2	2	2
3	1	3	3	3	3
4	1	4	4	4	4
5	2	1	2	3	4
6	2	2	1	4	3
7	2	3	4	1	2
8	2	4	3	2	1
9	3	1	3	4	2
10	3	2	4	3	1
11	3	3	1	2	4
12	3	4	2	1	3
13	4	1	4	2	3
14	4	2	3	1	4
15	4	3	2	4	1
16	4	4	1	3	2
Total	40	40	40	40	40

Figure 6.23 Modified L-16 array.

	Factors	Level 1	Level 2	Level 3	Level 4
1	Folic Acid	400 micro	500 micro gm	600 micro gm	700 micro gm
2	Phosphorus	60 micro	80 micro gm	100 micro gm	120 micro gm
3	Mineral - Magnesi	100 micro	200 micro gm	300 micro gm	400 micro gm
4	Calcium	50 micro	100 micro gm	150 micro gm	200 micro gm
5	Fibrous Food	500 micro	700 micro gm	900 micro gm	1100 micro gm

Array Type: M-16

Use <ctrl> + <arrows> to move cursor.

Inner Array Design   Print   Ok   Help   Cancel

Figure 6.24   Factor and level description for Example 6.4.

described in Figure 6.24. The tests as prescribed by the 16 experimental conditions were carried out with groups of five specimens (rat) in each condition. At the end of the test periods, the tumors in each specimen were examined for reduction in number, size, and percent malignancy. Upon examination, each specimen was given a numerical number ranging between 0 and 50, with 0 representing complete absence of tumor at the end of the test period. For analysis purposes, the average performances (QC = S) of the groups were considered as the results for the trial conditions, as shown in Figure 6.25. Based on the test results, determine:

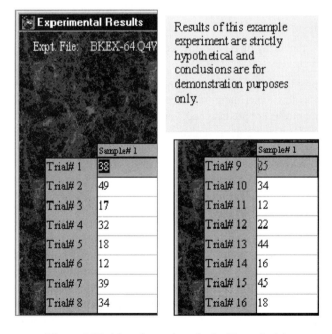

Experimental Results

Expt. File: BKEX-64.Q4V

Results of this example experiment are strictly hypothetical and conclusions are for demonstration purposes only.

	Sample# 1
Trial# 1	38
Trial# 2	49
Trial# 3	17
Trial# 4	32
Trial# 5	18
Trial# 6	12
Trial# 7	39
Trial# 8	34

	Sample# 1
Trial# 9	25
Trial# 10	34
Trial# 11	12
Trial# 12	22
Trial# 13	44
Trial# 14	16
Trial# 15	45
Trial# 16	18

Figure 6.25   Experimental results for Example 6.4.

(a) The trend of influence of fibrous food.

(b) The two factors most influential in reducing tumors.

(c) The number of points that can be gained by folic acid alone.

Solution: This example experiment has been designed using QT4 and saved as *BKEX-64.Q4W*. You may review the design and the results using the *Edit* menu option or practice the design process by executing the following steps.

QT4 Design Steps

1. Run QT4 and be in the main screen.

2. Select *Automatic design* from the *Design* menu item.

3. Click *Yes* at the prompt for *New experiment design.*

4. In the *Automatic design* screen, check the box for four-level factors and set the number of factors to 5. Click *OK* to proceed.

5. Click *Yes* when QT4 displays that an M-16 is selected for the experiment design.

6. Type in the descriptions of 5 four-level factors and their levels. Click *OK* to proceed.

7. Review the descriptions of factors and levels you just entered and the columns that QT4 has assigned them. Notice that QT4 has assigned the five factors you described to the five columns. You may edit any description, if necessary, here in this screen (see Figure 6.24). Click *OK* when done.

8. Enter the project title, objectives, and the quality characteristic (QC = S).

9. Review the array used for the design. Click *OK* to proceed to save the experiment file.

10. When prompted to save the experiment file, click *Cancel* and return to the main screen, as this experiment has already been saved as *BKEX-64.Q4W*. (If this experiment was not predesigned, you would need to name this experiment file.) Once you are in the main screen, load the experiment file *BKEX-64.Q4W* by selecting the *Open* option from the *File* menu item.

Experiment Review

1. Upon experiment design completion, you can review the 16 trial conditions that served as the condition/prescription of medication and diet for the various groups of test specimens. To review the trial condition, select *Trial conditions* from the *Conditions* menu item in the main screen. QT4 always starts with the first trial condition. You may now click on the *Next trial* button to view all other conditions, one at a time. Click *Return* when you are done.

2. Compare the experimental results displayed next to the array in the main screen with those shown in Figure 6.25. You will need to scroll down to view trial numbers beyond trial 8. You can also review factors, levels, and results by

selecting the appropriate options from the *Edit* menu item. You may edit any descriptions or results in these screens, but QT4 will not allow you to update this example experiment.

The answers to the questions posed in this example can be obtained by analyzing the results of the 16 trial conditions. Run QT4 and execute the following steps to determine the factor influences, their relative influence, and individual factor contributions.

QT4 Analysis Steps

1. Run QT4 and load the experiment file *BKEX-64.Q4W* from the *File* menu in the main screen. (You need not do this if you are already working with this file.)

2. Select *Standard analysis* from the *Analysis* menu item.

3. Check the box for *Smaller is better* in the *Quality characteristic selection* screen. Click *OK* to proceed.

4. Review the results of the experiments and check to see that they are the same as shown in Figure 6.25. Click *OK* to proceed.

5. In the *Main effects* screen, study the factor influences by selecting the *Plot* button. You can examine the nature of factor influences by reviewing the factor effects one at a time. You should click at the bottom-right scroll bar in the plot

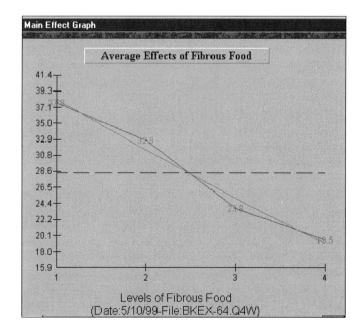

Figure 6.26   Main effect of fibrous food.

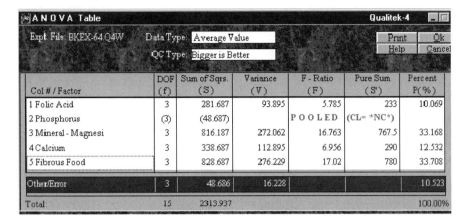

Expt. File: BKEX-64.Q4W		Data Type: Average Value					
		QC Type: Bigger is Better					
Col # / Factor	DOF (f)	Sum of Sqrs. (S)	Variance (V)	F - Ratio (F)	Pure Sum (S')	Percent P(%)	
1 Folic Acid	3	281.687	93.895	5.785	233	10.069	
2 Phosphorus	(3)	(48.687)		P O O L E D	(CL= *NC*)		
3 Mineral - Magnesi	3	816.187	272.062	16.763	767.5	33.168	
4 Calcium	3	338.687	112.895	6.956	290	12.532	
5 Fibrous Food	3	828.687	276.229	17.02	780	33.708	
Other/Error	3	48.686	16.228			10.523	
Total:	15	2313.937				100.00%	

Figure 6.27   ANOVA (Example 6.4).

screen to arrive at the average effect display of fibrous food (Figure 6.26) [answer to question (a)]. Click *Return* to go back to the main effect screen and then click *OK* or *ANOVA* to proceed.

6. Your ANOVA table should look like the one shown in Figure 6.27 after you have pooled one factor (phosphorus). Examine the right column of ANOVA, which represents the relative influence of the factors to the variation of result, and find that mineral-magnesium and fibrous food are the top two influential factors [answer to question (b)]. Click *OK* or *Optimum* to proceed.

7. The *Optimum condition* screen shows the contribution of individual factors to the most achievable performance as well as the condition of the factors. The effect of folic acid is found to lower 5.562 (first row, last column in Figure 6.28) the scale used to measure the various characteristics of the tumors in the laboratory specimens.

NINE WAYS TO USE A MODIFIED L-32 ORTHOGONAL ARRAY

Use a modified L-18 array to design experiments with:

- 1 two-level factor and 5, 6, 7, 8, or 9 four-level factors
- 6, 7, 8, or 9 four-level factors

Note: Assign a two-level factor to column 1 and four-level factors to the other columns arbitrarily.

The modified L-32 orthogonal array has 1 two-level and 9 four-level columns. It is an array that allows you to design experiments mixing 1 two-level factor with many four-level factors. The array is constructed such that the first column has sixteen 1's and sixteen 2's. The remaining columns all have four levels distributed in equal numbers

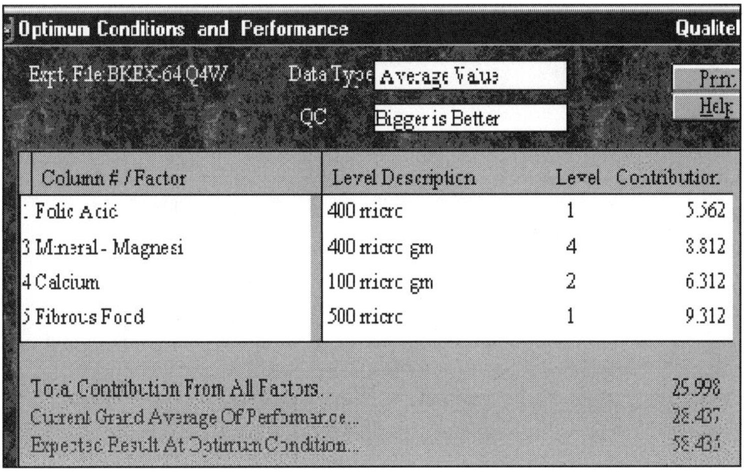

Optimum Conditions and Performance			Qualitel

Expt. File BKEX-64.Q4W   Data Type [Average Value]   [Prin:]
                                        QC [Bigger is Better]   [Help]

Column # / Factor	Level Description	Level	Contribution
2 Folic Acid	400 micrc	1	5.562
3 Mineral- Magnesi	400 micrc gm	4	8.812
4 Calcium	100 micrc gm	2	6.312
5 Fibrous Food	500 micrc	1	9.312

Total Contribution From All Factors..   25.998
Current Grand Average Of Performance...   28.437
Expected Result At Optimum Condition..   58.435

Figure 6.28   Factor contributions for Example 6.4.

along the depth of the column. This array will be useful to you when you have more than 5 four-level factors or you exceed the capacity of a modified L-16. Like the L-18 array, this is an array that allows you to design experiments with limited mixed levels.

You can review the content of the modified L-32 array by one of the example experiments provided with QT4. To review, run QT4 and go to the main screen. Select the *Open* option from the *File* menu item and load experiment file *REF-L32M.Q4W*. From the *Edit* menu item, select *Inner array*, which should bring up the array. Notice that the first column has only levels 1 and 2. All others have four levels. As mentioned before, the column totals shown at the bottom of each column are only for a quick review of column accuracy.

For experimental situations involving 6, 7, 8, or 9 four-level factors along with or without 1 two-level factor, QT4 always selects a modified L-32 array. You can quickly try some of these situations and check the design selection. To test the QT4 array selection, try designing an experiment with 1 two- and 6 four-level factors following these steps.

1. Run QT4 and go to the main screen.

2. Select *Automatic design* option from the *Design* menu item.

3. Click *Yes* to the prompt at the *Design new experiment* screen.

4. Check the box for two-level factors and set the number of factors to 1. Also, check the box for four-level factors and set the number of factors to 6 (Figure 6.29). Click *OK* to proceed.

5. QT4 shows that it has selected an M-32 (M for modified) for the design (see Figure 6.29). Click *Yes* to proceed to describe factors and levels for QT4 to complete the design.

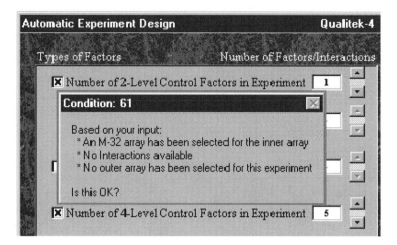

Figure 6.29 Automatic experiment design input.

You can repeat steps 1 to 3 and enter other factor combinations in step 4 to check that QT4 also selects the M-32 array for the design. In the *Automatic experiment design* screen you may input any of the following conditions and find that in each condition QT4 displays the same array selection (see Figure 6.10). To try different factor combinations, click *No* at the prompt for the array selected and enter a new input in the *Automatic experiment design* screen:

- 1 two- and 5 four-level factors
- 1 two- and 6 four-level factors
- 1 two- and 7 four-level factors
- 1 two- and 8 four-level factors
- 1 two- and 9 four-level factors
- 6 four-level factors only
- 7 four-level factors only
- 8 four-level factors only
- 9 four-level factors only

Example 6.5: Ride Comfort Study Using Analytical Model An analytical study group of an automobile manufacturer used a mathematical model to study the vibration experienced by a driver. The purpose of the study was to fine-tune the structural design parameters such that driver comfort is maximized. The model simulated the vehicle structural component in terms of weightless stiffness members connecting the distributed weights concentrated at rigid node points. Among the many input parameters for the model, 1 two- and 6 four-level factors, as shown in Figure 6.30, were considered for the study. For comparison performances at different structural combination, the peak amplitude of motion (displacement in millimeters) at the driver

	Factors	Level 1	Level 2	Level 3	Level 4
1	A:Engine Type	2.8 L Eng	3.1L Engine	-----------	-----------
2	B:Type of Tire	Brand G1	Brand G2	Brand F1	Brand F2
3	C:Engine Mount	Location	Location 2	Location 3	Location 4
4	D:Mount Stiffness	Forward	Present	Rear	Most Rear
5	E:Steering Column	Assembly	Assembly 2	Assembly 3	Assembly 4
6	F:Body Stiffners	None	Two - Cross	Two - Long.	One and One
7	G:Seat Mount	Inward	Outward	In -Out	Out - In
8	COLUMN UNUSED	*UNUSED*	--------------	--------------	--------------
9	COLUMN UNUSED	*UNUSED*	--------------	--------------	--------------
10	COLUMN UNUSED	*UNUSED*	--------------	--------------	--------------

Figure 6.30   Factor and level descriptions for Example 6.5.

seat location was calculated.

An L-32 modified array used to design the experiments provided 32 separate combinations of the input parameters for the computer runs. These runs produced calculated values of the magnitude of peak displacements at the driver seat location (Figure 6.31). For running the model, digitized excitation under standard driving and road conditions was used. From the results, is it possible to identify the two factors that will have the major influence on ride comfort?

Solution: This example experiment has already been designed using QT4 and saved as *BKEX-65.Q4W*. You may review the design and the results using the *Edit* menu option or practice the design process by executing the following steps.

QT4 Design Steps

1. Run QT4 and go to the main screen.
2. Select *Automatic design* from the *Design* menu item.

	Sample# 1		Sample# 1		Sample# 1		Sample# 1
Trial# 1	0.8	Trial# 9	2.1	Trial# 17	4.3	Trial# 25	2.8
Trial# 2	2.2	Trial# 10	1.1	Trial# 18	3.1	Trial# 26	2.52
Trial# 3	4.6	Trial# 11	1.7	Trial# 19	4.6	Trial# 27	1.7
Trial# 4	4.4	Trial# 12	3.4	Trial# 20	2.4	Trial# 28	4.8
Trial# 5	1.2	Trial# 13	3	Trial# 21	5.3	Trial# 29	3
Trial# 6	2.2	Trial# 14	1.4	Trial# 22	3	Trial# 30	2.1
Trial# 7	2.7	Trial# 15	2.1	Trial# 23	1.9	Trial# 31	1.3
Trial# 8	2.4	Trial# 16	4.6	Trial# 24	3.4	Trial# 32	3.4

Figure 6.31   Experimental results for Example 6.5.

3. Click *Yes* at the prompt for *New experiment design*.

4. In the *Automatic design* screen, check the box for two-level factors and set the number of factors to 1. Then check the box for four-level factors and set the number of factors to 6. Click *OK* to proceed.

5. Click *Yes* when QT4 displays that an M-32 is selected for the experiment design.

6. Type in the descriptions of 1 two-level factor and its levels. Click *OK* to proceed.

7. Type in the descriptions of 6 four-level factors and their levels. Click *OK* to proceed.

8. Review descriptions of factors and levels you just entered and the columns QT4 assigned them. Notice that QT4 has assigned the seven factors you described to the first seven columns of the array with the two-level factor in the first column. You may edit any description, if necessary, here in this screen (see Figure 6.30). Click *OK* when done.

9. Enter the project title, objectives, and the quality characteristic (QC = S).

10. Review the array used for the design. Click *OK* to proceed to save the experiment file.

11. When prompted to save the experiment file, click *Cancel* and return to the main screen, as this experiment has already been saved as *BKEX-65.Q4W*. (If this experiment were not predesigned, you would need to name this experiment file.) Once you are in the main screen, load the experiment file *BKEX-65.Q4W* by selecting the *Open* option from the *File* menu item.

Experiment Review

1. When you have finished designing the experiment, you can review the 32 trial conditions that describe the conditions of the input parameter (factor levels) for the model. To review the trial condition, select *Trial conditions* from the *Conditions* menu item in the main screen. QT4 always starts with the first trial condition. You may now click on the *Next trial* button to view all other conditions, one at a time. Click *Return* when you are done.

2. Compare the experimental results displayed next to the array in the main screen with that shown in Figure 6.31. You will need to scroll down to view trial numbers beyond trial 8. You can also review factors, levels, and results by selecting the appropriate options from the *Edit* menu item. You may edit any descriptions or results in these screens, but QT4 will not allow you to update this example experiment.

To determine which two among the factors studied are the most influential, you will need to perform analysis of variance (ANOVA) on the 32 results collected in the analytical model runs (see Figure 6.31). You can get ANOVA output as part of the

QT4 standard analysis screen. Run QT4 and execute the following steps to determine the relative factor influences on the variation of results.

QT4 Analysis Steps

1. Run QT4 and load the experiment file *BKEX-65.Q4W* from the *File* menu in the main screen (you need not do this if you are already working with this file).
2. Select *Standard analysis* from the *Analysis* menu item.
3. Check the box for *Smaller is better* in the *Quality characteristic selection* screen. Click *OK* to proceed.
4. Review the results of the experiments and check to see that they are as shown in Figure 6.31. Click *OK* to proceed.
5. In the *Main effects* screen, study the factor influences by selecting the *Plot* button. Click *OK* or *ANOVA* to proceed.
6. Study the last column of the *ANOVA* screen (see Figure 6.32) and observe that the factors engine mount (12.96%) and mount stiffness (26.10) have more influence on the variation of results than any of the other factors included in the study. Click *OK* twice to return to the main screen.

Additional Observations

1. Repeat steps 1 to 6 with the *Bigger is better* quality characteristic and observe that the ANOVA results do not change.
2. The last row of the ANOVA table (see Figure 6.32), labeled "All Other/Error" indicates the percentage influence due to factors not included in the experiments, and experimental error, if any. If experimental error is absent, 51.91% of the total influence is due to other controllable and uncontrollable factors.

A N O V A Table						Qualitek-4
Expt. File: BKEX-65.Q4W	Data Type: Average Value				Print	Ok
	QC Type: Smaller is Better				Help	Cancel

Col # / Factor	DOF (f)	Sum of Sqrs. (S)	Variance (V)	F - Ratio (F)	Pure Sum (S')	Percent P(%)
1 A:Engine Type	1	2.952	2.952	3.578	2.127	4.669
2 B:Type of Tire	3	2.942	.98	1.188	.467	1.025
3 C:Engine Mount	3	8.382	2.794	3.386	5.907	12.966
4 D:Mount Stiffness	3	14.368	4.789	5.805	11.893	26.105
5 E:Steering Column	3	.549	.183	.221	0	0
6 F:Body Stiffners	3	3.77	1.256	1.523	1.295	2.842
7 G:Seat Mount	3	2.694	.898	1.088	.219	.482
Other/Error	12	9.899	.824			51.911
Total:	31	45.558				100.00%

Figure 6.32 Relative influence of factors to ride comfort.

SUMMARY

In this step you have learned how to use the standard L-9, L-18, L-27, modified L-16, and modified L-32 arrays. You now know how to design experiments when several three- or four-level factors are mixed with 1 two-level factor. You also learned how to interpret factor influence plots for three- and four-level continuous factors in terms of an idealized smooth curve (least-squares fit).

REVIEW QUESTIONS

Q. Can an L-9 array be used for 3 three-level factors?

A. Yes. In this case you are testing one third of all possible factor combinations ($3^3 = 27$).

Q. Do you have to have a two-level factor to use L-18 array?

A. No. You should use an L-18 array whenever you have 5, 6, or 7 three-level factors whether or not you have an additional two-level factor to go with them.

Q. Should an L-18 be used to study 3 three- and 1 two-level factor?

A. You could use an L-18. But a better selection would be an L-9 with its one column downgraded to two-level. Downgrading of columns is covered in later steps.

Q. Are there general guidelines for selecting levels for a three-level factor?

A. Levels should always be established based on project needs and what are known about factor behavior. As general guidelines, applicable to continuous factors, you should take a level at the current working condition (say, level 2), one below (level 1), and one above (level 3). The levels 1 and 3 values should be selected as far away from level 2 as possible, making sure that the experiments with these conditions are still functional. For a new system, where current working factor levels are unknown, you may want to run separate tests with multiple levels of a single factor to establish its operating range.

Q. What considerations dictate the use of three- and four-level arrays?

A. Your primary considerations for selecting an array should be the number of factors and their levels that you select for the study. In order of sequence, you should prepare a long list of factors first. Based on time and budget available for the study, select the biggest array for the design. Since you do not yet know the levels of the factors, you should base the array selection on the assumption that all factors are at two levels unless you have prior knowledge about the nonlinearity and you are able to handle most factors at three levels. If your budget allows going after a three-level array (say an L-18 instead of L-16), you should do so, as it will provide you with more complete information about factor behavior.

EXERCISES

6.1 Indicate the orthogonal array that you would select to design experiments in the following situations:

(a) 3 three-level factors

(b) 1 two- and 5 three-level factors

(c) 7 three-level factors

(d) 12 three-level factors

6.2 In a study, 10 factors were identified as the primary influential factors. It was desirable to study most factors at three levels while keeping the number of experiments under 20. To design experiments for the study, what kind of adjustment would you need to make?

6.3 To determine the best recipe for a small-scale beer brewing process, an L-9 experiment was carried out to study the effects of four factors: fermentation time, brewing temperature, water condition, and amount of hops. The output from each trial was tested and the results, on a scale of 0 to 10 (QC = B), were recorded and saved in experiment file *BKEX-63E.Q4W*. Analyze the results using QT4 and determine the nature of the influence of temperature on taste.

6.4 An experiment was designed and conducted to study the effects of eight factors in molding an automobile hood using SMC plastics. The descriptions of 1 two- and 7 three-level factors included in the study are shown in Figure 6.33. The molded parts trial runs were evaluated by counting the number of defects (pops, cracks, pregel, etc.) in each part. From the results (Figure 6.34) of three samples tested in each of the 18 trial conditions, determine the two most significant factors affecting the defects in the parts. (*Hint*: Use experiment file *BKEX-64E.Q4W* to analyze the results using QT4.)

	Factors	Level 1	Level 2	Level 3
1	A:SMC Type	A-5 #971	#971	------------
2	B:GLASS CONTENT	27%	22%	18%
3	C:LOAD PATTERN	STANDARD	ALTERNATE 2	ALTERNATE 3
4	D:Mold Temperatur	er	Current	Higher
5	E:EDGE TREATMENT	NONE	SW IR PRIME	SIEBERT
6	F:PRIMER	STANDARD	100% POPSEA	POPSEA/2349
7	G:FINISHING TECHN	NONE	WET SAND 2	WET SAND
8	H:VACUUM FIRING	2 in.	2.5 in.	2.7 in.

Figure 6.33 Factor and level descriptions for Exercise 6.4.

	Sample# 1	Sample# 2	Sample# 3
Trial# 1	8	9	6
Trial# 2	12	14	11
Trial# 3	8	5	6
Trial# 4	4	5	5
Trial# 5	3	5	6
Trial# 6	5	6	8
Trial# 7	10	9	11
Trial# 8	8	8	9
Trial# 9	9	10	11
Trial# 10	10	8	11
Trial# 11	6	5	5
Trial# 12	6	8	5
Trial# 13	6	8	7
Trial# 14	6	6	8
Trial# 15	12	11	9
Trial# 16	9	12	10
Trial# 17	9	8	10
Trial# 18	9	11	12

Figure 6.34   Experimental results for Exercise 6.4.

EXERCISE ANSWERS

6.1   (a) L-9; (b) L-18; (c) L-18; (d) L-27.

6.2   Drop two factors and assign two levels to one factor.

6.3   None or very little effect. You will be able to draw this conclusion by examining the factor influence plot and ANOVA table.

6.4   Review experiment file *BKEX-64E.Q4W* for accuracy of factor descriptions and results. Analyze the results using the *Standard analysis* option and find answers from ANOVA. Factors: glass content and finishing technique.

Analysis of Variance

What You Will Learn in This Step

In this step you will learn how to determine the relative influence of factors under study in terms of discrete proportion. This step contains some simple methods and calculations for all terms in an ANOVA table. You will learn about principal concepts and interpretation of ANOVA results through a number of case examples you have worked out in earlier steps. From this step you will gain a working knowledge of ANOVA and feel comfortable interpreting the ANOVA results of your own experiments.

Thought for the Day

There is a great difference between knowing and understanding: You can know a lot about something and not really understand it.

—Charles F. Kettering

TWO PARTS OF THE ANALYSIS

Analysis in design of experiments (DOE) refers to the things that are done with results after experiments are carried out and test samples are evaluated. All calculations using the results are carried out to support the observations, conclusions, and recommendations made from the experiments. Depending on the complexities of the calculations involved, the analysis can be performed in two parts.

Part I: Simple Analysis

This part of the analysis is performed to produce a *grand average* of results and the *average effects* of factors. Calculations in this part involves only simple arithmetic (addition and division) operations but help us make observations and conclusions in the following areas:

1. Factor influence or *main effect*s
2. *Optimum condition* for a desired quality characteristic
3. Performance expected at the *optimum condition*

All you need is a simple four-function calculator to perform this analysis. You can always do this part when you are away from your computer or if you do not have analysis software. In QT4, item 1 is displayed in the *Main effects* screen and items 2 and 3 are contained in the *Optimum* screen as you proceed with analysis.

Part II: Analysis of Variance

Calculations involved here are not something most people feel comfortable doing by hand or would feel confident about. It involves squaring terms and taking the squared root of some of the terms. The analysis of variance (ANOVA) calculation is something that you would need as you begin to look for items beyond the three obtained from the simple analysis. Generally, ANOVA will support calculations, tests, and observations of the following nature:

1. Relative influence of factor and interaction to the variation of results
2. Test of significance of factor and interactions assigned to the columns
3. Confidence interval (C.I.) on optimum performance
4. Confidence interval on *main effect* of factors
5. Error factor/term, which includes the influence of all factors not included in the experiments and effects of experimental error

QT4 performs calculations and allows manipulations of all the items above in an *ANOVA* screen. As you have seen while using the QT4 *Analysis* option, it places ANOVA between *Main effect*s and *Optimum condition* screens.

WHY PERFORM ANOVA?

By now you know what an ANOVA table looks like and learned to make use of the right column of the table. We have not discussed where all the items in different columns in the table come from or how the items above are calculated. Depending on your needs, this part of the analysis may be beneficial to you. If learning about factor influences and how sensitive the result is to different factors were important to you, you would definitely need to perform ANOVA.

   No matter your application, soon after you run a few studies, you would want to learn to interpret ANOVA information and feel comfortable using it. With the ANOVA information about your experiments, you will be able to get a better insight into your product/process under study that would defy common sense and above and beyond what simple analysis offers you. One way to view ANOVA is to consider it as a snapshot of the status of your project performance, just the same way that an x-ray is taken of a human body to obtain insight about the condition of health. To a radiologist it contains a bundle of information. To a layman it is just a plastic slide. To learn how to interpret ANOVA very well is to become like a good radiologist reading an x-ray film.

ANOVA can be a vast statistical subject of study. You could spend a semester or two taking courses in statistical science. The lessons in this book are not to make you an expert in ANOVA, only to make you learn all the calculations. The purpose is simply to show you where all the numbers in the table come from and how the calculations are performed. The calculations are shown in detail to help you examine the interrelationships among the various ANOVA terms. Because it is only when you know the mathematical expression that you will understand how each term influences the conclusions being drawn. We will, however, not become involved in discussions of statistical theories and derivations of the terms involved in ANOVA.

ANOVA CALCULATION STRATEGY

The main objective of ANOVA is to extract from the results how much variation each factor (or interaction assigned to the column) causes relative to the total variation observed in the result. The term *variation* is indicated by several mathematical descriptions; we will make use of one soon. But, thinking graphically, the term *variation* brings images of population performance distributions shaped like an umbrella or a mountain. It is unimportant at this point whether it is a bell-shaped curve or a true normal distribution. What is important is the spread or width of the image, which is directly proportional to the variation within the data. Perhaps, for a study with factors A, B, C, and so on, the total variation in the results of experiments (all trail result) can be shown by a larger distribution and the individual factor influence distributions contained within it (Figure 7.1). To express the influence of an individual factor to the total amount, the influence by the individual factor is expressed as a fraction (%) of the total variation in the manner shown in Figure 7.2.

The calculation of individual factor influence is similar to finding the percentage of individual contributions in a group project when the total output is known. Suppose that a contingent of city road maintenance crews repairs 600 potholes in a shift. A fast-working member among the crews claims to have repaired 90 potholes alone. One way to express the contribution this person makes is by expressing his or her performance in terms of a fraction of the total as 90/600 = 0.15 or 15%. In the same manner, the variation caused by an individual factor can be expressed as a percentage of the total variation. The challenge therefore lies in being able to quantify the variation and express it in numerical terms such that we can divide one by the other.

For a set of data (results), Y_1, Y_2, . . . , Y_N, the total variation can be calculated by adding deviations of the individual data from the mean value. If the deviations were

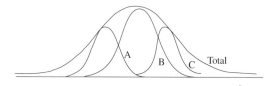

Figure 7.1 Total and individual factor influence distribution.

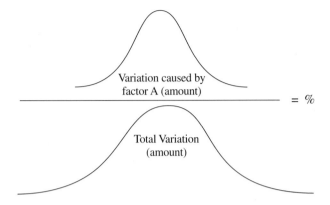

Figure 7.2   Factor influence as a fraction of total influence.

collected, as is, deviation from a data point that falls on the left of the average will be canceled by another equally away from the average on the right. To assure that all deviations are counted, the individual deviations are squared, which forces all deviation squared values to be positive (Figure 7.3).

$$S_T = \sum_{i=1}^{N} (Y_i - \overline{Y})^2$$

which can be reduced to the following form:

$$S_T = \sum_{i=1}^{N} Y_i^2 - \frac{T^2}{N}$$

Following a similar approach, the variation caused by an individual factor, say A, is obtained by an expression called the *factor sum of squares* as

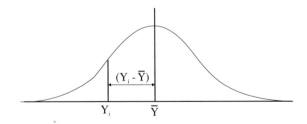

Figure 7.3   Calculation of total sum of squares.

$$S_A = \frac{A_1^2}{N_{A1}} + \frac{A_2^2}{N_{A2}} - \text{C.F.}$$

where C.F. is the correction factor ($= T^2/N$), N_{A1} the total number of experiments in which level 1 of factor A is present, and A_1 the total of results (y_i) that include factor A_1.

The total and factor sums of squares are the basic calculations needed for ANOVA. Four other quantities calculated as part of the ANOVA table information are all derived from the original sums of squares. For factor A, they are as follows:

Mean squares (or variance): $\quad V_A = \dfrac{S_A}{f_A}$

F-ratio: $\quad F_A = \dfrac{V_A}{V_e}$

Pure sum of squares: $\quad S'_A = S_A - (V_e \times f_A)$

Percent influence: $\quad P_A = \dfrac{S'_A}{S_T}$

where V_e is the variance for the *error term* (obtained by calculating error sum of squares and dividing by error *degrees of freedom*) and f_A is the *degrees of freedom* (DOFs) of factor A.

Degrees of Freedom

Degree of freedom (DOF) is an important number for analyses of results and you need to have a good working knowledge of it. DOF is a common term used in engineering and science. In some applications, such as vibration analysis, DOF has a physical meaning. For example, a vibrating mass–spring system with two *degrees of freedom* has two distinct mode of vibration which can be observed under appropriate excitation. Unfortunately, there is no such visible interpretation of DOFs applied to experimental data. When it comes to statistical analysis of experimental data, DOF is an indication of the amount of information contained in a data set. For example, if there are three numbers in a data set, the DOF = 3 – 1, which indicates the amount of additional information that can be derived by taking the differences among the three numbers. Looked at another way, if there are three people, you will need two comparisons to determine who is tallest.

In statistical analysis of experimental results, DOF is applied to characterize four separate items. You simply need to understand these four definitions.

DOF of a factor = number of level of the factor – 1

DOF of a column = number of level of the column – 1

$$\text{DOF of an array} = \text{total of all column DOFs for the array}$$

$$\text{DOF of an experiment} = \text{total number of results of all trials} - 1$$

From the definitions, it should follow that the DOFs of a two-level factor is 1 (= 2 − 1), for a three-level factor is 2, and for a four-level factor is 3 (= 4 − 1). Applied to the columns of an array, a column of an L-8 array has 1 (= 2 − 1) DOF, which will go to give 7 DOFs for the array, as it has seven columns with 1 DOF each.

The DOF of an experiment or the total DOFs of an experiment is calculated by taking into consideration the number of samples tested per trial condition. Thus, if an L-8 array is tested with five samples in each trial condition, the DOFs for the experiment will be 39 (= 8 × 5 − 1). Remember that each sample tested is expected to produce one result (single criterion or multiple criteria combined into one number). DOF plays important roles in the calculation of *confidence intervals* and in tests of significance using the calculated ANOVA table numbers. Several examples that follow in this step will demonstrate the use of DOF and other subtleties of ANOVA calculations. You should examine carefully the calculation of terms displayed by QT4 screens and their meaning in relation to the experiment under study.

Example 7.1: Microwave Oven–Popped Corn (This example demonstrates a procedure for calculations of factor influence, confidence interval, and pooling.) Three factors (power level, type of oil, and mixing/stirring) involved in the process of popping corn were studied using an L-4 experiment. A fixed amount of corn was popped several times in each of the four trial conditions. The unpopped kernels were weighed and the results were recorded in ounces (Table 7.1). The objective of the study was to determine the best condition for popping. (This experiment has been designed and is saved as *BKEX-71.Q4W* in QT4.) By performing ANOVA, determine the following items and show your computations:

(a) A factor that can be pooled

(b) The factor with the most influence in the variation of results

(c) The C.I. for the optimum performance

TABLE 7.1   Experimental Results

	Factor			
Experiment	A	B	C	Result
1	1	1	1	3
2	1	2	2	6
3	2	1	2	4
4	2	2	1	2

Solution: Hand calculation analysis of the results in this experiment will be compared with the calculated outputs of QT4. Run QT4, open file *BKEX-71.Q4W*, and proceed with the *Standard analysis* option from the *Analysis* menu item in the main screen. Compare and review the *Main effects* and *ANOVA* screens with Figures 7.4 and 7.5, respectively.

Part I: Simple analysis. The total of results is

$$T = 3 + 6 + 4 + 2 = 15$$

The total effects of factors are obtained by adding the results containing the effects of the factors at the level desired. Note that the average effects (Figure 7.4) are obtained by dividing the total effects calculated below by the number of results (two in this case) contained in the calculation.

$$A_1 = 3 + 6 = 9 \qquad A_2 = 4 + 2 = 6$$

$$B_1 = 3 + 4 = 7 \qquad B_2 = 6 + 2 = 8$$

$$C_1 = 3 + 2 = 5 \qquad C_2 = 6 + 4 = 10$$

Part II: ANOVA calculations. Let's now examine how ANOVA terms in Figure 7.5 are computed. The second column shows the number of DOFs of the items in the first column:

$$\text{total DOFs} = (\text{number of results}) - 1$$

$$= 4 - 1$$

$$= 3$$

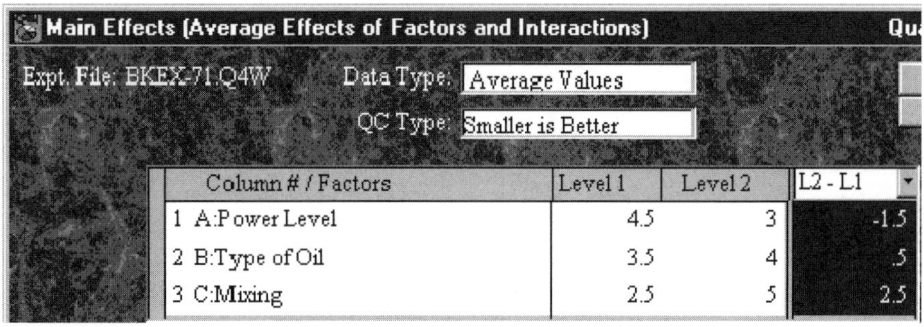

Main Effects (Average Effects of Factors and Interactions)			Qu
Expt. File: BKEX-71.Q4W Data Type: Average Values			
QC Type: Smaller is Better			

Column # / Factors	Level 1	Level 2	L2 - L1
1 A:Power Level	4.5	3	-1.5
2 B:Type of Oil	3.5	4	.5
3 C:Mixing	2.5	5	2.5

Figure 7.4 Factor average effects.

ANOVA Table — Qualitek-4

Expt. File: BKEX-71.Q4W Data Type: Average Value QC Type: Smaller is Better

Col # / Factor	DOF (f)	Sum of Sqrs. (S)	Variance (V)	F - Ratio (F)	Pure Sum (S')	Percent P(%)
1 A:Power Level	1	2.25	2.25	-----	2.25	25.713
2 B:Type of Oil	1	.25	.25	-----	.25	2.857
3 C:Mixing	1	6.25	6.25	-----	6.25	71.427
Other/Error	0					
Total:	3	8.75				100.00%

Figure 7.5 ANOVA table.

The total DOFs must be calculated as shown above. QT4 displays it at the bottom of the second column. Since all factors in this experiment are at two levels, the DOFs for the factors are all 1 (= 2 – 1). The DOFs for the error term must be calculated after independently calculating the DOFs for the factors and the total DOFs:

$$\begin{aligned} f_e &= (\text{total DOFs}) - (\text{total of all factor DOFs}) \\ &= 3 - (1 + 1 + 1) \\ &= 0 \end{aligned}$$

Make a note that the error DOF is zero in this case. The value of the error DOF will be more and more important in our discussions about many tasks with ANOVA in this and other examples. All column items for error terms such as DOF, S, and $P\%$ are calculated by subtracting the sum of all factor values from the corresponding totals.

The correction factor (C.F.) is used for calculation of all sums of squares. It remains constant for all factors, as it is composed of fixed quantities (T and N).

$$\text{C.F.} = \frac{T^2}{N} = \frac{15 \times 15}{4} = 56.25$$

The total sum of squares is calculated independently first using the formula shown earlier. This value is shown at the bottom of the third column in the ANOVA table.

$$\begin{aligned} S_T &= \sum_{i=1}^{N} Y_i^2 - \text{C.F.} \\ &= (3^2 + 6^2 + 4^2 + 2^2) - \text{C.F.} \\ &= 65 - 56.25 \\ &= 8.75 \end{aligned}$$

The factor sums of squares are calculated next, one factor at a time.

$$S_A = \frac{A_1^2}{N_{A1}} + \frac{A_2^2}{N_{A2}} - \text{C.F.}$$

$$= \frac{9^2}{2} + \frac{6^2}{2} - 56.25$$

$$= 40.5 + 18 - 56.25$$

$$= 2.25$$

$$S_B = \frac{B_1^2}{N_{B1}} + \frac{B_2^2}{N_{B2}} - \text{C.F.}$$

$$= \frac{7^2}{2} + \frac{8^2}{2} - 56.25$$

$$= 24.5 + 32 - 56.25$$

$$= 0.25$$

$$S_C = \frac{C_1^2}{N_{C1}} + \frac{C_2^2}{N_{C2}} - \text{C.F.}$$

$$= \frac{5^2}{2} + \frac{10^2}{2} - 56.25$$

$$= 12.5 + 50 - 56.25$$

$$= 6.25$$

These values are shown in the third column under the corresponding factor descriptions. The sum of squares for the error term is absent ($= 0$), as it should be. *If the error DOF is zero, the sum of squares for the error term must also be zero.* You can check this by calculating it from the known factor values.

$$S_e = S_T - (S_A + S_B + S_C)$$

$$= 8.75 - (2.25 + 0.25 + 6.25)$$

$$= 0$$

Mean squares (or variance) is simply the sum of squares per DOF. Since all factors have DOF as 1, the variance column in ANOVA shows the same number as S.

$$V_A = \frac{S_A}{f_A}$$

$$= \frac{2.25}{1}$$

$$= 2.25$$

Similarly, $V_B = 0.25$ and $V_C = 6.25$.

When both sum of squares and DOF for the error terms are zero, the variance for the error terms cannot be calculated as

$$V_e = \frac{S_e}{f_e}$$

$$= \frac{0}{0} \quad \text{which is indeterminate (neither large nor small)}$$

Consequently, F-ratios for the factor cannot be calculated when variance for the error term is an indeterminate. And of course, in absence of the variance for the error term F-ratios like

$$F_A = \frac{V_A}{V_e} \quad \text{cannot be calculated.}$$

With V_e absent, all pure sums of squares equal the corresponding sum of squares. This makes the S' column the same as the S column in an ANOVA table.

$$S'_A = S_A - (V_e \times f_A)$$

or

$$S'_A = S_A \qquad S'_B = S_B \qquad S'_C = S_C$$

Percent influence of the factors can now be calculated by comparing the *pure sums of squares* of the factors with respect to the *total sum of squares*.

$$P_A = \frac{S'_A}{S_T}$$

$$= \frac{100 \times 2.25}{8.75}$$

$$= 25.71\%$$

$$P_B = \frac{S'_B}{S_T}$$

$$= \frac{100 \times 0.25}{8.75}$$

$$= 2.86\%$$

$$P_C = \frac{S'_C}{S_T}$$

$$= \frac{100 \times 6.25}{8.75}$$

$$= 71.43\%$$

The total percent influence, shown at the bottom of the P column in the ANOVA table, is always set to 100%. The percent influence of the error term again is calculated in the same manner: the sum of all factor influences subtracted from the total. Since the error DOF is zero, the influence of the error term must also be zero, as found here. Also, since the total is always set to 100, the individual factor influences are properly referred as the *relative percentage influence* (the last column of the ANOVA table).

$$P_e = 100 - (P_A + P_B + P_C)$$

$$= 100 - (25.71 + 2.86 + 71.43)$$

$$= 0$$

What should you do when you have performed ANOVA for the first time? In QT4 analysis, what should you do when you are at the *ANOVA* screen? The first thing you should do is try to identify factors that are significant and those that are not. Whether a factor influence is significant or not is judged by subjecting the factor to the test of significance. If a factor passes the *test of significance*, it is considered significant. The factors that do not pass this test are considered insignificant and are usually treated as if they are not present. The process of ignoring a factor once it is deemed insignificant, called *pooling*, is done by combining the influence of the factor with that of the error term.

Pooling is a common practice of revising and reestimating ANOVA results. Pooling is strongly recommended, for two reasons. First, when a number of factors are included in an experiment, the laws of nature make it probable that half of them would be more influential than the rest. Second, in statistical predictions, we encounter two types of mistakes: alpha and beta mistakes. An *alpha mistake* is calling something important when it is not. A *beta mistake* is the opposite of an *alpha* mistake: Significant factors are mistakenly ignored. In deciding which factors are important and which are not, we should try to minimize the chance of committing an alpha mistake. Pooling factors that are not significant go toward reducing the chance of making this mistake. How is a factor pooled? How do we determine which factor to pool?

Pooling. A factor is *pooled* when it fails the test of significance. Unfortunately, the test of significance can be done only when the error term has nonzero DOF. So what should you do when the DOF of the error term is zero, as in this example experiment? Even if a test of significance is not possible, you should consider pooling. You should start with the factor that has the least influence (compare S values). As a rule, pool a factor if its influence (use S to compare instead of P) is less than 10% of the most influential factor. Refrain from pooling arbitrarily if this condition is not present and when a test of significance is not possible. Always rely on a test of significance, when possible, to determine which factor to pool. (Pooling by test of significance is discussed in the next example.)

By examining ANOVA, you can now answer the first two questions in the example. Since the error DOF is zero, you need to decide which factor to pool by comparing the sums of squares values of the factors.

(a) The factor type of oil has the smallest S (0.25) and it is also less than 10% of the highest S (6.25). Thus this factor is pooled. To pool, double-click on the factor description in the *ANOVA* screen (Figure 7.5). At the prompt for pooling shown in Figure 7.6, click *Yes* to proceed. After one or more factors are identified for pooling, ANOVA terms are recalculated. The recalculation of ANOVA always starts by establishing the new values for the error term by ignoring the presence of all factors identified for pooling (treat pooled factors as if they are not there).

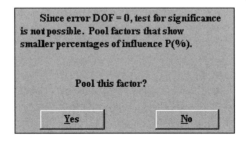

Figure 7.6 Pooling the prompt screen.

$$f_e = \text{total DOFs} - (\text{total of all significant factor DOFs})$$

$$= 3 - (1 + 1)$$

$$= 1$$

$$S_e = S_T - (S_A + S_C)$$

$$= 8.75 - (2.25 + 6.25)$$

$$= 0.25 \qquad (\text{which is same as } S \text{ for the factor pooled})$$

$$V_e = \frac{S_e}{f_e}$$

$$= \frac{0.25}{1}$$

$$= 0.25$$

With the revised error values, the factor F-ratios and pure sums can now be recalculated. Calculations for factor A is as shown here.

$$V_A = 2.25 \text{ (remains unchanged)}$$

$$F_A = \frac{V_A}{V_e}$$

$$= \frac{2.25}{0.25}$$

$$= 9.0$$

$$S'_A = S_A - (V_e \times f_A)$$

$$= 2.25 - (0.25 \times 1)$$

$$= 2.00$$

$$P_A = \frac{S'_A}{S_T}$$

$$= \frac{100 \times 2.00}{8.75}$$

$$= 22.86\%$$

Similarly, $P_B = 68.57$.

The influence of the error term is finally calculated from known values of the same for all surviving (significant) factors and the ANOVA revised as shown in Figure 7.7.

$$P_e = 100 - (P_A + P_C)$$

$$= 100 - (22.86 + 68.57)$$

$$= 8.57\%$$

Note that the new error percent influence (8.57%) is different from the percent influence (2.85%) of the factor pooled. The recalculated value of the error percent bears no direct relationship to the influences of the factors pooled and should not be expected to be of equal magnitude.

(b) Factor mixing has the most influence (68.57% after pooling), as shown in Figure 7.7.

(c) The confidence interval is calculated using ANOVA values. But since C.I. specifies the boundaries of the expected performance at the *optimum condition*, the option to calculate it is available from the *Optimum* screen. Click *OK* at the *ANOVA* screen to proceed to the *Optimum* screen shown in Figure 7.8. Note that the performance expected at the optimum condition is 1.75. The C.I. calculates the lower and upper limits of this value, within which the mean performance of the actual test samples should fall. The C.I. is calculated for a desired confidence level. This is user input and is a subjective choice. There is no fixed value with which everyone works. Most people use a confidence level between 80 and 95%. A desirable thing to do is to work with a fixed value (say, 90 or 95%) for all calculations and predictions in an experiment. Examine carefully how this subjective input affects your calculated values.

While at the *Optimum* screen, click C.I. to proceed to input values for the confidence level prompt shown in Figure 7.9. Type in 90 for 90% and click *OK*. Based on the confidence level entered by the user, QT4 calculates the C.I. values. There are

ANOVA Table					Qualitek-4	_ □
Expt. File: BKEX-71.Q4W	Data Type: Average Value				Print	Ok
	QC Type: Smaller is Better				Help	Cancel

Col # / Factor	DOF (f)	Sum of Sqrs. (S)	Variance (V)	F - Ratio (F)	Pure Sum (S')	Percent P(%)
1 A:Power Level	1	2.25	2.25	9	2	22.857
2 B:Type of Oil	(1)	(.25)		POOLED	(CL= *NC*)	
3 C:Mixing	1	6.25	6.25	25	6	68.571
Other/Error	1	.25	.25			8.572
Total:	3	8.75				100.00%

Figure 7.7 Pooled ANOVA.

Optimum Conditions and Performance			Qualitek-

Expt. File:BKEX-71.Q4W   Data Type Average Value   Print

QC   Smaller is Better   Help

Column # / Factor	Level Description	Level	Contribution
1 A:Power Level	Higher	2	-.75
3 C:Mixing	None	1	-1.25

Total Contribution From All Factors...   -2
Current Grand Average Of Performance...   3.75
Expected Result At Optimum Condition...   1.75

Figure 7.8 Optimum performance.

times when QT4 cannot calculate C.I., as it is unable to compute the value of the F-ratio for the applicable value of the error DOF. When C.I. is calculated by hand, the F-ratios are chosen from the standard tables (called F-tables). QT4 uses an approximate method to compute F-ratios instead of using a standard table. For values of error DOF smaller than 3, the computation is unreliable, as the F-ratio in this region is highly nonlinear. For this example (error DOF = 1) we will assume that the F-ratio computed by QT4, as shown in Figure 7.10, is correct.

The C.I. formula (same as QT4) shown below contains two new terms with which you need to be familiar. The first one is the standard value of F-ratios. The F value, as it is often referred to, is selected from a table of numbers [see references Ross (1988), Roy (1990), or Taguchi (1987) as listed in the Appendix] prepared based on two DOF numbers. In the case of C.I. on the optimum performance, these two DOFs for the F-ratio is $F(n_1, n_2)$, where n_1 (always 1) is the DOF of the mean performance and n_2 is the DOF for the error term, which is 1 in this case. The other term, the effective number of replications, is defined below.

User Supplied Confidence Level		

Confidence interval is the variation of the mean performance at the optimum condition.

Ok
Cancel
Help

Enter the confidence level : 90.00 %

Figure 7.9 Input for confidence level desired.

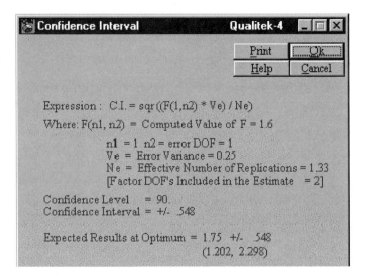

Figure 7.10 Confidence interval (C.I.).

- C.I. formula:

$$\text{Confidence interval (C.I.)} = \pm \left[\frac{F(1, n_2) \times V_e}{N_e} \right]^{0.5}$$

where $F(1, n_2)$ is the F value from the F table for factor DOF and error DOF at the confidence level desired, V_e the variance of the error term (from ANOVA), and N_e the effective number of replications:

$$N_e = \frac{\text{total number of results or } S/N}{\text{DOF of mean (always = 1) + DOF of all factors included in estimating the mean performance at optimum condition}}$$

- Effective number of replications:

$$N_e = \frac{4}{1 + 2}$$

$$= 1.33$$

From the experiment, $V_e = 0.25$, and the QT4-computed value of $F(1,1) = 1.6$. Remember that this F value differs from the standard table value for reasons explained earlier. The C.I. can now be calculated from these.

$$C.I. = \pm \left[\frac{F(1,n_2) \times V_e}{N_e} \right]^{0.5}$$

$$= \pm \left(\frac{1.6 \times 0.25}{1.33} \right)^{0.5}$$

$$= \pm 0.548$$

The upper and lower limits of estimated performance at the optimum condition at the 90% confidence level are found as follows:

$$\text{upper limit} = \text{expected result} + C.I.$$

$$= 1.75 + 0.548$$

$$= 2.298$$

$$\text{lower limit} = \text{expected result} - C.I.$$

$$= 1.75 - 0.548$$

$$= 1.202$$

Confidence Level and Confidence Interval

The confidence interval (C.I.) represents the boundaries on the expected results and is always calculated at a confidence level. When multiple samples are tested to confirm the results predicted from the experiment, the term *expected result* expresses an estimate of the mean (average) of the average performance. The term *confidence level* expressed as a percentage, say 90%, indicates how often the performance is expected to exceed the value estimated. In other words, 90% will imply 9 out of 10 times. Putting it all together, the performance expected, shown in Figure 7.11, means that the average results of a population of samples tested at the optimum condition is expected to be within 1.202 and 2.298. Since the C.I. was calculated at a 90% confidence level, if sev-

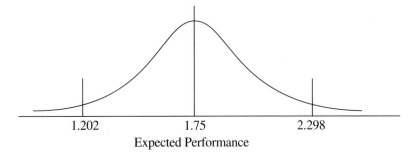

| 1.202 | 1.75 | 2.298 |

Expected Performance

Figure 7.11   C.I. for performance expected.

eral such sets are tested, 9 out of 10 times the averages of the sets are expected to fall within these limits.

ANOVA UTILITIES

In Example 7.1, you saw how the ANOVA table terms are calculated and how they are revised. In terms of the detailed math, that is as far as we will go in this book. Our focus now will be turned toward learning how ANOVA is used commonly. To explore its common function, we review a few of the earlier examples and learn to interpret what ANOVA presents in each experimental situation.

Error Term

The error term or error factor is the row above the total row in the ANOVA table. You already know (from Example 7.1) how it is calculated. Some discussions about what it means will be helpful. Most books on statistical experiments have labeled this term as experimental error. In reality this term is more than just experimental error. It represents the collective influence of all factors not included in the study, plus any experimental error if present. To call it experimental error would be to make this term indicate what went wrong when the experiment was run. Although this may happen to be the case once in a long while, often it is not a matter of how well the experiment was conducted. So what does the error term represent?

The error term combines the effects of three sources:

1. *Factors excluded from the experiment.* Generally, experiment planning sessions yield a larger number of factors than can be included in the experiment. Time and budget constraints would often prevent you from studying all factors that are identified. Depending on what is done with these factors when experiments are conducted, its effects will be combined with the error term.

2. *Uncontrollable factors (noise factors).* These are factors that are known to influence the results but are considered uncontrollable for cost or other reasons. For certain machining processes, the effects of change of room temperature, humidity, operator skill, tool wear, and so on, may fall in this category. Their influence on the result would also be part of the error term. In Taguchi experimental designs, these types of factors are called noise factors.

3. *Experimental error.* In an analysis of experimental results, error associated with the experimental setup is assumed to be due to uncontrollable factors. Furthermore, any error associated with a measuring instrument or repeatability is considered negligible. This experimental error refers strictly to any deliberate or unintentional error in collection, recording and manipulation of results, or incorrect settings of the factors under study.

Because it is not simply from the experimental setups, the error term in the ANOVA table of QT4 is labeled *All other/error*. This understanding of the constituents of the error term will play a major role in developing future experimental actions based on interpretations of ANOVA of the current experiments. To examine some characteristics of error terms, review Example 5.5. Run QT4 and open file *BKEX-55A.Q4W*. Select *Standard analysis* from the *Analysis* menu item in the main screen and proceed with the analysis. Stop and review the *ANOVA* screen, which should look like Figure 7.12.

The error term in this experiment is zero. Does it mean that there was no experimental error?   This experiment utilizes an L-4 array to study 3 two-level factors. The factors occupy all columns. Also, the experimental DOF (3) equals the factor DOF (3), which makes the error DOF equal to zero. The error term (S, V, P, etc. in ANOVA) will always be zero when the error DOF is zero. When the error term is zero because the error DOF is zero, it does not necessarily mean that there is no experimental error or that there are no effects from the factors not included in the experiment. It simply means that there was no provision in the experiment to capture the experimental error. The error term (S and P values whether zero and nonzero) is meaningful only when the error DOF is nonzero.

Which factor should you consider for pooling, and how would you go about it? Since the error DOF is zero, you should attempt to pool the weakest factor arbitrarily. Apply the 10% rule mentioned earlier to decide whether a factor is weak enough for pooling. Cutting speed has less than 10% of the influence of the strongest factor: feed rate, so it is a good target for pooling. Pool this factor by double-clicking on the factor description and revise the ANOVA.

From the pooled ANOVA shown in Figure 7.13, what can you say about the experimental error?   Remember that the error term not only represents any experimental error, it also includes the influences of two kinds of factors not included in the experiment. Whether or not it reflects any influence of control factors excluded from the experiment and the uncontrollable factors, depends on the experimental situation and what was done with these factors when you carried out the experiments. About the best explanation you can offer is that the experimental error, if present, will amount to less than 2.29%.

ANOVA Table					Qualitek-4	
Expt. File: BKEX-55A.Q4W	Data Type: Average Value				Print	Ok
	QC Type: Smaller is Better				Help	Cancel

Col # / Factor	DOF (f)	Sum of Sqrs. (S)	Variance (V)	F - Ratio (F)	Pure Sum (S')	Percent P(%)
1 A:Tool Type	1	12.25	12.25	-----	12.25	37.404
2 B:Cutting Speed	1	.25	.25	-----	.25	.763
3 C:Feed Rate	1	20.25	20.25	-----	20.25	61.831
Other/Error	0					
Total:	3	32.75				100.00%

Figure 7.12   ANOVA table.

A N O V A Table				Qualitek-4	

Expt. File: BKEX-55A.Q4W Data Type: Average Value   Print   Ok

QC Type: Smaller is Better   Help   Cance

Col # / Factor	DOF (f)	Sum of Sqrs. (S)	Variance (V)	F - Ratio (F)	Pure Sum (S')	Percent P(%)
1 A:Tool Type	1	12.25	12.25	49	12	36.641
2 B:Cutting Speed	(1)	(.25)		P O O L E D	(CL= *NC*)	
3 C:Feed Rate	1	20.25	20.25	81	20	61.068
Other/Error	1	.25	.25			2.291
Total:	3	32.75				100.00%

Figure 7.13   Pooled ANOVA.

Test of Significance

The term *test* here refers to the process of comparing a calculated factor F-ratio with the standard table or calculated reference value. The purpose of this test is simply to distinguish the significant factors from the insignificant ones. The judgment as to whether a factor influence is significant or not is done by comparing the calculated value of expected probability (confidence level) of occurrence with the desired level of confidence, which is subjective and supplied by the user. Consider, for example, that you decide to work with a minimum 90% confidence level. This means that you want an assurance that the influence (S or P in ANOVA) for the factor shown in the ANOVA table will happen more than 9 out of 10 times when such experiments are repeated. The assurance you want is a calculated value of the confidence level based on the ANOVA parameters (F, V, S, etc.) of the factor. If the confidence level calculated is higher than the desired value, the factor exceeds your expectation and is considered significant. By your own standard, then, this factor is important to you and should not be pooled. If this sounds a little complicated, it will become easier when you practice pooling a few factors in the examples that follow.

Many of your experiments will begin with zero error DOF. This will often be the case (there are exceptions) when you have filled all the columns of the array with factors and you have tested only one sample in each trial condition. With zero error DOF, you must attempt to pool the factor with the smallest sum of squares (use the 10% rule mentioned earlier). Soon after you pool one factor, and always when the error DOF is nonzero, you should pool factors only after performing the test of significance. There is only one key point you must remember: that when using QT4, the test of significance is not reliable for an error DOF below 3.

Pooling and revising ANOVA is one of the first tasks you should perform when you are at the *ANOVA* screen. To practice pooling by a test of significance, you will use one of the example experiments discussed earlier. Go to the main screen and Open the file *BKEX-63.Q4W*. Using the *Edit* menu, review the factors and the array used for this experiment. If you want to, review the example description in Step 6. Once you have made yourself familiar with this example experiment, select the *Standard analy-*

sis option from the *Analysis* menu in the main screen, proceed with the analysis, selecting QC = S, and go to the *ANOVA* screen shown in Figure 7.14.

Examine the ANOVA table and understand where the error DOF (4) comes from. This experiment has 27 results (L-27) from one sample tested in each trial condition. This makes the total DOF 26 (27 − 1). The experiment also includes 11 three-level factors, each of which has 2 DOFs. The error DOF is 4 (26 − 11 × 2), which is a good number to rely on using the test of significance in QT4. Also, notice that the error percent influence is 36.76%—keep an eye on this number as you pool the factors. This number would generally go up as you pool factors. You will of course, accept the factor influence values only after all insignificant factors are pooled.

Before you consider pooling, you must set your own level of expectation. As mentioned before, this is up to you, the experimenter. Your expectation level is expressed in terms of the minimum confidence level with which you want to work. You can work with a different level for every different experiment. But it is never a good idea to apply different standards for different factors in the same experiment. Most people will work with 80%, 85%, 90%, or 95% confidence levels. Let's suppose that you want to have a minimum of 85% confidence level before you consider a factor significant.

Look for a factor to pool. Find that factor *C*: toe plate has the smallest value of *S* (0.51). Double-click on the factor description and get the prompt about confidence levels shown in Figure 7.15. As you double-click, QT4 performs the test of significance in the background, which means that it calculates the standard confidence level based on the experimental data for the factor. The standard values of confidence level are obtained from a complex mathematical expression (gamma function) relating it to

	A N O V A Table					Qualitek-4	
Expt. File: BKEX-63.Q4W		Data Type:	Average Value			Print	Ok
		QC Type:	Smaller is Better			Help	Cancel
Col # / Factor	DOF (f)	Sum of Sqrs. (S)	Variance (V)	F - Ratio (F)	Pure Sum (S')	Percent P(%)	
1 A:Capsule Bracket	2	422.769	211.384	4.739	333.576	14.265	
2 B:Upper Jacket	2	146.732	73.366	1.645	57.54	2.46	
3 C:Toe Plate	2	.51	.255	.005	0	0	
4 D:Lower Jacket	2	87.657	43.828	.982	0	0	
5 E:Exten. Bracket	2	542.528	271.264	6.082	453.336	19.387	
6 F:Qtr. Beam-Upper	2	102.288	51.144	1.146	13.095	.56	
7 G:Qtr. Beam-Lower	2	12.066	6.033	.135	0	0	
8 H:Retainer Sleeve	2	601.158	300.579	6.739	511.965	21.894	
9 I:Rim Wire Dia.	2	26.768	13.384	.3	0	0	
10 J:Spoke Wire Dia	2	198.288	99.144	2.223	109.095	4.665	
11 K:Pencil Bracket	2	19.14	9.57	.214	0	0	
Other/Error	4	178.385	44.596			36.769	
Total:	26	2338.296				100.00%	

Figure 7.14   Unrolled ANOVA.

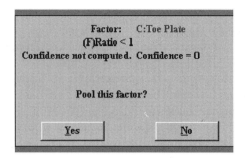

Figure 7.15 Prompt on calculated confidence level.

the factor F-ratio and error DOF (0.005 in this case, third row in Figure 7.15) obtained in the experiment. QT4 cannot calculate the confidence level for this example experiment since the F-ratio (0.005) is too small. The confidence level is arbitrarily set to zero (very small), which is less than the desired 85%. Click *Yes* to pool this factor. QT4 shows the revised *ANOVA* screen. Notice that the error DOF is increased by the DOF of the factor being pooled. In the same manner, pool factors G: qtr. beam-lower, K: pencil bracket, and I: rim wire dia. one at a time, since all these factors will have a confidence level below the value you desire. The revised ANOVA should now look as shown in Figure 7.16.

You would now wonder whether or not factor D: lower jacket, which is next in line for pooling, is significant. So double-click on this factor description and see that QT4 finds its confidence level as 85.2% (Figure 7.17). According to the minimum level (85%) you desired, this factor meets it. You now have all factors that are significant by your own standard. Click *No* in the prompt screen to return to ANOVA.

After pooling a few factors, if you find that you mistakenly pooled a factor or you did not maintain the sequence of pooling from the lowest one up, you can click the *Unpool* button to bring the ANOVA to its original form. Click *Unpool* and practice pooling the same set of factors again. You may also use the *Auto pool* button to pool a group of factors automatically, all at a time. In this option, QT4 pools all factors in the experiments that fall below the desired confidence level you enter. Generally, the revised ANOVA will look different if you automatically pool factors as opposed to doing one at a time. Pooling one factor at a time, always selecting the weakest among the surviving factors, is a better practice. You may practice *Unpool* and *Auto pool* options to see how the ANOVA differs. Return to the main screen when finished.

What does the percent influence of the error term mean? Consider another experiment with multiple results in each trial condition. From the main screen, Open file *BKEX-61.Q4W* and review the experimental details from the *Edit* menu. Perform standard analysis with the *Smaller is better* QC and be at the *ANOVA* screen shown in Figure 7.18. This is an L-9 experiment in which each trial condition was tested four times. The total DOF of the experiment is therefore 35 ($4 \times 9 - 1$), and the error DOF is 27 (35 − 4 factors × 2 DOF each).

Col # / Factor	DOF (f)	Sum of Sqrs. (S)	Variance (V)	F - Ratio (F)	Pure Sum (S')	Percent P(%)
1 A:Capsule Bracket	2	422.769	211.384	10.708	383.29	16.391
2 B:Upper Jacket	2	146.732	73.366	3.716	107.254	4.586
3 C:Toe Plate	(2)	(.51)		P O O L E D	(CL= *NC*)	
4 D:Lower Jacket	2	87.657	43.828	2.22	48.179	2.06
5 E:Exten. Bracket	2	542.528	271.264	13.742	503.05	21.513
6 F:Qtr. Beam-Upper	2	102.288	51.144	2.59	62.809	2.686
7 G:Qtr. Beam-Lower	(2)	(12.066)		P O O L E D	(CL= *NC*)	
8 H:Retainer Sleeve	2	601.158	300.579	15.227	561.679	24.02
9 I:Rim Wire Dia.	(2)	(26.768)		P O O L E D	(CL= *NC*)	
10 J:Spoke Wire Dia	2	198.288	99.144	5.022	158.809	6.791
11 K:Pencil Bracket	(2)	(19.14)		P O O L E D	(CL= *NC*)	
Other/Error	12	236.371	19.739			21.953
Total:	26	2338.296				100.00%

Figure 7.16 *pooled* ANOVA.

Suppose that you decide to work with 90% confidence level in this experiment. Check to see if any of your factors should be pooled. Click on the *Auto Pool* button. In the confidence level input screen shown in Figure 7.19, enter 90 (no need to reenter if it is already there) and click *OK* to proceed. QT4 shows the revised ANOVA as in Figure 7.20. Notice that the error DOF jumps to 31 (27 + 2 + 2) and error percent influence becomes 52.98 (≈ 53%). You could also have pooled this factor one at a time, each time checking to see if the confidence level for the factor meets your desired level.

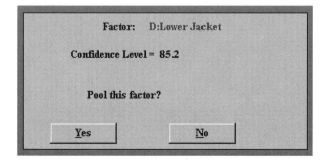

Figure 7.17 QT4 calculated confidence level.

A N O V A Table — Qualitek-4

Expt. File: BKEX-61.Q4W Data Type: Average Value QC Type: Smaller is Better

Col # / Factor	DOF (f)	Sum of Sqrs. (S)	Variance (V)	F - Ratio (F)	Pure Sum (S')	Percent P(%)
1 A: Lube levels	(2)	(93.05)		POOLED	(CL=82.89%)	
2 B: Engine Speed	2	148.383	74.191	3.004	98.994	5.35
3 C:Torque Applied	2	820.383	410.191	16.61	770.994	41.668
4 D: Surface Finish	(2)	(121.733)		POOLED	(CL=89.83%)	
Other/Error	31	881.537	132.085			52.982
Total:	35	1850.305				100.00%

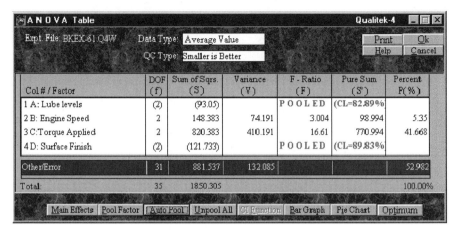

Figure 7.18 Pooled ANOVA.

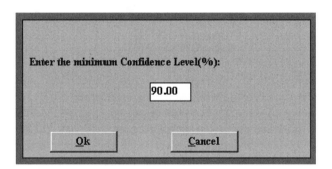

Figure 7.19 Confidence level desired for automatic pooling.

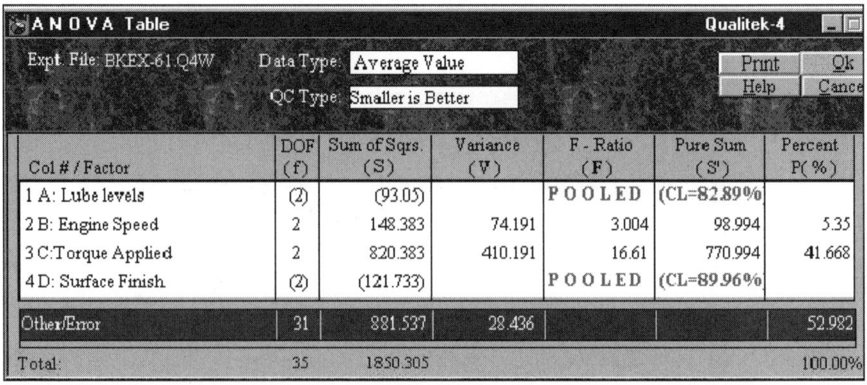

A N O V A Table — Qualitek-4

Expt. File: BKEX-61.Q4W Data Type: Average Value QC Type: Smaller is Better

Col # / Factor	DOF (f)	Sum of Sqrs. (S)	Variance (V)	F - Ratio (F)	Pure Sum (S')	Percent P(%)
1 A: Lube levels	(2)	(93.05)		POOLED	(CL=82.89%)	
2 B: Engine Speed	2	148.383	74.191	3.004	98.994	5.35
3 C:Torque Applied	2	820.383	410.191	16.61	770.994	41.668
4 D: Surface Finish	(2)	(121.733)		POOLED	(CL=89.96%)	
Other/Error	31	881.537	28.436			52.982
Total:	35	1850.305				100.00%

Figure 7.20 Revised ANOVA after pooling.

A high magnitude of influence of the error term (53% in the example) must be interpreted by the current state of the project. First, 53% error means that 47% (100 − 53) of the influence on the variation of results comes from the two significant factors. Does this mean that the results of the experiments are only 47% good? Unfortunately, this is likely to be a layperson's perception. You know how to interpret the results better now. You know that this (53%) influence is due to three things: experimental error, factors not included in the experiment, and uncontrollable factors. How much came from which source? This question will lead you to examine your project in further detail and review what went on during experiment planning. All you have to go by is this number. You need to examine each of the three sources of the error term carefully and eliminate one or more of them when possible.

Suppose that your review of the results shows that they are all correct and you carried out the experiment by setting it the way it is prescribed. Then the conduct of the experiment is not where the error term came from. Now, move on to looking into all other factors that were identified but not included in the experiment. If three other factors were identified, they could possibly be influential. The key data point is to determine the conditions of these factors when you ran the tests. If these factors were held fixed, they are not the source for the error term. If they were allowed to vary, it would definitely raise a question in your mind about their influence. Perhaps you will need to include these factors in your future studies.

What if the factors not included in the experiments were indeed held fixed? Then you need to look for the source of influence elsewhere. The last area is to think about all the things that you suspect have influence on the results but are uncontrollable, difficult to control, or too expensive to control. It is probably difficult cost-effectively to do something about them, and you have no plan to address them immediately. You may have to accept this portion of the influence and carry on with the improvement achievable despite the influence of the error term.

It should be clear to you from the discussions above that finding the reasons for influence represented by the error term in ANOVA is unique to the project you are studying. No single source or general trend can be established. What you should keep in mind is that just because the error influence is high, it does not necessarily imply that your experiment was bad or that it is a reflection of how precisely they were carried out. As long as you can ascertain that there was no deliberate or inadvertent error in setting up the experiments, you should have no reason to feel apprehensive about what your experimental data reveal. In that case, no matter what the magnitude of the error (70%, 80%, etc. are not uncommon), the result simply needs to be explained with facts about your project.

In industrial environment, many products and processes are too sensitive to the influence of uncontrollable and undetected factors. No matter how thoroughly all controllable factors are identified and how large an experiment is designed to study all factors, the error term may still show to have a large influence. This is typical of a system with natural or inherent variability. Expect a large error term (percent influence) in such systems. Regardless of the magnitude of the influence of the error term, the relative influence of the individual significant factors is useful information. When you are pooled on the *ANOVA* screen (Figure 7.20), click *OK* to proceed to the *Optimum* screen (Figure 7.21). Observe that customarily, the performance expected at the optimum condition is calculated by making use of significant factor contributions. Insignificant factors, those that are pooled, do not enter into specification of the optimum condition or its expected performance.

What can be done with factors that are pooled? Theoretically, the factors pooled (A: lube levels and D: surface finish in Figure 7.20) offer opportunities for you to treat them like uncontrollable factors. The optimum condition prescribed by analysis does not require any particular level. Which means that should you need to specify these factors, you are free to do so based on information (e.g., cost, availability, looks, etc.) other than that prescribed by the optimum condition.

When you can afford to, you should treat these factors as uncontrollable. Generally speaking, it will make your design more attractive since there are now fewer factors to adjust and control. Also remember that even if you end up specifying a nominal working level for these factors, you no longer need to specify tight tolerance on their specified levels. On the other hand, the relative influence of significant factors such as *B*: engine speed and *C*: torque applied (Figure 7.22) will dictate the degree of tightness you should impose on their specifications and tolerances. The higher the percent influence of a factor, the tighter the tolerance, and vice versa.

How does a higher error percentage of influence affect the confidence interval? The *Optimum* screen for most experiments captures the summary findings from the analysis of results. In summary, it tells you what the best design parameters are for the limits of expected performance. You can see how the error term affects the C.I. by calculating it for different ANOVA revision. From the *Optimum* screen (Figure 7.21), click the C.I. button and enter 90 for a 90% desired confidence level. QT4 calculates and displays C.I. as you click *OK* (see Figure 7.22). Review the numbers used for C.I. calculation (Figure 7.22) and where they came from. Return to ANOVA and repeat

Optimum Conditions and Performance				Qualitek
Expt. File:BKEX-61 Q4W	Data Type Average Value			Print
	QC Smaller is Better			Help

Column # / Factor	Level Description	Level	Contribution
2 B: Engine Speed	2000 rpm	2	-1.639
3 C:Torque Applied	100 ft lb	1	-5.139

Total Contribution From All Factors...	-6.778
Current Grand Average Of Performance...	88.638
Expected Result At Optimum Condition...	81.86

Figure 7.21 *Optimum conditions* with significant factors.

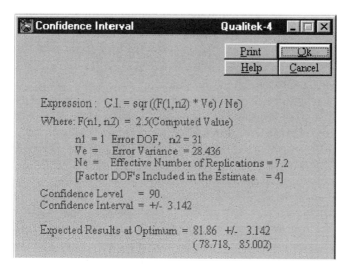

Figure 7.22   Calculated confidence interval.

the C.I. calculation with unpooled ANOVA. Observe the changes in C.I. value. Return to the *Optimum* screen and recalculate C.I. with different confidence levels (say, 70, 80, 85, etc.). Note how C.I. changes with changes in desired confidence level.

Directly or indirectly, error terms in ANOVA influence the optimum condition. First, pooling in ANOVA dictates the number of factors for the optimum condition. Second, the error term directly affects the C.I. by changing values of F_e and V_e in the relationship shown in Figure 7.22. Although it is difficult to draw a general trend of relationship of C.I. with respect to a change in error terms, the C.I. always becomes larger as the desired level of confidence demanded is higher.

Note that the effect of the process of pooling on the test of significance performed by QT4 is opposite to the manual way (textbook method) of doing it. Conventionally, the experimentally obtained factor F-ratio is compared with the standard table value (called an F-table) for the desired confidence level. If the experimental F-ratio exceeds the standard value, the factor is considered significant. QT4 software follows the converse approach (compare the confidence level instead of the F-ratio), as the confidence level is much more easily computed from the same relationship for a given value of F-ratio.

EXPERIMENTERS: BE AWARE

When should you accept the ANOVA results (percent of influence)?   Your acceptance of a factor influence will depend on your desired level of confidence. The test of significance (also called an F-test) provides the criteria for making such decisions. The F-test determines statistically whether the constituents (factor sums of squares) of the total sum of squares are significant with respect to the components in the error variance. In cases where the error DOF is zero and you cannot perform an *F*-test, accept all factors having influences greater than 10% of the highest factor influence.

What does absence of error DOF and percent of influence mean?   When *degrees of freedom* (DOF) of error is absent, $f_e = 0$ (which results in $P_e = 0$), it does not mean that there is no error but that information concerning factors contributing to the error term cannot be determined specifically. In these instances, you should consider pooling factors starting with a lower percentage influence, one at a time, to generate error DOF.

How many factors should you pool?   There is no fixed number. Should you meet the guidelines discussed in this step, the recommended practice is to continue pooling factors until the error DOF is approximately half the total DOF of the experiment.

What if high values of error term (80%, 90%, etc.) are present?   A high value of error term is not alone a reflection of the "quality" of the experiment. If *P_e is high, it does not mean that the experiment is bad or that the factor influences are unreliable. Instead, such a conclusion must be made only after testing for significance.*

Why should you pool?   Pooling generally enlarges the error DOF and its contribution. When the error DOF is greater, the confidence level on the factors is lower. That makes the results look bad. So, why pool?

- To make a conservative estimate. The higher probability allows us to identify true contributors.
- To reduce the chance of identifying less significant factors as being significant.

When must you test multiple samples in each trial condition?   When samples are repeated (i.e., multiple samples are tested in each trial condition), the total DOF of the experiment is much larger than the DOF of the array used for the experiment. In this case, if the results are analyzed by standard analysis techniques (as has been done so far), ANOVA will show a high error DOF. Even though it presents a definite advantage, it should not be a primary criterion in deciding whether or not to run multiple samples. Whether to repeat experiments or how many samples to test in each trial condition are determined by such considerations as cost of sample, variability in results from sample to sample, available time, and so on.

SUMMARY

In this step you have learned more about ANOVA and where all the terms in the table come from. You have also learned how the factor influences are tested for significance and how factors are pooled based on the test of significance. Additionally, you have practiced utilizing ANOVA and understanding how it affects the optimum condition and the expected performance in the optimum condition.

REVIEW QUESTIONS

Q.   Should one or more columns of the array used for the experiment be left empty such that experimental error can be captured even before pooling?

A. Not necessarily. If there are more factors than the number of columns available, you should fill it with factors. Remember, you would probably be able to pool one or more factors to make room for error DOF.

Q. If the error influence is greater than 50%, should you scrap the results and redo experiments?

A. Not necessarily. The information you obtain from the experiment can be good regardless of the magnitude of the Error influence. As long as you can assure that the experiments were carried out properly and the results analyzed were accurate, the ANOVA carries good information about your product or process.

Q. What indicates that the experiment is not satisfactory and should be rerun?

A. The definite indication that you should repeat your experiment is when you do not confirm the prediction of performance at the optimum condition. Failing to confirm means that the average performance from a number of samples tested at the optimum condition is not within the limits specified by the C.I.

Q. What would happen if I choose not to utilize ANOVA at all?

A. If you skip ANOVA as part of your analysis, you will not be able to make a conservative estimate about the optimum condition. You can make a conservative estimate of performance when you are able to specify and control the significant factors. In the absence of ANOVA, it is difficult for you to distinguish all factors that are statistically significant. Further, without the help of ANOVA terms, you will not be able to compute the range of expected results and be able to confirm the improvement you predict.

Q. Is there a minimum percentage influence below which a factor can be considered insignificant?

A. There is no such number. Use the 10% rule, which is to consider a factor insignificant when its influence is less than 10% of the highest factor influence.

Q. If the influence of the error term is 90% (i.e., the combined factor influence is 10%), can we still make use of the factor influences?

A. The relative factor influences, which are what the last column of ANOVA shows, are still good as long as you confirm your results. The high percentage of error may indicate that your system (product/process) has a high degree of inherent variability caused by unknown or uncontrollable factors.

EXERCISES

7.1 What does it mean when ANOVA indicates a high percentage of influence ($P_e = 30\%$, 60%, or 80%; $f_e > 0$)? Select all appropriate answers.

    (a) It shows that the experiment was poorly conducted and should be repeated.

    (b) It does not necessarily indicate that the experiment was bad.

    (c) It indicates that the performance of the product/process is largely influenced by factors not included in the study.

(d) It means that the result at the same trial condition varies significantly when repeated.

(e) It means that the results vary extensively from trial condition to trial condition.

7.2 How do you determine when and what factors to pool?   Select all appropriate answers.

(a) Pool all factors that fail the test for significance (say, 95% confidence).

(b) Start pooling with the factor that has the least percent of influence (P_e).

(c) There is no general percentage guideline for pooling. It depends on several factors.

(d) You should always attempt to pool if error DOF = 0.

(e) In general, try to pool until the error DOF is about half the total DOFs.

7.3 ANOVA is performed to determine which of the following?   Select all appropriate answers.

(a) Optimum performance

(b) Relative influence of individual factor on the variation of results

(c) Significance of influence of individual factor

(d) Best design condition

(e) Confidence level of expected performance at optimum

(f) Confidence level of the main effects of a factor.

(g) Performance at conditions other than those covered by the experiment

7.4 Which of the following are basic parts in analysis? Select all appropriate answers.

(a) Average column effects.

(b) ANOVA.

(c) Optimum performance calculations.

7.5 Refer to Table 7.2 to answer the following questions. (Experiment used an L-8 array and two samples per trial condition.)

(a) What are the total degrees of freedom for this experiment?

(b) What influence do factors other than those studied have on the final result?

(c) If you wanted to study only those factors that contribute more than 10% to the outcome, which factors would you pool?

(d) If the factors *chocolate chips* and *sugar* were pooled, what will be the DOFs of the error term?

(e) What does the percentage influence of the error term indicate? Discuss how this quantity can influence your follow-up experimental strategies.

TABLE 7.2 ANOVA for Exercise 7.5

Factor	f	S	V	F	$P(\%)$
Egg	1	33.063	33.063	25.190	35.11
Chocolate chips	1	1.563	1.563	1.190	0.28
Sugar	1	0.563	0.563	0.429	0.00
Butter	1	14.063	14.063	10.714	14.10
Flour	1	5.063	5.063	3.857	4.15
Time	1	18.063	18.063	13.762	18.52
Nuts	1	7.563	7.563	5.762	6.91
All other/error	8	10.500	1.310	20.94	20.94
	15	90.440			100.00

(f) If the trial conditions were repeated five times (instead of twice) and the results were analyzed by standard analysis, what will be the degrees of freedom of the error term?

7.6 Consider the factors of Table 7.3. Calculate the ANOVA statistics for the experimental results shown in Table 7.4.

(a) Complete the following calculations for the relative influence of factor A.

$$T = 12 + 10 + 15 + 16 = 53 \quad \text{or} \quad \overline{T} = 13.25 \qquad \text{C.F.} = \frac{53 \times 53}{4} = 702.25$$

$$A_1 = 12 + 10 = 22 \qquad A_2 = 15 + 16 = 31$$

$$S_T = (12^2 + 10^2 + 15^2 + 16^2) - 702.25 = 22.75$$

$$S_A = \frac{(22-31)^2}{2+2} = 20.25 \qquad \text{DOFs of } A, f_A = 2 - 1 = 1$$

$$V_A = \qquad P_A = \qquad = 89.01\%$$

TABLE 7.3 Factors and Effects for Exercise 7.6

	Average Effects	
Factor	L_1	L_2
A	11	15.5
B	13.5	12.5
C	14	12.5

TABLE 7.4 Array for Exercise 7.6

| Trial | Column | | | Results |
	A	B	C	
1	1	1	1	12
2	1	2	2	10
3	2	1	2	15
4	2	2	1	16

    (b) Verify that the ANOVA terms for factors B and C are as shown in Table 7.5.

7.7 For an experiment whose pooled ANOVA is as shown in Table 7.6, the optimum condition is A_2, B_1 (QC = B), and $Y_{opt} = 16.25$.
    (a) Determine the effective number of replications.
    (b) If $F(1, 1) = 1.60$ at the 60% confidence level, determine the confidence interval.

$$\text{C.I.} = \left(\frac{1.6 \times}{1.33} \right)^{0.5} = \pm 0.548$$

7.8 Analyze the results of the experiment in Example 6.2 (file *BKEX-62.Q4W*) using QT4, and determine which factors are insignificant if you desire 85% as a minimum level of confidence.

TABLE 7.5 ANOVA for Exercise 7.6

Column	Factor	f	S	V	F	S'	P (%)
1	A	1	20.25	20.25	—	20.25	89.01
2	B	1	0.25	0.25	—	0.25	1.09
3	C	1	2.25	2.25	—	2.25	9.89
Error factor		0					
		3	22.75				100

TABLE 7.6 Pooled ANOVA for Exercise 7.7

Column	Factor	f	S	V	F	S'	$P(\%)$
1	A	1	20.25	81	20		87.91
2	B	(1)	Pooled				
3	C	1	2.25	2.25	9	2.0	8.79
Error factor		1	0.25	0.25			3.30
Total		3	22.75				100

EXERCISE ANSWERS

7.1 Answers (b), (c), and (d) are correct.

7.2 All answers are correct.

7.3 Answers (b), (c), (e), (f), and (g) are correct.

7.4 Answers (a) and (c) are correct.

7.5 (a) 15; (b) 20.94; (c) chocolate chips, sugar, flour, and nuts; (d) 10; (e) answers will vary; (f) 32.

7.6 Partially worked out.

7.7 (a) 1.33; (b) ±3.02.

7.8 Three factors: D: inner panel, F: hinge reinforcement, and G: striker reinforcement.

Experimental Design for Studying Factor Interaction

What You Will Learn in This Step

Interaction among factors is quite common in industrial experiments. You can be highly effective in interpreting the experimental results if you have a good understanding of interaction between two factors. In this step, you will learn how to design experiments to include interactions and how to analyze results to determine if interaction is present, whether or not it is significant, or which factor levels are most desirable. Through a number of example experiments, you will have the opportunity to learn about how to detect the presence of interactions between factor pairs even when you are not able to design them into your experiments. By completing this step, you will be able to develop a more practical strategy for experimentation.

Thought for the Day

When I was a boy of 14, my father was so ignorant I could hardly stand to have the old man around. But when I got to be 21, I was astonished at how much the old man had learned in seven years.

—Mark Twain

When you plan and run experiments as a project team, the simpler the experiment, the better it is for all to understand. Simple experiments are those that you can design using the standard orthogonal array (the 10 arrays you have already used). Indeed, a vast majority of your experiments can be of this type. But there will be times when you will need to learn some new technique to design your experiments more cleverly to satisfy certain experimental needs. You will need to seek advance experiment design and analysis methods primarily for two reasons: either your factors have factor levels that cannot be accommodated with the standard array or because you were unable to confirm the predicted optimum performance from your simple experiments. The second reason is a likely scenario for experiments with factors of all kinds and is encountered more commonly. What do you do when your experiment is not satisfactory?

A good indication of whether or not an experiment is satisfactory is obtained when the result of a confirmation test matches the predicted value. A confirmation test is run at the conclusion of an experiment and at the optimum condition predicted by analysis of the results. If the average of the sample results (mean of population) falls within the range of expected performance as specified by the confidence interval (C.I.), the experiment is considered satisfactory. If, on the other hand, the average is outside the C.I. limits, you have reason to suspect that the model you used to make such predictions about performance is not quite adequate. The analytical model here refers to the method by which the optimum factor levels are identified and the performance at the optimum condition is estimated. Of course, the model bases such determination strictly on the behavior of the factors included in the study. If the confirmation test result is off, it is most likely that the model is unaware of actual factor behavior. It is often due to the fact that a factor behaves differently in the presence of other factors such that its trend of influence changes when levels of other factors change. This phenomenon is known as the *interaction effect*.

In this step you will find a detailed explanation of how to prepare your design for study of interactions, how to test for its presence, how to find out whether or not it is significant, and finally, how to make corrections in the prediction of optimum performance due to interactions. The explanation of theory is for your awareness; QT4 takes care of rules you need to follow and designs and analyzes the experiment automatically. If you are familiar with the theory of interaction design and analysis, you may skip to the first examples in this step. Otherwise, for a good grasp of the interaction studies presented in this step, keep the following checklist in mind:

1. Understand what interaction is and what indicates that it exists.
2. Learn how to design experiments that allow studies of preselected interactions.
3. Know how to test for the presence and the significance of interaction.
4. Ascertain what to do when interaction is found to be significant. Discover that all possible factor interactions can be tested for their presence only, even if none are included in the experimental design.

WHAT IS INTERACTION, ANYWAY?

Interaction means *interdependence*. Most factors you study have an influence on results. This is what we obtain as factor average effects (a plot of main effects). If these factors all behave independent of each other, a plot of the main effect (shows the trend of influence) will not change no matter which other factor it is with. A single experiment will yield information about the factor that could be used at all times regardless of where the factor appears. Unfortunately, most factors you would deal with have some kind of interaction with other factors in the group. To get an understanding of what interaction is, consider the following common situations.

Aspirin and Alcohol. To cure a severe headache, Johnny usually takes two aspirin tablets. This seems to work well to reduce his pain most of the time. Occasionally, however, when he has such headaches after he has consumed a can of beer (16% alcohol) and takes aspirin, the headache becomes worse. Johnny has learned the hard way that aspirin and alcohol don't mix.

Now, put on your experimenter hat and examine the effects on the user of the two factors *aspirin* and *beer*. The criterion of measurement or result is the level of headache. If you were to plot the effect of the aspirin, you will get a trend of influence that looks like the lower line in Figure 8.1. You will then be tempted to conclude how well aspirin cures headaches. Perhaps you would feel confident about making the conclusion if you were not aware of how the drug fails to act the way you expected it to work when there is liquor in the stomach. You are reminded of the other behavior—that is, how the same dose (level) of the drug aggravated the pain—as shown by the top line in the same graph. Faced with this situation, what should you say about the effect of aspirin? Because the two factors interact, you will have to modify your conclusions to reflect the possible effects when the two factors (alcohol and aspirin) are used together.

Temperature and Humidity.   A change in summertime temperature from 70°F to 90°F is welcome news to most residents of Michigan and many other states in the United States but may not be good news to people living in southern states, especially those near the ocean, where the air is likely to contain more moisture. Why does the same level of temperature change affect us in different ways?

Let's look at the two factors, temperature and humidity, and how they influence our comfort level. This is what you would measure as the results (criterion of evaluation). In just the same way as you would for any factor, the effects of temperature can be plotted for residents of these two areas as shown in Figure 8.2. For residents in northern states, the temperature may have an effect like that of the lower line (h_1), which would indicate that the comfort level goes higher with higher temperature. The trend of influence is opposite for residents of the southern states—the line (h_2) goes down with increased temperature. Here again the effect of temperature depends on what the level of the other factor, humidity, is. This behavior indicates that there is interaction

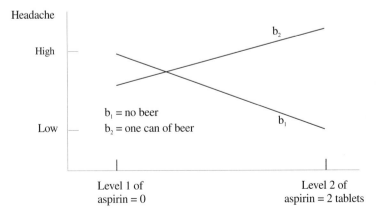

Figure 8.1   Plot of interaction between aspirin and beer.

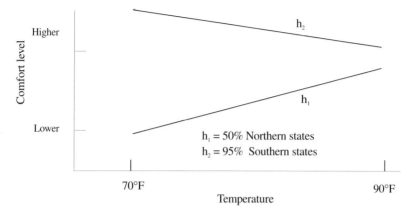

Figure 8.2   Plot of interaction between temperature and humidity.

between temperature and humidity. Symbolically, an interaction between two factors, A and B, will be depicted as $A \times B$.

Whereas interactions between known factors such as aspirin and alcohol and temperature and humidity are commonly experienced, interactions between factors in industrial experiments are not, although they are quite likely to be present. To detect and correct for the presence of interactions, you must take special care to design your experiments and analyze the results.

The two situations above show how the plots (two lines for each interacting pair) of the factor are expected to look if there is interaction between them. When there is interaction, the two lines will be nonparallel. It is unimportant whether they intersect within the range of the plot or what the magnitude of the effect (y-axis value) is. What is important are the trends (slopes) of the two lines and whether there is a difference in the trends, indicating a nonparallel relationship. The angle between the lines is the key indicator. Conversely, if from the experimental results you can generate data to plot two lines of effects for one of the interacting factors, the angle between the lines will indicate the presence of interaction.

Is Interaction a Factor or a Result?   An interaction (e.g., $A \times B$, $B \times C$, $D \times E$, etc.) between factors is something that changes the way the factors involved influence the results. In that respect it is an effect, similar to that for factors, which goes to increase or decrease the net effects of the factors on the result. It is not something you can see, touch, feel, or otherwise control. After all, if you had any control over it, you would rather not have it. So an interaction is not a factor, as you cannot control it, nor is it a result, since it has an effect on the result. The best way to relate to interaction is to view it as an effect, just like a factor effect. Like factors, interaction will occupy columns in an array. The effects are calculated in the same way as one would calculate that for factors. Since it is not an input you can control, unlike factors, interactions do not enter into descriptions of trial conditions. This will be clearer to you later in this step, when you design experiments to study interactions.

FORMS OF INTERACTIONS

When your experiments include a number of factors, the possibility of each factor interacting with another one or more factors increases quadratically. Suppose that there are seven factors, A, B, C, D, E, F, and G, in an experiment. As defined earlier, interaction is the influence of levels of one factor (say, B) on the effect of another (say, A). If A only had influence from B, there will be just $A \times B$ to consider. But while A is somewhat influenced by B, it can also be influenced by C. This means that there could possibly be three-factor interaction ($A \times B \times C$), in the same manner as we have seen is true of two-factor interactions. And, of course, these three-factor interactions may in turn be influenced by the level of D, which then will produce four-factor interactions ($A \times B \times C \times D$), and so on. For a number of factors, the possible two-factor combination ($A \times B$) is quite large. The number of combinations of interactions among three or more factors is even greater.

For the moment, let us consider the number of possible interactions between two factors and how many pairs can be formed. For example, if there were only two factors, A and B, there could be only one interacting pair, $A \times B$. But for seven factors (A, B, ..., G), there could be 21 possible pairs. If N is the number of factors, the possible number of pairs is calculated by the formula $N \times (N - 1)/2$. Finding the interacting pairs from a group of factors is similar to determining the number of single games you need to arrange to find the best one among N players. The interacting pairs for groups of up to seven factors are shown in Table 8.1.

The table shows only the possibilities of two-factor interactions. The possible interactions among three, four, or a higher number of factors are too numerous to list. For instance, when 7 two-level factors are studied by designing an experiment with an L-8 array, only eight of the possible total of 128 ($= 2^7$) combinations are tested. In this case, statistics about only the average effects of the factors can be determined. As shown above, this group of seven factors has 21 two-factor interactions, 35 three-factor interactions, 35 four-factor interactions, and few more. If you were to study all such effects, you will need to carry out experiments with all possible combinations (the full factorial). But that's too many to do, and often the number is prohibitively large. Orthogonal arrays (the 10 arrays introduced earlier) can be used to study interactions between two factors only. For this reason, and to keep it simple, our discussions will be about two-factor interactions only. You also need to be aware, that even if we study interactions between two factors, the factors involved may have two, three, or four levels. In the two situations discussed earlier, interactions between aspirin and alcohol and between temperature and humidity, all factors were at two levels. What would the interaction plot (graph of two lines) look like if the interacting factors (A and B) were at different levels?

Suppose that factors A and B are both at three levels. In creating an interaction plot, the effect of any one factor (A or B) can be plotted on the x-axis showing three locations for level effects. Suppose that you decide to plot A along the x-axis (Figure 8.3). The effect of factor A can now be plotted for a fixed level of factor B. Any such plot will now be composed of two straight-line segments. Bear in mind that factor B has

TABLE 8.1 Interaction Possibilities

Number of Factors	Number of Interacting Factor Pairs
(2) A, B	(1) $A \times B$
(3) A, B, C	(3) $A \times B, A \times C, B \times C$
(4) A, B, C, D	(6) $A \times B, A \times C, A \times D, B \times C, B \times D, C \times D$
(5) A, B, C, D, E	(10) $A \times B, A \times C, A \times D, A \times E,$ $B \times C, B \times D, B \times E, C \times D, C \times E, D \times E$
(6) A, B, C, D, E, F	(15) $A \times B, A \times C, A \times D, A \times E, A \times F, C \times D, C \times E,$ $C \times F, B \times C, B \times D, B \times E, B \times F, D \times E, D \times F, E \times F$
(7) A, B, C, D, E, F, G	(21) $A \times B, A \times C, A \times D, A \times E, A \times F, A \times G,$ $B \times C, B \times D, B \times E, B \times F, B \times G,$ $C \times D, C \times E, C \times F, C \times G,$ $D \times E, D \times F, D \times G, E \times F, E \times G,$ $F \times G$

three levels. Thus there will be three such plots, as shown in Figure 8.3. Recall the decision criteria for interactions. Interaction exists when the lines are nonparallel. In this case you have too many lines to compare. Since each plot has two line segments, there are many possible ways to compare the angle between the lines. As you can see, the task of deciphering where the two factors interact and where they do not becomes quite complicated. Interactions between factors at three and four levels are supported by the corresponding arrays and it is also partially supported by QT4. Again, in this book, for the sake of simplicity, we discuss only the interactions between 2 two-level factors and make some related observations about interactions between higher-level factors. Henceforth, unless otherwise defined, the word *interaction* will mean interaction between 2 two-level factors.

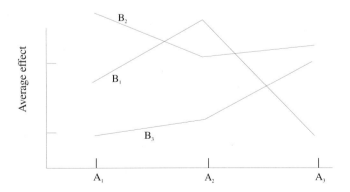

Figure 8.3 Plot of interaction between 2 three-level factors.

SORTING OUT INTERACTIONS BETWEEN 2 TWO-LEVEL FACTORS

Even if you only wish to study interactions between 2 two-level factors, there are always too many combinations to consider. The number of interacting pairs increases rapidly as the number of factors increase. For example, if there are 7 two-level factors considered, 21 pairs of interactions could be studied. Could you possibly study them all? Addressing all interactions is sure to increase the size of your experiment. Perhaps you would like to study a selected few. But which among these 21 will you study? How would you know (before experiment) which ones are more important than others? You will learn the answer to those questions in the discussion below.

Interaction effects between 2 two-level factors are studied by experiments designed using two-level orthogonal arrays such as L-4, L-8, L-16, and L-32 (all two-level arrays except L-12). The effect of an interaction is obtained (calculated) in the same manner as a factor when columns of the array are reserved to study the effect. Generally, prior knowledge and group discussions help determine the interactions of interest. Once interactions to be studied are selected, reserving the appropriate column for them can complete the experiment design. The number of columns necessary to gather information about interactions $(A \times B)$ that is between 2 two-level factors is dictated by the degrees of freedom (DOFs) of interaction. The DOF value of an interaction is calculated simply by multiplying the factor DOFs:

$$\text{DOFs of interaction } A \times B = (\text{DOF of } A) \times (\text{DOF of } B)$$

$$= (2 - 1) \times (2 - 1)$$

$$= 1$$

Since an interaction such as $A \times B$ above has 1 DOF, it can be studied by keeping one column of the two-level array empty. Keeping a column empty would mean that no factor is assigned to this column. Instead, the column is reserved for the interaction. For example, if factors A and B are assigned to columns 1 and 2, respectively, column 3 must be reserved to study the interaction effect $(A \times B)$ between them. This is because the interaction effect between factors placed in columns 1 and 2 is found to be mixed with the effects of another factor if assigned to column 3. If you know this information and you want to study the interaction $A \times B$, you would keep the third column empty. Otherwise, perhaps you would normally assign the next factor instead (Figure 8.4).

The location of the column for interaction that is reserved (left empty) to study the effect of interaction depends on where the interacting factors are located. Since many pairs of interacting factors are assigned everywhere in the array, there could be hundreds of questions about where their interaction effects will show up. The task of finding which columns receive (are confounded or mixed with) interaction effects from which pair of factors is extensive and time consuming. Fortunately, along with the development of the set of orthogonal arrays, Taguchi has provided the interaction col-

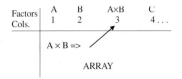

Figure 8.4 Mixing of interaction effects.

umn information applicable to each array. The interaction confounding information for any arbitrary location of the interacting factors is provided in a table of numbers called the *triangular table* (see Table 8.2).

Using the triangular table (TT), you can determine the interaction column where interaction effects between factors will be mixed. The column of interaction naturally depends on the location of the interacting factors. While designing your experiment, suppose that you assign factor A in column 2 and factor B in column 4 of the array (L-4, L-8, etc.). If you also want to study the effect of interaction $A \times B$, you would need to know which column you should leave empty. You will use the triangular table (see the Appendix) to determine the columns you should reserve to study interactions between two factors assigned to any two columns.

There are separate triangular tables (TTs) for two-, three-, and four-level factors. We will only make use of the table for two-level arrays, which is needed to make provisions to study interactions between 2 two-level factors. The number across the top of a TT corresponds to column numbers in the array ("2" for column 2). The numbers in parentheses at the beginning of each row of a TT also represent column numbers, such as (4) in the fourth row of the TT. The number in the TT identified by the intersection of these two numbers (2 and 4)—that is, the intersection point between a ver-

TABLE 8.2   Triangular Table for Two-Level Factor Interactions

1	2	3	4	5	6	7	8	9	10	11	12	13	14	15
(1)	3	2	5	4	7	6	9	8	11	10	13	12	15	14
	(2)	1	6	7	4	15	10	11	8	9	14	15	12	13
		(3)	7	6	5	4	11	10	9	8	15	14	13	12
			(4)	1	2	3	12	13	14	15	8	9	10	11
				(5)	3	2	13	12	15	14	9	8	11	10
					(6)	1	14	15	12	13	10	11	8	9
						(7)	15	14	13	12	11	10	9	8
							(8)	1	2	3	4	5	6	7
								(9)	3	2	5	4	7	6
									(10)	1	6	7	4	5
										(11)	7	6	5	4
											(12)	1	2	3
												(13)	3	2
													(14)	1
														(15)

tical line through 4 and a horizontal line through 2—is the column (6), with which interaction (of factors in 2 and 4) is confounded. You will always use a TT while designing experiments that include interaction studies. For convenience, the readings from the TT are presented in the following notation:

$$2 \times 4 \Rightarrow 6, \quad 1 \times 2 \Rightarrow 3, \quad 4 \times 8 \Rightarrow 12 \quad \text{etc.}$$

The numbers at the left of each expression indicate the column location of the factors. The number on the right of the $\Rightarrow$ sign indicates the column to which the interaction effect goes. (Note that the sign $\Rightarrow$ is used instead of an =, as it is not an algebraic equation.)

Use the TT (Table 8.2) to review the following readings:

$$1 \times 2 \Rightarrow 3, \quad 2 \times 3 \Rightarrow 1, \quad 3 \times 1 \Rightarrow 2$$

$$2 \times 4 \Rightarrow 6, \quad 4 \times 6 \Rightarrow 2, \quad 6 \times 2 \Rightarrow 4$$

$$2 \times 7 \Rightarrow 5, \quad 1 \times 5 \Rightarrow 4, \quad 3 \times 4 \Rightarrow 7 \quad \text{etc.}$$

Here are some key observations for you to remember:

1. Any valid reading of the TT forms an interacting group of columns (IGC) (e.g., 1, 2, and 3; 2, 4, and 6, etc.).
2. All IGCs are commutative. Refer to the first two groups above. If you know that 1×2 points to 3, then without reading the table, because of the commutative property, you know what 2×3 and 3×1 are.
3. Each array has a fixed number of independent IGCs. An L-8 array has only one and L-16 has five IGCs. One common IGC for L-8 array is columns 1, 2, and 3. See if you can find another reading for two factors placed between columns 1 and 7 from the TT that does not include column 1, 2, or 3. (You will not be able to find it.)

It is a simple matter to read the TT and find where the interaction effects go when you have a few factors and limited number of interactions (two or three) to include in your design. It is a different situation when you need to lay out an experiment to study many factors and a large number of interactions among them. To facilitate the process of column assignment, certain common readings of the TT have been represented in graphical form, called a *linear graph*. In a linear graph the number on the line shows the interaction column between factors placed in a column identified by numbers at the ends on the line. You can conveniently use them when necessary or always refer back to the TT. A common set of linear graphs is shown in Figure 8.5.

The triangular table and linear graphs shown above are used when designing experiments with L-4, L-8, and L-16 arrays. Expanded (up to 31 columns) triangular tables and linear graphs for L-16 and larger arrays are provided in the Appendix. Keep

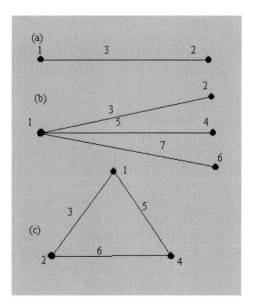

Figure 8.5   Common linear graphs.

in mind that these tools for interaction design applies only to all two-level orthogonal arrays except L-12. The L-12 is a specially designed array in which, unlike all other two-level arrays, the interaction effect between factors assigned to 2 two-level column is distributed equally to all 11 columns. This makes an L-12 array unable to identify the effects of interaction between factors assigned to any column. To put it another way, the L-12 does not have any interacting group of columns and must not be used to study interactions. However, for the same reasons, it is an excellent choice when interactions between factors is known to be present but are deliberately ignored. For example, when you want to study up to 11 two-level factors, and you suspect that interactions are present between many pairs of factors not all of which you can afford to study, your best bet will be to use an L-12 array for the experiment design.

INTERACTION DESIGN, ANALYSIS, AND CORRECTION

As you know by now, your experiment needs to be specially designed to study interactions. Once you performed such experiments, you will also need to perform special analysis to test for the presence and significance of such interactions. Subsequently, when interactions are found significant, you will be required to modify the optimum design and performance as dictated by the significant interactions. You will also now learn all the tasks involved in interaction studies.

Example 8.1: Heat Treatment Process Study Supplier of a small steel machine part conducted a study of the heat treatment process used for the production process.

The part, which is manufactured at several facilities of the company, was of warranty concern to the company. A team was formed comprised of people from all participating plants for the purpose of performing design of experiments. Collectively, the team identified a number of factors and interactions for the study (Table 8.3). Participants from the three plants (plants I, II, and III) volunteered to design and carry out separate experiments to study some or all of the factors identified. The available budget for the project dictated that initial study be limited to use of an L-8 array for the experiment design.

In the heat treatment process for steel, when steel is quenched at a sufficiently rapid cooling rate from above the upper critical temperature, a microstructure called *martensite* is formed. A martensitic layer of steel which is formed on the surface of the part is *nonductile* and is very hard. The objective of the study was to determine heat treatment process parameters that produce the highest surface hardness number obtained by a Shore scleroscope hardness measuring instrument.

During the planning session, the group agreed on the seven factors listed in Table 8.3 in the order of their importance. All team members were made aware of the fact that $21 (= 7 \times 6/2)$ pairs of interactions are possible. Although no one had prior knowledge about the degree of influence of all the interactions, a few participants had stronger feelings about the six interactions listed below. Staying within the limits of L-8 experiments, each plant chose to study a different combination of factors and interactions. Realizing that the L-8 array has seven columns and that each interaction requires one column, plant I selected two interactions and five factors for the study (Table 8.4). The factors and interactions groups selected by plants II and III are also shown in the table.

(a) Plant I designed experiments by assigning factors A, B, C, D, and E in columns 1, 2, 4, 5, and 7, respectively. The eight experiments were carried out and the results (average hardness) obtained were recorded as 66, 75, 54, 62, 52, 82, 52, and 78 (Shore hardness).

(1) Determine the optimum condition and the estimated range of performance at the 90% confidence level.

TABLE 8.3   Factors and Interactions Identified for Example 8.1

Factors	Interactions
A: temperature	A: temperature $\times B$: time duration
B: time duration	B: time duration $\times C$: gas concentration
C: gas concentration	C: gas concentration $\times D$: quenching oil
D: quenching oil	C: gas concentration $\times A$: temperature
E: quench rate	A: temperature $\times C$: gas concentration
F: carbon content (part material)	A: temperature $\times D$: quenching oil
G: part orientation (heating feeder)	

(2) If the experiments were to be repeated, which factors and interactions should be included?

(b) For the factor and interactions plant II wants to include in the study, design the experiment and prescribe a better choice of factors or interactions for their study.

(c) Design experiments for the two combinations of factors and interactions that plant III proposes to study:

Solution: **(a)** This example experiment has been designed using QT4 and the results saved under experiment file *BKEX-81A.Q4W*. You may review the design and results using the *Edit* menu option or practice the design process by executing the steps shown here. If you were new to interaction design and analysis, you would benefit by practicing to design experiments using both the *Automatic design* and *Manual design* QT4 options described below:

QT4 Design Steps

1. Run QT4 and go to the main screen.
2. Select *Automatic design* from the *Design* menu item.
3. Click *Yes* at the prompt for new experiment design.
4. In the *Automatic design* screen, check the box at the bottom of the screen. You would always select this box (Figure 8.6) when all your factors are at two levels, with or without interactions. Click *OK* to proceed.
5. Type in the descriptions of the factors and their levels you want to study (Figure 8.7). It is a good idea to list the factors in order of their importance. At this point QT4 is unaware of the number of factors you would list and obviously does not know which array to select for the design or to which columns to assign the factors. You can list 63 factors or less. Remember that the level description can be both numbers and characters. Click *OK* to proceed.
6. QT4 confirms the number of factors you entered in the prompt message. Click *Yes* in this prompt if the factor count is correct.

TABLE 8.4 Factors and Interactions for Each Plant in Example 8.1

Plant I	Plant II	Plant III
A: temperature	A: temperature	A: temperature
B: time duration	B: time duration	B: time duration
C: gas concentration	C: gas concentration	C: gas concentration
D: quenching oil	D: quenching oil	D: quenching oil
E: quench rate	E: quench rate	E: quench rate
$A \times B$	$A \times B$	$A \times B$
$B \times C$	$C \times D$	$A \times C$
		$B \times C$ or $A \times D$

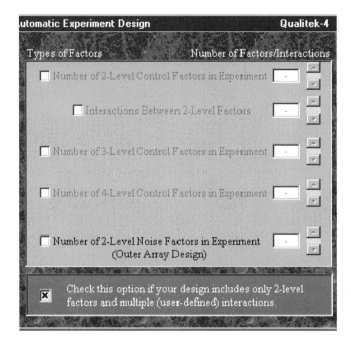

Figure 8.6 Automatic design input for interaction studies.

7. In this screen QT4 shows the factors you described in the box at the left. You can now indicate the interactions you want to study by double-clicking the factor of your choice. QT4 assumes that the pair of factors you select first is the most important (if known) one in your mind. To select $A \times B$, double-click on factor A first. Notice that factor A is removed temporarily from the box at the left.

	Factor	Level 1	Level 2
1	A:Temperature	800 Deg F	925 Deg F
2	B:Time Duration	20 Sec	40 Sec
3	C:Gas Concentration	20% Carbon	40% Carbon
4	D:Quanching Oil	Type 1	Type 2
5	E:Quaench Rate	Rapid	Slower
6			
7			

Input of 2-Level Factors   Qualite

Enter Dummy Da

Factor Descriptions

Figure 8.7 Factor and level descriptions for automatic interaction design.

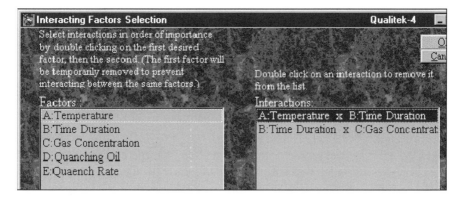

Figure 8.8   Interaction selection screen.

Double-click on factor B next. The left box is now revised to show all factors, and the interaction $A \times B$ is shown in the right box. In the same way, you can select interactions $B \times C$ by double-clicking on factors B and C in the left box one at a time. Your screen should look as in Figure 8.8. If you want to practice again or are not satisfied with your selection, click *Cancel* to restart from the *Factor description* screen.

8. In the *Interaction placement* screen, which comes next, QT4 shows you the array it has selected for the design and the column assignments for the factors and interactions. You should carefully review this screen for interactions you wish to study. Click *OK* to display the complete factors and their level descriptions and the column assignments as shown in Figure 8.9.

Inner Array Design

Array Type: L-8

Use <ctrl> + <arrows> to move cursor.

	Factors	Level 1	Level 2
1	A:Temperature	800 Deg F	925 Deg F
2	B:Time Duration	40 Sec.	60 Sec.
3	Interaction A x B	*INTER*	-----------
4	C:Gas Concentrati	20 % Carb	40% Carbon
5	D:Quenching Oil	Type 1	Type 2
6	Interaction B xC	*INTER*	-----------
7	E:Quench Rate	Rapid	Slower

Figure 8.9   Factor and level descriptions with interaction column assignments.

9. You have the option to edit any descriptions you like in this screen. Notice which factor is assigned where and which column is reserved for interactions. In *Automatic design*, QT4 is not burdened with experimental design rules. It designs the experiment automatically when it can. Click *OK* and proceed to the next few screens in the design process, with which you are already familiar.

10. Provide the description of experiments in the next screen. Indicate the *Bigger is better* quality characteristic for this experiment, as the higher Shore hardness value is indicative of a harder surface.

11. Review the array used for the design. Note that the columns reserved for interactions are left unchanged. Click *OK* to save the file.

12. When prompted to save the experiment file, type a file name you like (say, *MyExpt81.Q4W*). Using QT4, you will be able to save this and other experiment files that utilize L-8 arrays. So you should practice saving files using names of your choice. The file name you enter should be eight characters or less. The extension .Q4W is optional. QT4 will add it if you forgot. Click *OK* and return to the main screen.

Experiment Review

1. Your experimental file should be the same as *BKEX-81A.Q4W*, provided with QT4. Once you are at the main screen, load this file, selecting the *Open* option from the *File* menu item. Now select *Trial condition* from the *Condition* menu item and review the eight trial conditions. Examine them and see that the interaction column has no effect on the description of the trial condition. In other words, there is nothing to do because of the interaction. As discussed earlier, interaction is not input or anything you can control. Thus, when describing the trial condition, the columns, which are reserved for interaction studies, are ignored. The first two trial conditions are as shown in Figure 8.10.

2. In the main screen, compare the experimental results displayed next to the array with those prescribed in the problem statement. If the results look different from those specified, edit and revise the file by selecting *Results* from the *Edit* menu item.

At this point you are ready to analyze the results and find answers to the questions in the example. But before you proceed with the analysis, perhaps you should review the methods for interaction design. A good way to be proficient in interaction design is to use the *Manual design* option of QT4 to design your experiment. Skip the alternative design steps if you want to move on to analysis, or review the manual design method for a detailed explanation of guidelines for interaction design.

Alternative Design (QT4 Manual Design)

1. From the main screen, choose the *Manual design–Inner array* option from the *Design* menu item and proceed to design the experiment selecting an L-8 array. Here you are in charge. You must select which array to use for the design. Since there are 5 two-level factors and two interactions, you will need an array that has

Descriptions Trial Conditions		Qual

Trial Condition 1   (Random order for running this Trial is 5)

Factors	Level Description	Level #
A:Temperature	800 Deg F	1
B:Time Duration	40 Sec.	1
C:Gas Concentrati	20 % Carb	1
D:Quenching Oil	Type 1	1
E:Quench Rate	Rapid	1

Trial Condition 2   (Random order for running this Trial is 2)

Factors	Level Description	Level #
A:Temperature	800 Deg F	1
B:Time Duration	40 Sec.	1
C:Gas Concentrati	40% Carbon	2
D:Quenching Oil	Type 2	2
E:Quench Rate	Slower	2

Figure 8.10   Trial conditions unaffected by interactions.

at least seven columns. An L-8 has seven columns. Although just having a sufficient number of columns does guarantee that the array will work, it is a good start. (In later examples you will see where just enough columns will not be a sufficient condition.) A rule of thumb for selecting an array for the design is to find one that has DOFs equal more than the DOFs of all the factors and interactions included in the study. Again, matching DOFs alone is not sufficient to assure that you will have a design. All you would conclude at this point is that the array has the DOFs necessary. There are just two guidelines for designing an experiment that includes interaction:

(a) In the process of design, start assigning factors that are part of the interacting group. In case of multiple pairs of interacting factors, start with the pair known to have the stronger interactions. Select interacting factor pairs arbitrarily if their relative strengths are unknown.

(b) Always reserve interaction columns as prescribed by the triangular table (see Table 8.2). Assign the remaining factors arbitrarily to the available columns.

In manual design you need to make a decision about which factor is to be assigned to which column. It may help to work out a scheme of assignment on a separate page before you start typing the factor description in QT4. From the first rule above you can select the more important (if you know) of the two interaction pairs $A \times B$ and $B \times C$. Suppose that you decide to take care of factors A and B first. This means that you will assign two columns for them. Since at the start, all seven columns are available, you may decide to assign A to column 1 and B to column 2, as shown in Figure 8.11. Any two columns (numbers shown refer to the column of the array used) may be chosen. However, once the factors are assigned, you do not have a choice about which column you must reserve to study the interaction between the two factors. It must be done following the TT. QT4 has built-in TTs and helps you with the column assignment.

2. QT4 helps you identify and reserve columns for interactions in two ways. In the first you know the column—for example, in this case, column 3 should be reserved for interaction between factors A and B ($1 \times 2 \Rightarrow 3$). Place the cursor on column 3 (row in Figure 8.11) and click on the *Col inter* button. The *Interacting columns selection* screen (Figure 8.12) will show all possible column interactions that become mixed with column 3. Click on 1×2 (which highlights it) to select the interacting factor columns of your choice and click *OK*. You will notice that the *Inner array design* screen now shows that it has been reserved for interaction between columns 1 and 2. It is important that you do not edit the text under level 1 of the interaction column. If you wish to practice interaction column assignment more, put your cursor on the column and click the *Reset col* button to repeat the process. You may also modify the description of interaction to, say, $A \times B$ instead of *Inter cols* 1×2, which is the default notation.

The second method of identifying the interaction column will become clear as you proceed with the design and treat $B \times C$ interaction.

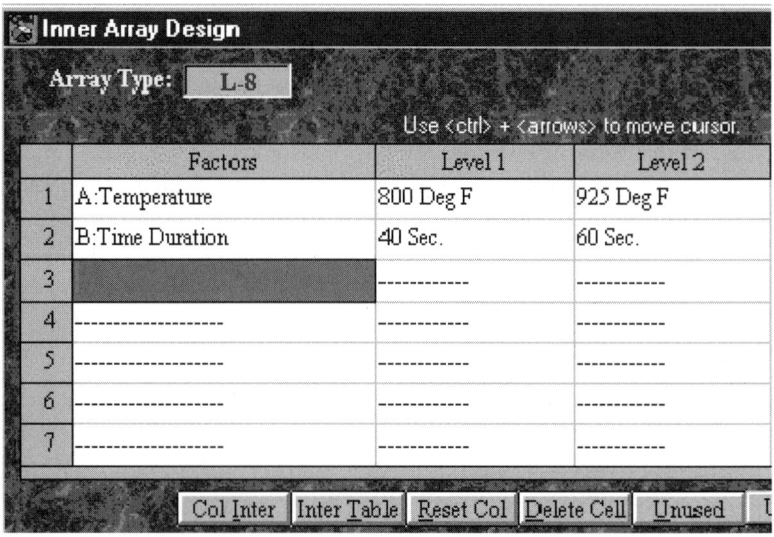

Figure 8.11 Manual experiment design input.

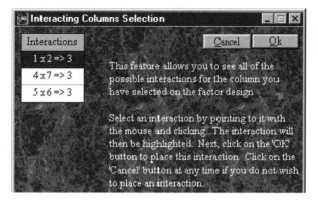

Figure 8.12   Interaction selection screen.

3. To continue with your design, you need to treat the second interacting pair, $B \times C$, next. But factor B has already been assigned to a column. You need to assign factor C to a column such that you are able find another column, as prescribed by the TT, to be reserved for $B \times C$. This is what you need to do carefully. Since column 4 is available, assign factor C to column 4 and click the *Inter table* button for interaction column selection as shown in Figure 8.13. In this screen, enter column numbers 2 and 4 in boxes under A and B, and check the box to set the interaction. As you check the box, QT4 instantly shows the interaction column from the TT. Click *OK* to return to the design screen. Notice that column 6 has been reserved for interaction between columns 2 and 4 (Figure 8.14), where factors B and C are assigned.

4. Complete the experiment design by entering descriptions of factors D and E in columns 5 and 7, respectively. Then click *OK* to proceed. You may continue with the design input and save this file under a different file name if you like.

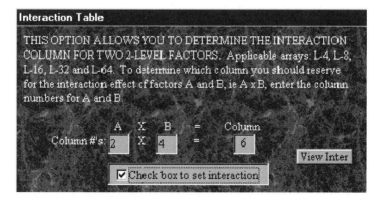

Figure 8.13   Automatic interaction column reading from triangular table.

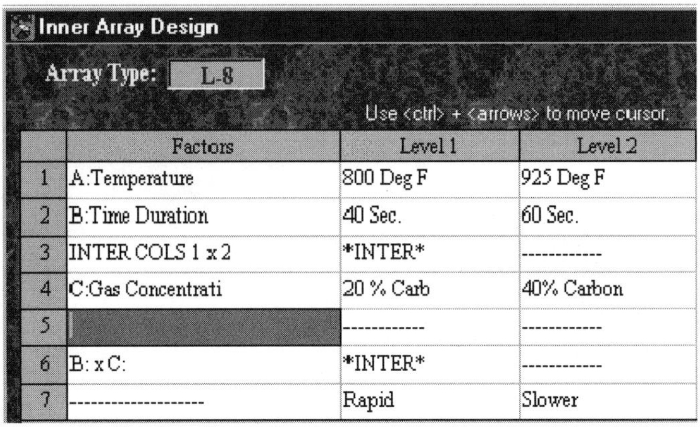
Inner Array Design

Array Type: L-8

Use <ctrl> + <arrows> to move cursor.

	Factors	Level 1	Level 2
1	A:Temperature	800 Deg F	925 Deg F
2	B:Time Duration	40 Sec.	60 Sec.
3	INTER COLS 1 x 2	*INTER*	------------
4	C:Gas Concentrati	20 % Carb	40% Carbon
5		------------	------------
6	B: x C:	*INTER*	------------
7	-------------------	Rapid	Slower

Figure 8.14 Automatic interaction column identification.

In manual design you have the option to assign factors to any columns of your choice. For instance, you could have assigned factor B to column 1 instead of A, or you could have started with the $B \times C$ interacting pair of factors first. Obviously, experiments designed for the same group of factors and interactions by different people are likely to have different column assignments. As long as the two design guidelines described above are not violated, the experiment designs are correct. In the automatic design option, QT4 will always produce the same design based on minimum mixing (highest resolution) of interaction effects.

Analysis of Results

1. Run QT4 and load the experiment file *BKEX-81A.Q4W* from the *File* menu in the main screen. (You need not load again if you are already working with this file.)
2. Select *Standard analysis* from the *Analysis* menu item.
3. Check the box for *Bigger is better* in the *Quality characteristic selection* screen. Click *OK* to proceed.
4. Review the results of the experiments and check to see that they are as shown in part (a) of the problem description. Edit results if they are not the same. Click *OK* to proceed.
5. The *Main effects* screen shown in Figure 8.15 lists factor and interaction average effects (referred to as *column effects*) for each. In analyses of column effects, the interaction is treated just as the factors. The levels of the interaction have no physical interpretation, but analytical values. For example, $(A \times B)_1$, the level 1 effect of column 3, which is reserved for interaction $A \times B$, has an average value of 67.75 (third row in Figure 8.15). The average effect of $A \times B$ for level 2, that is, $(A \times B)_2$, equals 62.5. These numbers reveal the same kind of information as they do for factors. The difference between the two level averages shows how

Main Effects (Average Effects of Factors and Interactions)   Qua

Expt. File: AJUNK1.Q4W   Data Type: Average Values

QC Type: Bigger is Better

Column # / Factors	Level 1	Level 2	L2 - L1
1 A:Temperature	64.25	66	1.75
2 B:Time Duration	68.75	61.5	-7.25
3 INTER COLS 1 x 2	67.75	62.5	-5.25
4 C:Gas Concentrati	56	74.25	18.25
5 D:Quenching Oil	70	60.25	-9.75
6 B: x C:	64.5	65.75	1.25
7 E:Quench Rate	65.5	64.75	-.75

| Noise Effects | Interactions | Plot | Multi Plot | Anova |

Figure 8.15   Factor and interaction average effects.

sensitive the result is to the interaction. Their relative value also helps you identify which one to include in the computation of performance at the optimum condition.

The column average effects are more useful when they are presented in the graphical form (column effects) shown in Figure 8.16, which is obtained when you click the *Multi plot* button. The slope in each graph is directly proportional to the difference of the two-level average effects shown in the last column of Figure 8.15. If the line is close to being flat (i.e., the difference is small), the effect of the factor or interaction placed in this column is smaller. The influence is absent when the line is perfectly horizontal. The slopes of the plots of column effects (always a single straight line for a two-level column) also allows you to determine the relative influences of each. Bear in mind, though, that this comparison indicates only their relative order of influence, not the exact magnitude (e.g., 25%, 45%, etc.) of influence. For that you must perform ANOVA.

What do the average effect plots tell you about the two interactions included in the study?   The average effect plots of the interactions $A \times B$ and $B \times C$ (see Figure 8.16) shows that $A \times B$ is stronger than $B \times C$. But whether $A \times B$ is significant, or whether $B \times C$ is negligible (insignificant), cannot be determined from here. It must await ANOVA calculations.

If the pairs of factor interactions included in your experiment were the only ones present, the observations made above are all you would need. Unfortunately, there are more interacting pairs present: namely, 10 pairs from the five factors ($= 5 \times 4/2$). Recall that the decision about which interactions to study was a matter of prior knowledge or guesses and also that inclusion of interaction meant sacrificing a column in

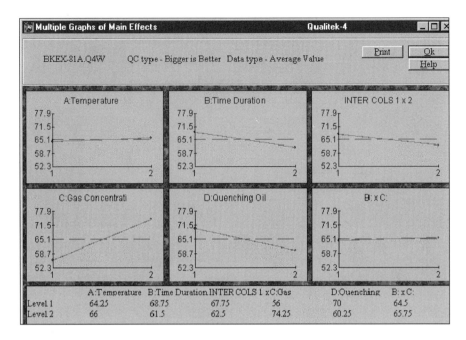

Figure 8.16 Factor and interaction average effects.

the array, which could have been used to study a factor. Only a limited number of interactions were selected. What should you do about the other eight of the 10 possible interactions? Are these interactions present, and if so, which are more severe than others?

Generally, for complete information about interactions, you need to find two characteristics for each: Is it present; and if present, is it significant? These two inquiries require different calculations and tests to find the answer. For those interactions that are included in the study (columns reserved), the ANOVA calculations recognize their presence and their relative influence. The ANOVA can calculate influence and performed tests of significance only on those interactions that occupy a column of the array. It cannot make any calculation on interactions that are not included in the study. You will soon see how ANOVA handles the two interactions in this example experiment.

Just because you were not able to include all possible interactions in your study, you are not completely in the dark about all of them. You can still determine partial information, that is, whether or not they are present, by performing some additional calculation. You will not be able to determine which are significant, but you will be able to rank them in the order of their strength of presence. Since all interactions that are found to be present are not necessarily significant, you will only use this information for future repeat experiments.

To perform a test for the presence of interaction of all possible interacting factor pairs, click the *Interactions* button from the bottom of the *Main effects* screen (Figure

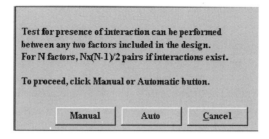

Figure 8.17   Test of presence of interaction selection.

8.15). At the prompt shown in Figure 8.17, click on *Auto* for QT4 to test all possible interactions automatically (all 10 among 5 factors).

As part of the automatic test of the presence of interactions, QT4 makes calculations for all possible (10 in this case) pairs of interactions and lists them in the order of their strength of presence. The term *severity* is used to express the degree of strength of presence numerically, with the most severe interaction on top of the list, as shown in Figure 8.18. This screen contains information about the interactions that are valuable for adjusting the optimum condition because of the interactions. But to understand the information in different columns (Figure 8.18), you will first need to examine carefully how the presence of an interaction test is performed and how the items for this test are calculated.

Methods for test of presence of interactions.   The presence of interactions is tested by comparing the slopes of the interaction plot shown for the factors *aspirin* and *beer* in Figure 8.1. The plot for the presence of interactions consists of two lines, repre-

Automatic Test for Presence of Interaction   Qu

Number of interactions between two factors calculated = 10

#	Interacting Factor Pairs (in Order of Severity)	Columns		SI(%)	Col	Opt.
1	A:Temperature x D:Quenching Oil	1 x	5	65.17	4	[2,1]
2	B:Time Duration x E:Quench Rate	2 x	7	57.35	5	[1,1]
3	D:Quenching Oil x E:Quench Rate	5 x	7	42.64	2	[1,1]
4	A:Temperature x B:Time Duration	1 x	2	42	3	[1,1]
5	A:Temperature x E:Quench Rate	1 x	7	41.66	6	[2,1]
6	A:Temperature x C:Gas Concentrati	1 x	4	34.82	5	[2,2]
7	C:Gas Concentrati x E:Quench Rate	4 x	7	22.34	3	[2,2]
8	C:Gas Concentrati x D:Quenching Oil	4 x	5	6.25	1	[2,1]
9	B:Time Duration x C:Gas Concentrati	2 x	4	4.9	6	[1,2]
10	B:Time Duration x D:Quenching Oil	2 x	5	4.41	7	[1,1]

Figure 8.18   Results of test of interactions.

senting the trends of influences of one factor for two levels of the other factor. The choice of the *x*-axis can be for any of the interacting factors. In Figure 8.1, the effect of the factor *aspirin* was first plotted for a fixed value of *beer* (level 1). The second line is then drawn showing the trend of influence of the factor (beer) at another level. The decision about the presence of interactions is then made by strictly comparing the slopes of the two lines. The strength of presence of interaction is also calculated by the magnitude of the angle, which ranges between 0 and 90°. The term *severity index* (SI) is defined such that SI = 100% when the angle between the lines is 90° and SI = 0 when the angle is zero. After all interactions are calculated, QT4 uses the SI values to order them (see Figure 8.18, SI column). Among the 10 interactions tested, the SI values range between 65.17 and 4.41. To see the most severe interaction plot for $A \times D$ (between temperature and quenching oil), double-click on the description and see that the plot is as shown in Figure 8.19.

The interaction plots are obtained by graphing the combined effects of the pair of factors involved. For example, to test for the presence of interaction between factors A and D ($A \times D$), you will need to calculate four numbers, such as the average effects of A_1D_1, A_1D_2, A_2D_1, and A_2D_2. The average effect of A_1D_1 is found by averaging results that contain effects A_1 and D_1 together. Refer to Table 8.5 and see that factor A is assigned to column 1, in which A_1 is located in the first four rows. Hence the first four trial results contain the influence of A_1. Considering column 5, where factor D is assigned, observe that the influence of D_1 is contained in

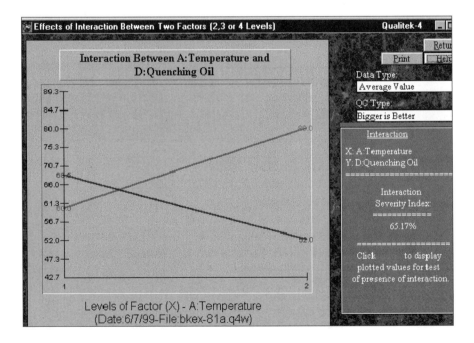

Figure 8.19 Plot for the most severe interaction.

TABLE 8.5 Orthogonal L-8 (2^7) Array with Results

				Factor (Column)				
Trial	A (1)	B (2)	$A \times B$ (3)	C (4)	D (5)	$B \times C$ (6)	E (7)	Results
1	1	1	1	1	1	1	1	66
2	1	1	1	2	2	2	2	75
3	1	2	2	1	1	2	2	54
4	1	2	2	2	2	1	1	62
5	2	1	2	1	2	1	2	52
6	2	1	2	2	1	2	1	82
7	2	2	1	1	2	2	1	52
8	2	2	1	2	1	1	2	78

the first, third, sixth, and eighth trial conditions. Therefore, the results of the first and third trial conditions contain the influence of A_1 and D_1 together. Thus the average effect of the combined factor effects is calculated as

$$\overline{(A_1 D_1)} = \frac{66 + 54}{2} = 60 \text{ (first and the third trial results only)}$$

Similarly,

$$\overline{(A_1 D_2)} = \frac{75 + 62}{2} = 68.5 \qquad \overline{(A_2 D_1)} = \frac{78 + 82}{2} = 80 \qquad \overline{(A_2 D_2)} = \frac{52 + 52}{2} = 52$$

The interaction plot lines in Figure 8.19 are graphed using the numbers calculated above. Depending on the factor selected for plotting along the x-axis, the four average effects are grouped as shown below.

$\overline{(A_2 D_1)} - \overline{(A_1 D_1)} = 80 - 60$, which shows the effect of factor A (as it changes from A_1 to A_2) for a fixed value of D (D_1).

$\overline{(A_2 D_2)} - \overline{(A_1 D_2)} = 52 - 68.5$, which shows the effect of factor A (as it changes from A_1 to A_2) for a fixed value of D (D_2).

For all interaction plots, QT4 labels the factors involved in X and Y. The actual descriptions of X and Y ($X = A$ and $Y = D$ in this case) are always displayed on the right-hand side of the plot (see Figure 8.19). You can also see the values of the combined factor average effects that QT4 uses by clicking on the word *Here*. The numbers used for graphing interaction plots for $A \times D$ interaction is shown in Figure 8.20. When you are done reviewing the numbers used for the plot, click *OK* to return to the interaction

```
┌─────────────────────────────────────┐
│         Plotted Values              │
│  =============================      │
│                                     │
│   X1Y1: 60        X3Y1:             │
│   X2Y1: 80        X4Y1:             │
│                                     │
│   X1Y2: 68.5      X3Y2:             │
│   X2Y2: 52        X4Y2:             │
│                                     │
│   X1Y3:           X3Y3:             │
│   X2Y3:           X4Y3:             │
│                                     │
│   X1Y4:           X3Y4:             │
│   X2Y4:           X4Y4:             │
│                                     │
│            [  Ok  ]                 │
└─────────────────────────────────────┘
```

Figure 8.20 Level average values calculated for interactions.

plot. Click *Return* to go back to the *Automatic test of presence of interaction* screen (see Figure 8.18).

As you realize, interaction $A \times D$ was not one of the interactions included in the study. The two interactions that were included ($A \times B$ and $B \times C$) rank fourth and ninth in the list. Also understand that calculations for tests of the presence of interactions are done by examining the columns where the factors are assigned. For example, calculations of A_1D_1, A_2D_1, and so on, make use of the columns where factors A and D are assigned and do not need a separate column reserved for the interaction $A \times D$ (does not exist in this example). Thus the average effects needed for the test of interaction for all possible factors, whether or not they were included in the experiment, can be calculated.

QT4 uses the following expression to calculate the severity index (SI):

$$SI = \frac{[(\overline{A_2D_1}) - (\overline{A_1D_1})] - [(\overline{A_2D_2}) - (\overline{A_1D_2})]}{2 \times \text{constant}}$$

For this example,

constant = difference between maximum and minimum combined factor average effects for all pairs of interaction

$$= 80 - 52$$

$$= 28$$

and

$$SI = \frac{(80 - 60) - (52 - 68.5)}{2 \times 28}$$

$$= \frac{36.5}{56}$$

$$= 65.17\%$$

Now that you know how SI for each interaction pair is calculated, you will easily follow the rest of the information in Figure 8.18. The information under *Columns* indicates the columns where the two factors are located. Thus, for the first row ($A \times D$), 1×5 indicates the columns where factors A and D are located. The number under the *SI* column shows the calculated value of the severity index (SI). The column number (4) indicated under the column labeled *Col* represents the column where the interaction effect of $A \times D$ will be present. Unless this column is left empty, the interaction effect of $A \times D$ will be mixed with the effect of the factor assigned to the column. If you were to repeat the experiment and study interaction $A \times D$ (assuming that A and D are as assigned), you will need to keep column 4 empty so that you can study the effect of $A \times D$.

The last column, *Opt*, can be very useful information for your project. This column lists the desirable levels for the interacting factors. The numbers [2,1] for $A \times D$ in the first row indicates that level 2 of A and level 1 of D are the desirable levels. If $A \times D$ interaction was included in your study (the only way you can test for significance is by reserving the column) and it is found to be significant, these levels for the factors should be adjusted to compensate for interaction effects. Later as you proceed with analyses in this example), you will see how interaction analysis dictates change in levels of some factors which interact with others.

The desirable levels for the factors, such as [2,1] for factors A and D, are selected from the interaction plot (Figure 8.19) you just reviewed. The factor levels are obtained by identifying the desirable point (magnitude of average effects) in the graph. The desirable point, of course, depends on the quality characteristic (QC). For QC, *bigger is better* in this case, the top right point has the highest magnitude (80) and represents the level description $A_2 D_1$ ($= 80$), which you can check by examining the corresponding interaction plots. Obviously, the desirable level for a factor may be different when its interactions with several other factors are considered (A interacting with B, C, D, etc.). For example, level A_1 could be desirable from $A \times B$, but A_2 could be preferred when $A \times C$ interaction is considered. What should you do with this information? How can you make a compromise in selecting the levels of the factor for best result? For the most part, you keep it for future studies. Information about those factors whose interactions have been included in the experiment may come to your needs immediately.

The interaction plot for any two factors can also be obtained for any two factors singly. From the *Main effects* screen (see Figure 8.15), you can click the *Manual* button and follow the screen instruction. Following the *Manual* option, QT4 displays the interaction plot between factors A and D, which should look the same as the automatic option.

After you are done reviewing interactions, return to the *Main effects* screen and click *OK* to proceed to the *ANOVA* screen (Figure 8.21). It is a common practice to revise the ANOVA by pooling factors or interactions assigned to the columns. The pooling refers to the process of ignoring column effects (factor or interactions occupying the column) that are considered insignificant. For a pooling process, you may want to review examples in Step 6.

The procedure for deciding which column effects are significant and which are not need not be an arbitrary decision. Whether or not a column effect is significant is judged by testing for significance. The test for significance and the supporting theory was covered in Step 7 under discussions of ANOVA. Unfortunately, the test of significance can only be done when error DOE is nonzero. In this case, error DOF is zero. We will then need to attempt to pool some factors and interactions, if possible, arbitrarily. A common practice is to consider a column effect insignificant and pool when its influence is less than 10% of the highest percentage of influence for any factor in the experiment. Following this guideline, factor E, interaction $B \times C$, and factor A are pooled one at a time in the order described. The revised ANOVA is shown in Figure 8.22.

What does ANOVA say about the experiment? ANOVA provides information in the following three areas:

1. *Which factors are significant, and the relative percentages of influence.* From the revised ANOVA, factor C: gas concentration has the most influence, followed by factors D and B. Factors A and E were considered insignificant. These two insignificant factors can be set at the peak of their tested level, allowing the opportunity to save cost. Interaction $B \times C$ is found insignificant now and was weakly present (listed ninth from the test of the presence of interaction; see Figure 8.18).

Col # / Factor	DOF (f)	Sum of Sqrs. (S)	Variance (V)	F - Ratio (F)	Pure Sum (S')	Percent P(%)
1 A:Temperature	1	6.125	6.125	-----	6.125	.596
2 B:Time Duration	1	105.125	105.125	-----	105.125	10.237
3 INTER COLS 1 x 2	1	55.125	55.125	-----	55.125	5.368
4 C:Gas Concentrati	1	666.125	666.125	-----	666.125	64.869
5 D:Quenching Oil	1	190.125	190.125	-----	190.125	18.514
6 B: x C:	1	3.125	3.125	-----	3.125	.304
7 E:Quench Rate	1	1.125	1.125	-----	1.125	.109
Other/Error	0					
Total:	7	1026.875				100.00%

| Main Effects | Pool Factor | Auto Pool | Unpool All | CI Function | Bar Graph | Pie Chart | Optimum |

Figure 8.21 Unpooled ANOVA table.

Col # / Factor	DOF (f)	Sum of Sqrs. (S)	Variance (V)	F - Ratio (F)	Pure Sum (S')	Percent P(%)
1 A:Temperature	(1)	(6.125)		POOLED	(CL=81.38%	
2 B:Time Duration	1	105.125	105.125	30.397	101.666	9.9
3 INTER COLS 1 x 2	1	55.125	55.125	15.939	51.666	5.031
4 C:Gas Concentrati	1	666.125	666.125	192.614	662.666	64.532
5 D:Quenching Oil	1	190.125	190.125	54.975	186.666	18.178
6 B: x C:	(1)	(3.125)		POOLED	(CL=100%)	
7 E:Quench Rate	(1)	(1.125)		POOLED	(CL= *NC*)	
Other/Error	3	10.375	3.458			2.359
Total:	7	1026.875				100.00%

Figure 8.22 Pooled ANOVA table.

2. *Which interactions are significant.* Interactions influence our conclusions from the experiment only when it is found to be present and determined to be significant. Thus $B \times C$ interaction is not of concern to us, but we need to see how $A \times B$ interaction (interaction columns 1×2), which is found to be significant, influences the factor-level selection.

3. *What the influence from all other factors is.* This is the percent influence in the row labeled *Other/Error*, which is 2.359% for this experiment. This means that the influence from the pool factors, factors not included in the experiment (when these factors are not held fixed), uncontrollable factors (noise factors), and the experimental error, if any, amounts to only 2.359% of the total influence. The magnitude of this number helps to decide about the future course of action in the project. No immediate action is called for because of the magnitude of this number, which offers some room for speculations, as the number is comprised of several things.

Click *OK* or the *Optimum* button in a revised *ANOVA* screen to display the *Optimum* screen (Figure 8.23). The optimum screen shows two pieces of information of immediate needs: the levels of the significant factors (level 1 for factor B, level 2 for C, and level 1 for D) and the expected performance when the factor levels are set as prescribed. The estimated performance, 85.375, already includes the effect of significant interaction. No further adjustment is necessary for the expected performance because of the interaction. But levels of the factors identified may need to be adjusted because of the interaction.

How are factor levels adjusted because of the interaction? An interaction found significant in this case is $A \times B$, which dictates what the levels of these two factors should be. If interaction were not significant, the levels of A could be any level, as it has been pooled, and that of B is 1, as selected by the program based on the average effect (see Figure 8.15, second row). Now that $A \times B$ is found significant, what will be their levels? This question can be answered by reviewing the $A \times B$ interaction plot obtained by clicking on $A \times B$ (fourth line) in the *Interaction* screen or obtained from

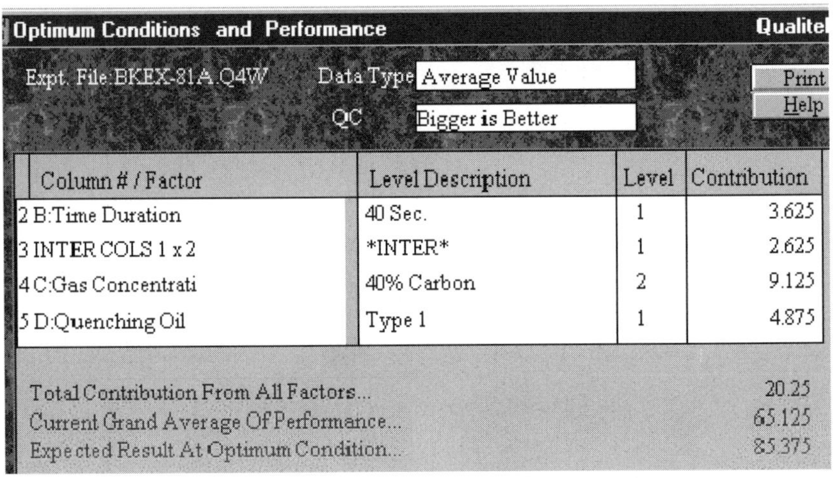

Optimum Conditions and Performance			Qualitek

Expt. File BKEX-81A.Q4W   Data Type Average Value   Print / Help

QC   Bigger is Better

Column # / Factor	Level Description	Level	Contribution
2 B:Time Duration	40 Sec.	1	3.625
3 INTER COLS 1 x 2	*INTER*	1	2.625
4 C:Gas Concentrati	40% Carbon	2	9.125
5 D:Quenching Oil	Type 1	1	4.875

Total Contribution From All Factors...   20.25
Current Grand Average Of Performance...   65.125
Expected Result At Optimum Condition...   85.375

Figure 8.23 Optimum condition and performance.

the last column under $A \times B$ interaction (levels 1, 1, Figure 8.18). In this case the levels for A and B are 1 and 1. No adjustment of level of factor B is called for, as it is already 1. The final specification for the confirmation test will thus be as shown in Table 8.6.

To determine the boundaries of expected performance, click on the *C.I.* button while at the *Optimum* screen and enter 80 in the input box for confidence interval. When a number of samples are tested at the optimum condition, the average performance is expected to fall within 83.469 and 87.281, as shown in Figure 8.24.

Recall that Example 8.1 has two parts. So far you have found a solution for part (a). Solutions for parts (b) and (c) follow.

(b) This project also has 5 two-level factors and two interactions. But because the interactions $A \times B$ and $C \times D$ are two independent pairs (have no common factor), the experiment design is accomplished using an L-16 array. Follow the automatic design option as in part (a) and proceed to design the experiment. Describe the five factors and indicate the interactions as shown in Figure 8.25. (Review the file *BKEX-81B.Q4W* after designing the experiment.)

TABLE 8.6   Optimum Condition

Factor	Level
A	1
B	1
C	2
D	1
E	Either 1 or 2

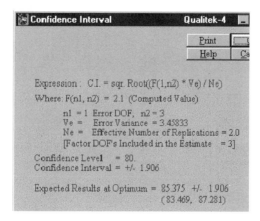

Figure 8.24 Confidence interval on expected performance.

As you proceed, QT4 shows you the placement of the factor and interactions in the L-16 array used for the design (Figure 8.26).

Notice that the experiment utilizes only 7 of the 15 columns of the array, leaving 8 columns unused. If it is essential that the two interactions be studied, plant II could study eight additional factors or study more interactions without increasing the size of the array. A better choice may be to keep the same five factors but select two dependent interactions with a common factor between the two (e.g., $A \times B$ and $A \times C$) such that an L-8 array could be used for the design.

(c) The experiment designs in this case will be different depending on the choice of the last interaction. For factors and interactions A: temperature, B: time duration, C: gas concentration, D: quenching oil, E: quench rate, $A \times B$, $A \times C$, and $B \times C$, the experiment is designed to require eight columns of a two-level array. Since L-8 has only seven columns, and L-12 cannot be used for interaction studies, the design is ac-

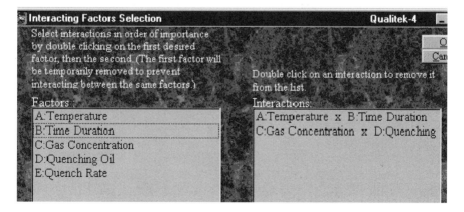

Figure 8.25. Interaction selection for automatic design

Inner Array Design

Array Type: L-16

Use <ctrl> + <arrows> to move cursor.

	Factors	Level 1	Level 2
1	A:Temperature	800 deg. F	925 deg. F
2	B:Time Duration	40 sec.	60 sec.
3	A:Temp. x B:Time Duration	*INTER*	------------
4	C:Gas Concentration	20% carbon	40% Carbon
5	E:Quench Rate	Rapid	Slower
6	COLUMN UNUSED	*UNUSED*	------------
7	COLUMN UNUSED	*UNUSED*	------------
8	D:Quenching Oil	Type 1	Type 2
9	COLUMN UNUSED	*UNUSED*	------------
10	COLUMN UNUSED	*UNUSED*	------------
11	COLUMN UNUSED	*UNUSED*	------------
12	C:Gas Conc. x D:Quenching Oil	*INTER*	------------
13	COLUMN UNUSED	*UNUSED*	------------
14	COLUMN UNUSED	*UNUSED*	------------
15	COLUMN UNUSED	*UNUSED*	------------

Figure 8.26 Factor and interaction placement in inner array

Inner Array Design

Array Type: L-16

Print   Ok
Help   Cancel

Use <ctrl> + <arrows> to move cursor.

	Factors	Level 1	Level 2	Level 3	Level 4
1	A:Temperature	800 deg.	925 deg. F	------------	------------
2	B:Time Duration	40 sec.	60 sec.	------------	------------
3	A:Temp x B:Time D	*INTER*	------------	------------	------------
4	C:Gas Concentrati	20% carbo	40% Carbon	------------	------------
5	A:Temp x C:Gas Co	*INTER*	------------	------------	------------
6	B:Time Dur.x C:Ga	*INTER*	------------	------------	------------
7	D:Quenching Oil	Type 1	Type 2	------------	------------
8	E:Quench Rate	Rapid	Slower	------------	------------
9	COLUMN UNUSED	*UNUSED*	------------	------------	------------
10	COLUMN UNUSED	*UNUSED*	------------	------------	------------
11	COLUMN UNUSED	*UNUSED*	------------	------------	------------
12	COLUMN UNUSED	*UNUSED*	------------	------------	------------
13	COLUMN UNUSED	*UNUSED*	------------	------------	------------
14	COLUMN UNUSED	*UNUSED*	------------	------------	------------

Figure 8.27 Factor descriptions and column assignments.

complished using an L-16 array. Follow the automatic design option as in part (a) and proceed to design the experiment. Describe the five factors, indicate the interactions, and obtain an experimental design as shown in Figure 8.27. (Review the file *BKEX-81C.Q4W* after designing the experiment.)

If, on the other hand, the interaction $B \times C$ is replaced by $A \times D$ (*A*: temperature, *B*: time duration, *C*: gas concentration, *D*: quenching oil, *E*: quench rate, $A \times B$, $A \times C$, and $A \times D$), this experiment will require an L-16 array. Follow the automatic design option as in part (a) and proceed to design the experiment. Describe the five factors, indicate the interactions, and obtain the experimental design as shown in Figure 8.28. (Review the file *BKEX-81D.Q4W* after designing the experiment.)

Example 8.2: Electrostatic Powder Coating Process Optimization A process improvement team in a plant manufacturing a dc motor (armature) lists factors suspected to have a major influence on the winding operation (Table 8.7). Two separate studies to investigate two combinations of interactions were proposed:

(a) $A \times B$, $B \times C$, and $A \times C$
(b) $A \times B$, $A \times C$, and $A \times D$

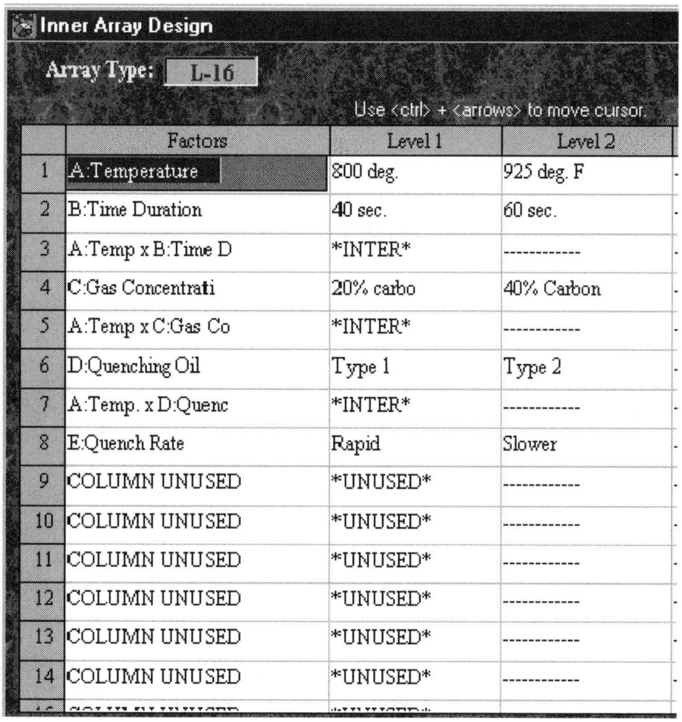

Figure 8.28 Factor descriptions and column assignments.

TABLE 8.7   Factors and Levels for Example 8.2

Factors	Level	
	1	2
A: air box pressure (lb/in^2)	200	400
B: brushing speed (rev/min)	300	500
C: voltage (kilovolts)	40	70
D: cure temperature (°C)	200	300

The objective in both experiments was to determine the process parameters needed for minimum defects in the coil winding.

Solution: (a) From the main screen of QT4, use the *Automatic design* option to describe the factors and their levels. [See the detailed steps in the solution of Example 8.1(a).] Be sure to indicate the interactions appropriately and obtain the design as shown in Figures 8.29 and 8.30. You may need to edit the interaction description to fit into the available character spaces. Three samples in each experimental condition were tested and results evaluated in terms of the number of defects from a batch size of 1000 products (Figure 8.31). You can design and save this experiment using a file name of your choice or review file *BKEX-82A.Q4W.*

Proceeding with analysis using *Standard analysis* from the main screen and performing *Interaction analysis* from the *Main effects* screen, the six possible interactions are arranged in order of their severity of presence in Figure 8.32. Realize that even though only three interactions ($A \times B$, $B \times C$, and $C \times A$) were included in the study, there are six ($4 \times 3/2$) possible interactions among the four factors. From the last column in Figure 8.32 or by studying the corresponding interaction plot, you can determine the levels of the factors suitable for lowering the defects. For example, if $A \times B$ were found to be significant, level 1 of factor A and level 1 of factor B would be de-

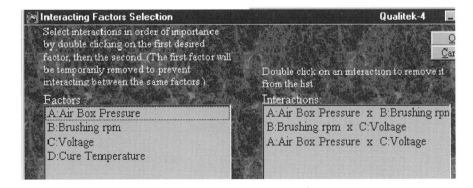

Figure 8.29   Factor and interaction descriptions.

Figure 8.30 Factor descriptions and column assignments.

	Factors	Level 1	Level 2
1	A:Air Box Pressure	200 psi	400 psi
2	B:Brushing rpm	300 rpm	500 rpm
3	A:Air Box Pressure x B:Brus	*INTER*	------------
4	C:Voltage	40 KV	70 KV
5	A:Air Box Pressure x C:Voltage	*INTER*	------------
6	B:Brushing rpm x C:Voltage	*INTER*	------------
7	D:Cure Temperature	200 deg.C	300 deg. C

Inner Array Design. Array Type: L-8. Use <ctrl> + <arrows> to move cursor.

Figure 8.31 Experimental results.

	Sample# 1	Sample # 2	Sample# 3	Sample # 4	Sample# 5	Sample# 6	Averages
Trial# 1	30	28	34				30.666
Trial# 2	26	32	29				29
Trial# 3	67	75	82				74.666
Trial# 4	44	57	53				51.333
Trial# 5	56	45	38				46.333
Trial# 6	84	72	79				78.333
Trial# 7	77	65	69				70.333
Trial# 8	43	44	52				46.333
							53.375

Automatic Test for Presence of Interaction

Number of interactions between two factors calculated = 6

#	Interacting Factor Pairs (in Order of Severity)	Columns	SI(%)	Col	Opt.
1	C:Voltage x D:Cure Temperatur	4 x 7	68.4	3	[2,2]
2	A:Air Box Pressur x D:Cure Temperatur	1 x 7	58.24	6	[1,1]
3	B:Brushing rpm x C:Voltage	2 x 4	57.1	6	[1,1]
4	A:Air Box Pressur x B:Brushing rpm	1 x 2	56.03	3	[1,1]
5	A:Air Box Pressur x C:Voltage	1 x 4	37.21	5	[1,2]
6	B:Brushing rpm x D:Cure Temperatur	2 x 7	35.61	5	[1,2]

Figure 8.32 Interaction severity for all possible interactions.

Col # / Factor	DOF (f)	Sum of Sqrs. (S)	Variance (V)	F - Ratio (F)	Pure Sum (S')	Percent P(%)
ANOVA Table — Qualitek-4						
Expt. File: BKEX-82A.Q4W Data Type: Average Value QC Type: Smaller is Better						
1 A:Air Box Pressur	1	1162.037	1162.037	31.056	1124.62	13.501
2 B:Brushing rpm	1	1276.046	1276.046	34.103	1238.629	14.87
3 A:Air Box x B:Bru	1	2072.037	2072.037	55.376	2034.62	24.426
4 C:Voltage	1	108.37	108.37	2.896	70.953	.851
5 A:Air Box x C:Vol	1	408.37	408.37	10.914	370.953	4.453
6 B:Brushing x C:Vo	1	2262.043	2262.043	60.455	2224.626	26.707
7 D:Cure Temperatur	1	442.047	442.047	11.814	404.63	4.857
Other/Error	16	598.67	37.416			10.335
Total:	23	8329.625				100.00%

Figure 8.33   ANOVA showing factor and interaction influence.

sirable (fourth row, last column). Similarly, $B \times C$ indicates level 1 for B and level 1 for C (third row, last column). But $A \times C$ (fifth row, last column) dictates that C should have level 2. In such a situation, you must decide on the level that produces the most favorable result, which can be calculated from the *Estimate* button in the *Optimum* screen (Figure 8.34).

Optimum Conditions and Performance — Qualitek

Expt. File: BKEX-82A.Q4W   Data Type: Average Value   QC: Smaller is Better

Column # / Factor	Level Description	Level	Contribution
1 A:Air Box Pressur	200 psi	1	-6.959
2 B:Brushing rpm	300 rpm	1	-7.292
3 A:Air Box x B:Bru	*INTER*	1	-9.292
4 C:Voltage	70 KV	2	-2.126
5 A:Air Box x C:Vol	------------	2	-4.125
6 B:Brushing x C:Vo	*INTER*	1	-9.709
7 D:Cure Temperatur	300 deg. C	2	-4.292

Total Contribution From All Factors...	-43.796
Current Grand Average Of Performance...	53.375
Expected Result At Optimum Condition...	9.579

Figure 8.34   Optimum condition and performance.

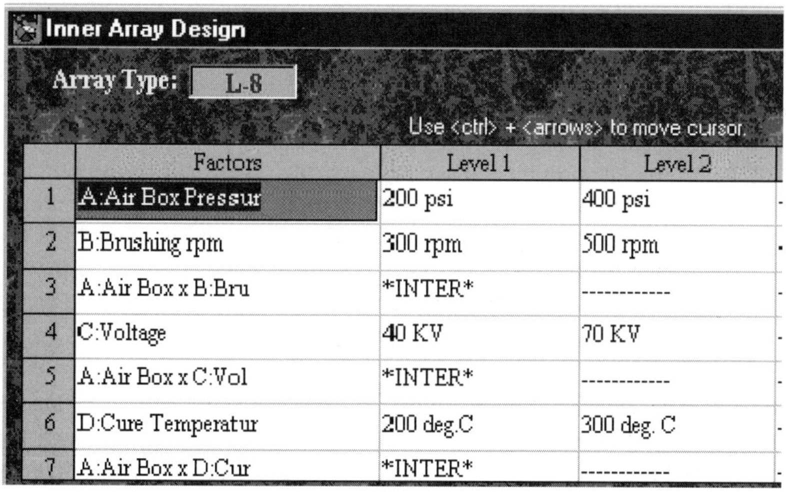

Figure 8.35   Factor descriptions and column assignments.

As you proceed with analysis, ANOVA (Figure 8.33) shows that all factors and interactions have a significant influence (over the 90% confidence level). This means that levels of the interacting factors (A, B, and C) may have to be adjusted for the interactions. The factor levels selected for the optimum condition, based on factor average effects, as shown in Figure 8.34, indicate level 1 for factor A, level 1 for factor B, and level 2 for factor C. Since these levels are also what was identified earlier from the interactions, no level adjustment is needed for the interacting factors. For confirmation tests, use the following combination and expected results, which includes the effects of interactions:

Optimum condition: $A_1B_1C_2D_2$

Performance expected = 5.579

(b)   If you were to study interaction $A \times B$, $A \times C$, and $A \times D$ instead of $A \times B$, $B \times C$ and $C \times A$, how would your design change? Review the solution. From the main screen of QT4, use the *Automatic design* option to describe the factors and their levels. [See the detailed steps in the solution of Example 8.1(a).] Be sure to indicate the interactions appropriately and obtain the design as shown in Figure 8.35 (Review the experiment file *BKEX-82B*.)

STRATEGY FOR EXPERIMENTAL STUDY WITH LARGER NUMBER OF FACTORS

When your experiment involves a larger number of factors, the number of possible interactions becomes even larger. In this case you may not be able to address all the interactions between 2 two-level factors you wish to study. You may have to choose between factors and their interactions. Since the number of columns is limited by the array selected, every interaction you select comes at the cost of a factor that could have been studied instead. The general guideline is to fill all columns with factors first. Then if there are columns available, reserve them for the interactions. Remember that even when you fill all columns of the array with factors only, you can still determine the presence of interactions (the severity of the presence of interactions and not their significance), which allows you to identify interaction pairs for future study. When you complete your experiment following these guidelines, if your experimental results are not confirmed, you risk repeating the experiment with the top few interactions included in the study. The guidelines can be summarized in these two steps:

1. Conduct experiments first by taking as many factors $(A, B, \ldots, G, $ L-8 array) as possible. Determine more severe interactions $(A \times C$ and $B \times C)$ among the possible 21 $(= 7 \times 6/2)$ from an analysis of interactions. Also, identify the factors that are less significant than others (say, D and F). Find optimum conditions and attempt to confirm results. If confirmed, you need not repeat the experiment.

2. If the results are not confirmed (the confirmation test result is not within the confidence interval), redesign and rerun the experiment, which includes the top interactions $(A \times C$ and $B \times C)$ in place of insignificant factors $(D$ and $F)$.

INTERACTIONS BETWEEN THREE- AND FOUR-LEVEL FACTORS

Interactions between factors that have three or four levels are extremely complicated to analyze and for which to derive design corrections. QT4 routinely displays such interactions, but no specific recommendation about how to select the factor levels for the optimum condition are dependable.

SUMMARY

In this step you have learned how to design experiments to include interactions and how to analyze the results to determine if interaction is present, and if found significant, how to determine the most desirable condition. You have also learned that when a number of factors is large, the number of interactions becomes even larger. Thus preference should be given to the study of factors.

REVIEW QUESTIONS

Q.   What is an interacting group of columns (IGC)?

A. For two-level arrays, these are any three columns (e.g., 1, 2, and 3) that form a valid reading of the triangular table.

Q. What constitutes an independent interacting group of columns?

A. These are IGCs without any common columns among them. For example, 1 2 3, 4 8 12, 7 9 14 are three independent groups. But 1 2 3, 2 4 6 are two dependent groups, as column 2 is common among the two.

Q. How many independent IGCs are there in an L-8 array?

A. There is only one independent IGC in an L-8 array. This is why an L-8 array can be used to design an experiment with only one independent interaction.

Q. What are linear graphs?

A. Linear graphs (stick model representations) are graphical representations of readings of the triangular table. They are made strictly for convenience in designing common experiments.

Q. How many independent interactions can be studied using an L-16?

A. An L-16 has five independent IGCs. Therefore, five independent pairs of interactions (e.g., $A \times B$, $C \times D$, $E \times F$, $G \times H$, and $I \times J$) can be studied together with 10 two-level factors.

Q. Is it possible to learn about interaction between factors after the experiments are carried out?

A. Information about interactions is of two types: whether it is present, and if present, whether it is significant. The information about how strongly an interaction is present can be found for all possible combinations of interactions, even if no column in the array was reserved for interactions. However, the significance of an interaction can only be found in ANOVA when an appropriate column is reserved for the study.

Q. When columns are all filled with factors only, how can the information about the presence of interaction be utilized?

A. The presence of interaction information (severity index) is not of immediate value without supporting information about the significance. Such information will, however, be useful when the results are not confirmed and experiments are repeated.

Q. When should you make an adjustment in the design because of interactions?

A. Only when an interaction is found significant in ANOVA. The design adjustment is dictated by the desirable performance point in the interaction plot (with two lines).

Q. How safe should you feel ignoring interactions?

A. It is highly risky with L-4 experiments. Your chances of being safe increase dramatically when you use larger arrays such as L-8, L-16, or L-32.

Q. Why is L-12 not suitable for interaction studies?

A. An L-12 array is a specially designed array where interaction effects are distributed to all columns instead of one. This characteristic makes L-12 unsuitable for interaction but qualifies it strongly for studies of factors alone.

EXERCISES

8.1 Design experiments to study the following situations. Select the orthogonal array and identify the column assignments. Indicate the array used and the column assignment of the factors and interactions.

    (a) 2 two-level factors (A and B) and the interaction ($A \times B$) between them

    (b) 6 two-level factors (A, B, C, D, E, and F) and interaction $A \times C$

    (c) 4 two-level factors (A, B, C, and D) and two interactions $A \times B$ and $C \times D$

    (d) 5 two-level factors (A, B, C, D, and E) and two interactions $A \times B$ and $B \times C$

    (e) 10 two-level factors (A, B, C, etc.) and three interactions $A \times B$, $B \times C$, and $C \times D$

8.2 Design an experiment to study four factors: A, B, C, and D (at two levels each) and interactions $B \times C$ and $B \times D$.

    (a) Can an L-8 array be used? If so, assign factors and interactions to the appropriate columns (Table 8.8).

    (b) If we wanted to study $A \times B$ and $C \times D$, instead of the interactions $B \times C$ and $B \times D$, which orthogonal array would we use: L-8, L-12, or L-16?

8.3 From results of the experiment described in Table 8.9, determine the following effects:

    (a) Average effect of B_2

    (b) Average effect of interaction $(A \times B)_1$

    (c) Determine the interaction between A and B by calculating $\overline{A_1B_1}$, $\overline{A_1B_2}$, $\overline{A_2B_1}$, and $\overline{A_2B_2}$.

TABLE 8.8   L-8 (2^7) Array for Exercise 8.2

Trial	Column						
	1	2	3	4	5	6	7
1	1	1	1	1	1	1	1
2	1	1	1	2	2	2	2
3	1	2	2	1	1	2	2
4	1	2	2	2	2	1	1
5	2	1	2	1	2	1	2
6	2	1	2	2	1	2	1
7	2	2	1	1	2	2	1
8	2	2	1	2	1	1	2

TABLE 8.9 Array for Exercise 8.3

| | Factor (Column) | | | | | | | |
Trial	A (1)	B (2)	$A \times B$ (3)	C (4)	D (5)	E (6)	F (7)	Results
1	1	1	1	1	1	1	1	3
2	1	1	1	2	2	2	2	2
3	1	2	2	1	1	2	2	6
4	1	2	2	2	2	1	1	8
5	2	1	2	1	2	1	2	4
6	2	1	2	2	1	2	1	5
7	2	2	1	1	2	2	1	6
8	2	2	1	2	1	1	2	3

8.4 In an experiment involving two factors A and B, each at two levels, the following readings were recorded:

$$\overline{A_1B_1} = 35 \qquad \overline{A_1B_2} = 55$$

$$\overline{A_2B_1} = 45 \qquad \overline{A_2B_2} = 60$$

Do factors A and B interact with each other?

8.5 When interactions between factors under investigation are suspected, explain why it is a common practice to reserve special columns of the orthogonal array. Select all appropriate answers.
  (a) To test for the presence of interaction
  (b) To determine the relative influence of interaction
  (c) To determine the optimum condition

8.6 In an experiment involving 7 two-level factors, an L-8 array was used. Is it possible to determine if there is interaction between any two factors?

8.7 If A and B are 2 two-level factors in your experiment, which column will you reserve to study interactions between them when:
  (a) A is in column 2 and B is in column 4
  (b) A is in column 8 and B is in column 4
  (c) A is in column 7 and B is in column 9
  (d) A is in column 11 and B is in column 13
  (e) A is in column 1 and B is in column 3

8.8   Can an L-8 array be used to study the following two-level factors and interactions? Answer yes, no, or maybe.
   (a)   6 factors and 1 interaction
   (b)   5 factors and 2 interactions
   (c)   4 factors and 3 interactions
   (d)   3 factors and 1 interaction
   (e)   3 factors and 3 interactions

8.9   Design an experiment to study two-level factors A, B, C, D, and E and
interactions $A \times B$ and $C \times D$. Indicate the array and the column assignments.

8.10   If you were to study 4 two-level factors (A, B, C, and D) and interactions $A \times
B$, $B \times C$, and $C \times D$, what is the smallest array that you will use for the design?
Indicate the column assignment.

8.11   When interaction is determined to be present and is considered significant, how
does it influence the outcome of your study? Select all appropriate answers.
   (a)   It may change the optimum condition.
   (b)   It influences the estimate performance at optimum.
   (c)   It alters the main effects of some factors.
   (d)   It changes the grand average of performance.

8.12   If you fail to include some interaction that may indeed be present, will it affect
the description of the trial conditions?

8.13   Recognizing that some levels of interactions between factors included in the
experiment are always present, with limited experimental scopes and a large
number of factors to study, which of the following should be your experimental strategy?
   (a)   Select the largest orthogonal array possible and include several of the interactions you suspect. Fill the remaining columns with factors.
   (b)   Fill all columns with factors. Test for the presence of interactions after
the experiments are done. Study interactions in the future experiments.

8.14   **(a)**   Can you study two-level factors A, B, C, and D and interactions $A \times B$,
$C \times A$, and $B \times C$ using an L-8 array?
   (b)   Which array will you use to design an experiment involving two-level
factors A, B, C, and D and interactions $A \times B$ and $C \times D$?
   (c)   For the experiment shown in Table 8.10,
      (1)   Determine if the influence of interaction $A \times B$ is more significant
than the influence of factor A.

TABLE 8.10   L-8 Array for Exercise 8.14

			Factor (Column)					
	A	B	$A \times B$	C	D	E	F	
Trial	(1)	(2)	(3)	(4)	(5)	(6)	(7)	Result
1	1	1	1	1	1	1	1	24
2	1	1	1	2	2	2	2	30
3	1	2	2	1	1	2	2	16
4	1	2	2	2	2	1	1	22
5	2	1	2	1	2	1	2	32
6	2	1	2	2	1	2	1	22
7	2	2	1	1	2	2	1	26
8	2	2	1	2	1	1	2	18

(2)   Complete the following calculations to determine whether interaction between factors C and D exists.

$$\bar{A}_1 = \frac{24 + 30 + 16 + 22}{4} = \qquad \overline{(A \times B)_2} = \frac{16 + 22 +}{4} =$$

$$\bar{A}_2 = \frac{32 + 22 + 26 + 28}{4} = \qquad \overline{(A \times B)_1} = \frac{24 + 30 +}{4} =$$

$$\overline{C_1 D_1} = \frac{24 + 16}{2} = 20 \qquad \overline{C_2 D_1} = \frac{22 + 18}{2} =$$

$$\overline{C_1 D_2} = \frac{32 + 26}{2} = \qquad \overline{C_2 D_2} = \qquad =$$

8.15   Engineers and production specialists in a supplier plant wish to optimize the production of foam seats. The improvement project has been undertaken as a result of complaints from the customer about the quality of the delivered parts. The main defects found in the foam parts are (1) excessive shrinkage, (2) too many voids, (3) inconsistent compression set, and (4) varying tensile strength. There appears to be general agreement that these are the primary objectives; however, there is no consensus as to their relative importance (weighting). Most people involved are aware that just satisfying one of the criteria may not always satisfy the others. It is believed that a process design that produces parts within the acceptable ranges of all the objective criteria would be preferable.

Conventional wisdom will dictate that a designed experiment be analyzed separately using the readings for each of the objectives (criteria of evaluations). In this way, four separate analyses will have to be performed and optimum design conditions determined. Since each of these optimums is based on only one objective, there is no guarantee that they all will prescribe the same factor levels. To release the design, however, we need only one combination of factor levels. Such design must also satisfy all objectives in a manner consistent with the consensus priority of the project team members.

Combining all the evaluation criteria into a single index (OEC) which combines the subjective as well as the objective evaluations and also incorporates the relative weightings of the criteria may produce the design you are looking for. Of course, even if you analyze your experiment using the *overall evaluation criterion* (OEC), you may still perform separate analysis for individual objectives.

Discussions and investigations into possible causes of the subquality parts revealed many variables (not all are necessarily *factors*): (1) chemical ratio, (2) mold temperature, (3) lid close time, (4) pour weight, (5) discoloration of surface, (6) humidity, (7) indexing, (8) flow rate, (9) flow pressure, (10) nozzle cleaning time, (11) type of cleaning agent, and others. Most team members suspect that there are interactions between chemical ratio and pour weight and between chemical ratio and flow rate. Past studies indicated possible nonlinearity in the influence of chemical ratio, and thus four levels of this factor are also desirable for the experiment. But since no scientific studies have been done in the recent past, no objective evidence of interaction or nonlinearity is available. Because of the variability from part to part, it is a common practice to study a minimum of three samples for any measurements. The funding and available time for the project are such that only 30 to 35 samples can be molded. Discuss the problem, including one or all of the following considerations.

(a) Determine the evaluation criteria, their ranges of evaluations, QC, and their relative weights.

(b) Determine the OEC equation in terms of the unknown evaluations (x_1, x_2, x_3, etc.) and QC.

(c) Assuming evaluations for a test sample, show calculation of the OEC.

(d) List all factors and interactions that you want to include in the experiments, and why.

(e) Indicate the array and the column assignments for your experiment design.

8.16 An experiment was designed to study five factors (A, B, C, D, and E in columns 1, 2, 4, 5, and 6 of an L-8 array) and an interaction between A and B ($A \times B$ in column 3). The objective of the study was to determine the best design for minimum tool wear. The results of tests with one sample at each of the trial conditions are given by 30, 60, 45, 55, 25, 40, 50, and 35.

(a) Determine the optimum condition with interaction.

(b) Determine the expected performance at the optimum condition.

(c) Determine the main effect of factor D.

EXERCISE ANSWERS

8.1 You may use the QT4 *Automatic design* option to answer these questions.

   (a) L-4 array. Assign A in column 1, B in column 2, and $A \times B$ in column 3.

   (b) L-8 array. Assign A in column 1, C in column 2, and $A \times C$ in column 3. Fill other columns with the remaining factors in arbitrary order.

   (c) L-16 array. Assign A in column 1, B in column 2, $A \times B$ in column 3, C in column 4, D in column 8, and $C \times D$ in column 12. Discard all other columns (unused) of the L-16 array.

   (d) L-8 array. Assign A in column 1, B in column 2, $A \times B$ in column 3, C in column 4, and $B \times C$ in column 6. Assign D to column 5 and E to column 7.

   (e) L-16 array. Assign A in column 1, B in column 2, $A \times B$ in column 3, C in column 4, $B \times C$ in column 6, D in column 8, and $C \times D$ in column 12. Assign the rest of the factors to the remaining column. Discard the two unused columns.

8.2 (a) L-8 can be used. Assign B in column 1, C in column 2, $B \times C$ in column 3, D in column 4, A in column 5, and $C \times D$ in column 6.

   (b) L-16 array.

8.3 (a) $(6 + 8 + 6 + 3)/4 = 5.75$; (b) $(3 + 2 + 6 + 3)/4 = 3.5$; (c) 2.5, 7, 4.5, and 4.5.

8.4 Yes.

8.5 Answer (b) is correct.

8.6 Yes. The presence of interactions between all possible combinations of the seven factors can be found even though no columns were reserved for them.

8.7 Check the triangular table.

8.8 (a) Yes; (b) maybe; (c) maybe; (d) yes; (e) yes (satisfy common factor requirements).

8.9 L-16.

8.10 L-16.

8.11 Answers (a) and (b) are correct.

8.12 No.

8.13 Answer (b) is correct.

8.14 **(a)** Yes; **(b)** L-16; **(c)** calculate average effects.

8.15 Answers will vary.

8.16 **(a)** $A_2B_1C_1D_1E_1$; **(b)** 17.5; **(c)** 47.5 – 37.5.

Experimental Design with Mixed-Level Factors

What You Will Learn in This Step

Experiment Designs with all factors at one level are accomplished easily using the available standard orthogonal arrays. But there are not many orthogonal arrays with mixed factor levels to suit experiment designs with larger numbers of mixed-level factors that are commonly confronted in the industrial environment. Most mixed-level designs, however, can be accomplished by modifying the standard orthogonal arrays. In this step you will learn methods of upgrading and downgrading columns of orthogonal arrays to accommodate factors at two, three, and four levels.

Thought for the Day

As nuclear and other technological achievements continue to mount, the normal life span will continue to climb. The hourly productivity of the worker will increase.

—Dwight D. Eisenhower

Designing experiments with all factors in the same level are easier to accomplish. Generally, one of the standard arrays can be used as is. Frequently, however, factors included in the study have different levels (two, three, and four levels) and require arrays with columns suitable to accommodate the levels of the factors included in the study. Most often, you can create the array suitable for the experiment by modifying the standard orthogonal arrays with which you are already familiar.

Suppose that you have situations where you need to study:

1. 1 four-level factor and 4 two-level factors
2. 3 three-level factors and 1 two-level factor
3. 2 four-level factors, 1 three-level factor, and 6 two-level factors

Based on what you know so far, you will find that none of the standard arrays can be used to design these experiments. Consider situation 1. You may think of using an L-8 array where you will be able to put the 4 two-level factors in any four of the 7 two-level columns. But what will you do about the four-level factor? The array has no column that has four levels. Try to find an array for situation 2. An L-9 array would almost work, as it has 4 three-level columns. The 3 three-level factors will go in these columns. But what about the two-level factor? How should you assign this factor to a column that has three levels? What about using an L-18 array? Perhaps you could use this one. But it would mean you would waste 4 of the 7 three-level columns. Is there a smaller array that could do the job? If you were to modify an array to make it suitable for the design, which array should you modify?

HOW TO DETERMINE WHICH ARRAY TO MODIFY

The array you will modify depends on the needs of your experiment. The size of the array your experiment will require is determined by counting *degrees of freedom* (DOFs) of the factors included in the study. The array you will select for modification must equal or exceed the DOFs your experiment requires. This is only the necessary condition. There are situations when even though an array has sufficient DOFs, it may not be suitable for the design.

Degrees of freedom (DOFs) is a common concept in engineering and statistical science. In vibration studies and in mechanical engineering, the DOFs are defined as the number of independent variables needed to define completely the motion of the system under observation. In statistical science, DOFs can be defined as the number of observations (data available) made in excess of the theoretically minimum number necessary to estimate a statistical parameter, which we are after. For the purposes of experimental designs, the DOFs applicable to different items are as redefined (defined earlier in pages 211 and 212 in Step 7) below:

$$\text{DOF of a factor} = (\text{number of levels of the factor}) - 1$$

$$\text{DOF of a column} = (\text{number of levels in the column}) - 1$$

$$\text{DOF of an array} = \text{total of DOFs all of columns in the array}$$

$$\text{DOF an experiment} = (\text{total number of results obtained}) - 1$$

By definition, then, a two-level factor has 1 DOF and a factor with four levels has 3 DOFs. Similarly, a column of an L-18 array has 2 DOFs. Since there are 7 two-level columns in an L-8 array, it has 7 DOFs. You will need to compare the DOFs of the experimental factors with the DOFs of the standard array.

In a situation where you need to study 4 two-level factors and 1 four-level factor (*a*), 7 DOFs are needed for the experiment:

$$\begin{aligned} &3 \text{ DOFs} &&\text{for 1 four-level factor } (4 - 1) \\ &\underline{+\,4 \text{ DOFs}} &&\text{for 4 two-level factors } [4 \times (2 - 1)] \end{aligned}$$

Total: 7 DOFs

Now you have to look for an array that has at least 7 DOFs. Considering the DOFs of all available arrays, you will find that an L-8 array has 7 DOFs. It will therefore be a good array to consider modifying and utilizing for the design.

The DOFs of an experiment are determined by considering the number of samples tested and results recorded. For example, if an L-8 array were used for the design and three samples were tested in each of the eight experimental conditions (also called trial conditions), the DOFs will be 23 (24–1). For better understanding of the ANOVA of your experimental results, you will need to understand how the DOFs of the experiment are calculated.

Example 9.1: Generator Noise Study Development engineers in an automotive component manufacturing facility carried out designed experiments to study the audible performance noise of a generator. The factors selected for the study included one at four levels and four at two levels (Table 9.1). The factor with four levels has discrete (fixed) levels and it was important that all levels be tested. The object of the study was to determine the design specifications for reduced performance noise. The experiments utilizing an L-8 array produced the results 50, 62, 70, 75, 68, 65, 65, and 74 for the eight trial conditions. Determine the best design condition for results measured in decibels (*smaller is better*).

Solution: This project required study of 1 four-level factor and 4 two-level factors. The degrees of freedom (DOFs) needed for the experiment are 7 (3 DOFs for the four-level factor + 4 DOFs for the 4 two-Level factors). The smallest array with 7 DOFs that would be suitable for the design is an L-8. An L-8 has 7 two-level columns, of which four columns can be used for the 4 two-level factors A, B, C, and D). As it has no columns to accommodate the four-level factor (X), its column must be upgraded before this factor can be assigned.

Method of upgrading column levels. A four-level column will possess 3 DOFs. Each column of an L-8 array has 1 DOF. Thus 3 two-level columns of an L-8 array must be combined to make it into a four-level column (the procedure will also apply to L-16, L-32, and L-64 arrays). Which 3 two-level columns should you combine to make 1 four-level column?

TABLE 9.1 Factors and Levels for Example 9.1

Factor	Level			
	1	2	3	4
X: casement structure	Present design	Texture	Bolted	Press-fit
A: air gap	Present gap	Increased gap		
B: impregnation	Present type	Harder type		
C: contact brush	Type 1	Type 2		
D: stator structure	Present design	Epoxy coated		

The columns you would combine are not chosen arbitrarily; they must be an interacting group. The three columns you would combine must form a valid and independent reading from the triangular table (TT). Possible TT readings applicable to an L-8 array are [1 2 3], [2 4 6], [1 4 5], and so on. Any one of these groups of columns can be used to upgrade into four-level columns. Notice, however, that an L-8 has only one independent interaction group of columns (L-16 has five), which allows it to be upgraded only once.

The procedure for column modifications calls for selecting a set of interacting columns (1, 2, and 3 is one set), discarding any one of the three columns (say, column 3), and combining the two to form a new column. While combining the two columns, each of which has 1's and 2's, it is important to follow a consistent set of rules, such as combining levels 1 and 1 into a new 1, levels 1 and 2 into a new 2, and so on.

Conversion rules: $1\ 1 \Rightarrow 1$ $1\ 2 \Rightarrow 2$ $2\ 1 \Rightarrow 3$ $2\ 2 \Rightarrow 4$

Since the two levels in each of columns 1 and 2 can only be combined in four different ways, the conversion rules above may be applied to upgrading all two-level columns. The procedure for upgrading 3 two-level columns into a four-level column for the L-8 array for the example experiment is shown in Table 9.2.

To prepare a new column, examine the level numbers in columns 1 and 2. If the numbers involve 1 and 1, replace them with a 1; if 1 and 2, replace them with a 2; and so on. Continue the process for all eight rows. Stick to the rules you established for all rows. After combining columns 1 and 2 into a new column, replace the three columns with the new column as shown in Table 9.3 (QT4 places it in column 1 and considers columns 2 and 3 as unused). To complete the experiment design, assign the four-level factor (X) to column 1 and the remaining 4 two-level factors A, B, C, and D to columns 4, 5, 6, and 7, respectively.

You will have two options in handling this experimental design using QT4: *Manual design* and *Automatic design*. When you select the *Manual design* option, you will

TABLE 9.2   Method of Column Upgrading (L-8 Array)

				Column			
Experiment	1	2	3	4	5	6	7
1	1	1 > 1[a]	1	1	1	1	1
2	1	1 > 1	1	2	2	2	2
3	1	2 > 2	2	1	1	2	2
4	1	2 > 2	2	2	2	1	1
5	2	1 > 3	2	1	2	1	2
6	2	1 > 3	2	2	1	2	1
7	2	2 > 4	1	1	2	2	1
8	2	2 > 4	1	2	1	1	2

[a]New four-level column.

TABLE 9.3   L-8 Array with One Upgraded Four-level Column

	Factor (Column)				
Experiment	X (new col.)	A (4)	B (5)	C (6)	D (7)
1	1	1	1	1	1
2	1	2	2	2	2
3	2	1	1	2	2
4	2	2	2	1	1
5	3	1	2	1	2
6	3	2	1	2	1
7	4	1	2	2	1
8	4	2	1	1	2

need to modify the column by editing column 1 (in the case of this example) and setting columns 1 and 2 as unused. In the *Automatic Design* option, the array will be modified and factors assigned to the appropriate column automatically. The *Automatic design* option will be used for designing this experiment.

Experiment design: Run QT4 and be in the main screen. From the *Design* menu item, select the *Automatic design* option and check the factor-type boxes as shown in Figure 9.1. Once you have indicated the types of factors involved in your study, QT4 is able to select and show you the array it will modify for the design, as shown in Fig-

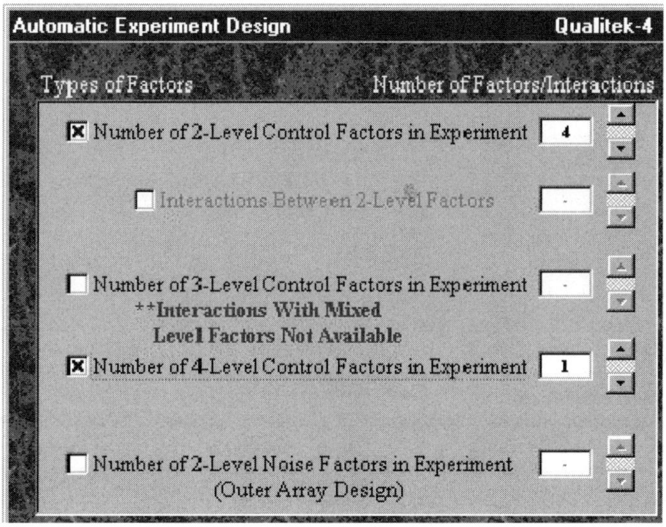

Figure 9.1   *Automatic design* input.

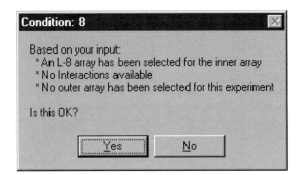

Figure 9.2 Array selection for the design.

ure 9.2. Click *Yes* at this prompt. QT4 then asks you to describe each type of factor separately, starting with the two-level factors first. Describe the 4 two-level factors as shown in Figure 9.3. Describe the 1 four-level factor next, as shown in Figure 9.4.

As you proceed, QT4 shows you the descriptions of all factors and the locations of the column to which they are assigned (Figure 9.5). You may alter any description of factor and level in this screen. Based on the factor levels described, QT4 automatically modifies the columns of the array. The modified array, which should be the same as the one shown in Table 9.3, is shown in Figure 9.6. Notice that column 1 is replaced by the newly created column, and columns 2 and 3 are set to contain all zeros (unused columns).

This experimental file has already been created and saved as *BKEX-91A.Q4W*, but you can save your experiment under any name, as you like. After saving your experiment file, return to the main screen. As usual, the experimental descriptions are obtained by reading the rows of the array after factors are assigned. You can review the

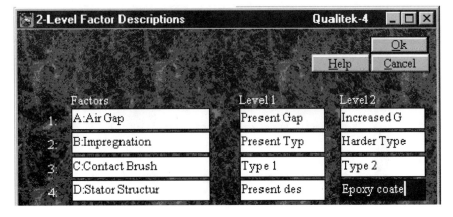

Figure 9.3 Descriptions of two-level factors.

Figure 9.4   Description of four-level factor.

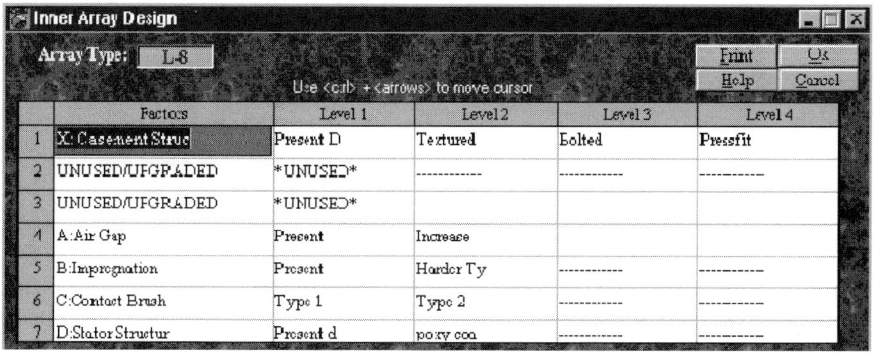

	Factors	Level 1	Level 2	Level 3	Level 4
1	X: Casement Struc	Present D	Textured	Bolted	Pressfit
2	UNUSED/UPGRADED	*UNUSED*	-----------	-----------	----------
3	UNUSED/UPGRADED	*UNUSED*			
4	A :Air Gap	Present	Increase		
5	B:Impregnation	Present	Harder Ty	-----------	----------
6	C:Contact Brush	Type 1	Type 2	-----------	----------
7	D:Stator Structur	Present d	po xy coa	-----------	----------

Array Type: L-8

Figure 9.5   Factor descriptions and column assignment.

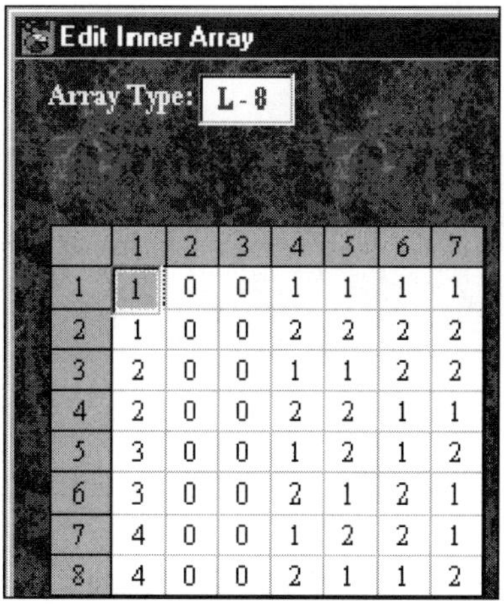

Edit Inner Array

Array Type: L - 8

	1	2	3	4	5	6	7
1	1	0	0	1	1	1	1
2	1	0	0	2	2	2	2
3	2	0	0	1	1	2	2
4	2	0	0	2	2	1	1
5	3	0	0	1	2	1	2
6	3	0	0	2	1	2	1
7	4	0	0	1	2	2	1
8	4	0	0	2	1	1	2

Figure 9.6   Modified array.

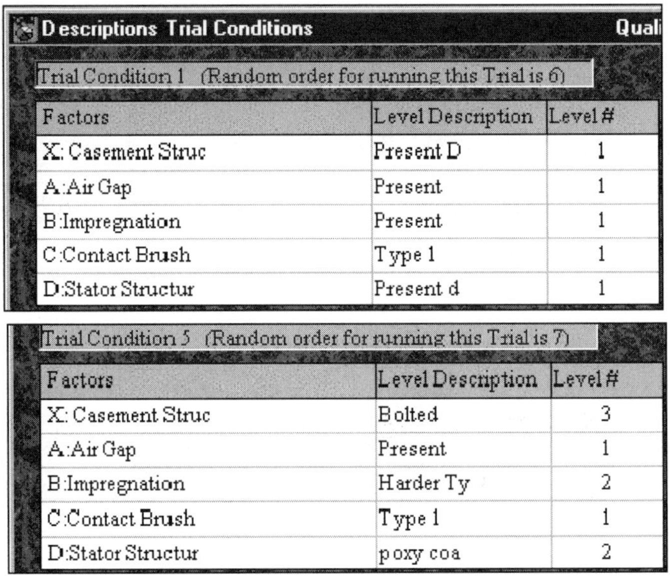

Descriptions Trial Conditions   **Quali**

Trial Condition 1   (Random order for running this Trial is 6)

Factors	Level Description	Level #
X: Casement Struc	Present D	1
A: Air Gap	Present	1
B: Impregnation	Present	1
C: Contact Brush	Type 1	1
D: Stator Structur	Present d	1

Trial Condition 5   (Random order for running this Trial is 7)

Factors	Level Description	Level #
X: Casement Struc	Bolted	3
A: Air Gap	Present	1
B: Impregnation	Harder Ty	2
C: Contact Brush	Type 1	1
D: Stator Structur	poxy coa	2

Figure 9.7   Trial conditions 1 and 5.

eight experimental conditions in this case from the *Condition* menu item from the main screen. Two of the eight experimental conditions are shown in Figure 9.7.

Analysis.   After experiments are carried out, the results are entered using the *Enter results* from the *Results* menu from the main screen. To perform analysis, select *Standard analysis* from the *Analysis* menu. Indicate the *Smaller is better* quality characteristic and proceed with the analysis. Your main effects screen should be the same as shown in Figure 9.8. Notice that only the first factor has four average effects, corresponding to its four levels. All other factors have only two average effects, as ex-

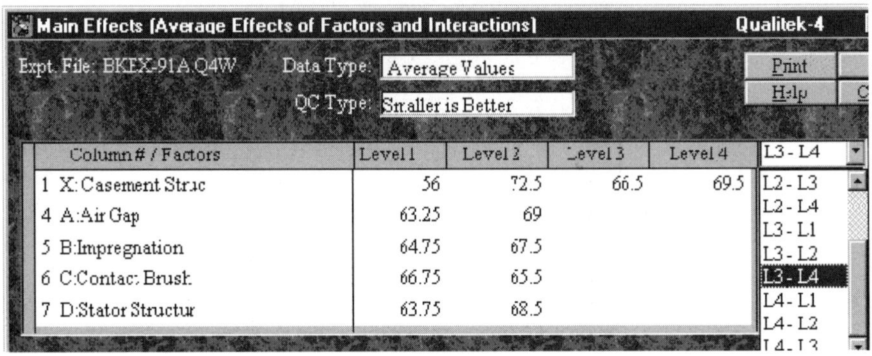

Main Effects (Average Effects of Factors and Interactions)   **Qualitek-4**

Expt. File: BKEX-91A.Q4W   Data Type: Average Values   Print
                              QC Type: Smaller is Better   Help

Column # / Factors	Level 1	Level 2	Level 3	Level 4	L3 - L4
1 X: Casement Struc	56	72.5	66.5	69.5	L2 - L3
4 A: Air Gap	63.25	69			L2 - L4
					L3 - L1
5 B: Impregnation	64.75	67.5			L3 - L2
6 C: Contact Brush	66.75	65.5			L3 - L4
					L4 - L1
7 D: Stator Structur	63.75	68.5			L4 - L2
					L4 - L3

Figure 9.8   Main effects of factors.

pected. Also notice that, by default, the difference in average effects (last column) shows the difference between levels 2 and 1. To see differences between other levels, when applicable, you need to open the box and select the option you desire.

When there are more than two factor levels, QT4 always attempts to draw a least-squares quadratic curve (first plot in Figure 9.9) through the calculated average effects, which should represent a more realistic trend than those of the straight-line segments. Beware that the least-squares fit is more meaningful when the factor is of a continuous type. There can be 10 possible two-factor interactions among the five factors included in the experiments (Figure 9.10). Although interactions were not of interest, information about the presence of interactions can be obtained by selecting the *Interaction* button from the *Main effects* screen (Figure 9.8). If you select the automatic option, QT4 calculates interactions between all possible pairs of factors, regardless of the number of levels of the interacting pairs.

Interaction studies between factors that are other than two-levels are beyond the scope of this book. However, you can review the nature of interaction plots in QT4 between two mixed-level factors. For example, you can review the plot between four- and two-level factors by selecting the seventh interaction in the list (Figure 9.11). ANOVA of the results (Figure 9.12) identifies factors that are significant and their relative influences. Note that the error DOF was zero to start with, which forces you to pool the factors arbitrarily. Also notice that the four-level factor (X) has 3 DOFs, which in turn affects the numbers for the variance and the F-ratio.

The optimum condition and the expected performance are obtained as you proceed with analysis (Figure 9.13). You can obtain a conservative estimate when only the significant factors are used to make the prediction. You can now have QT4 calculate the boundaries of the performance expected (confidence interval) by clicking on the *C.I.*

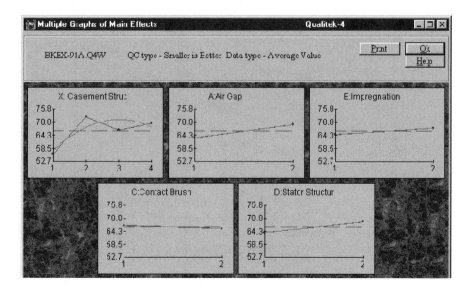

Figure 9.9 Plot of main effects.

Automatic Test for Presence of Interaction   Q

Number of interactions between two factors calculated = 10

#	Interacting Factor Pairs (in Order of Severity)	Columns	SI(%)	Col	Opt.
1	B:Impregnation x C:Contact Brush	5 x 6	71.05	3	[1,1]
2	B:Impregnation x D:Stator Structur	5 x 7	67.24	2	[1,1]
3	A:Air Gap x C:Contact Brush	4 x 6	62.9	2	[1,1]
4	A:Air Gap x D:Stator Structur	4 x 7	54	3	[1,1]
5	C:Contact Brush x D:Stator Structur	6 x 7	44.11	1	[1,1]
6	A:Air Gap x B:Impregnation	4 x 5	39.47	1	[1,1]
7	X:Casement Struc x C:Contact Brush	1 x 6	34	7	[1,1]
8	X:Casement Struc x D:Stator Structur	1 x 7	34	6	[1,1]
9	X:Casement Struc x A:Air Gap	1 x 4	14	5	[1,1]
10	X:Casement Struc x B:Impregnation	1 x 5	14	4	[1,1]

Figure 9.10   Interaction between factors.

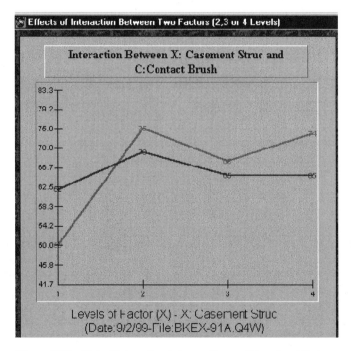

Figure 9.11   Plot of interaction between four- and two-level factors.

ANOVA Table						Qualitek-4	

Expt. File: EKEX-91A.Q4W   Data Type: Average Value   QC Type: Smaller is Better   Print Ok Help Cance

Col # / Factor	DOF (f)	Sum of Sqrs. (S)	Variance (V)	F - Ratio (F)	Pure Sum (S)	Percent P(%)
1 X: Casement Struc	3	309.375	103.125	11.301	282	64.255
4 A:Air Gap	1	66.125	66.125	7.246	57	12.987
5 B:Impregnation	(1)	(15.125)		POOLED	(CL=100%)	
6 C:Contact Brush	(1)	(3.125)		POOLED	(CL= *NC*)	
7 D:Stator Structur	1	45.125	45.125	4.945	36	8.202
Other/Error	2	18.25	9.125			14.556
Total:	7	438.875				100.00%

Figure 9.12   Pooled ANOVA.

button and specifying the 80% confidence level. (QT4 is unable to calculate at a higher percentage in this case.) As shown in Figure 9.14, the performance expected at the optimum condition is between 46.5 and 54.9 decibels. The result expected at the optimum condition represents the average performance of a number of samples tested at the optimum condition. No sample may perform exactly as the estimated optimum value, but all samples (80 out of 100 for 80% confidence level) are expected to perform within the boundary specified.

Example 9.1 demonstrated how 3 two-level columns are combined to form a four-level column. The method applies to columns of all lengths. Other methods of com-

Optimum Conditions and Performance				Qualitel

Expt. File:BKEX-91A.Q4W   Data Type Average Value   QC Smaller is Better   Print Help

Column # / Factor	Level Description	Level	Contribution
1 X: Casement Struc	Present D	1	-10.125
4 A:Air Gap	Present	1	-2.875
7 D:Stator Structur	Present d	1	-2.375
Total Contribution From All Factors...			-15.375
Current Grand Average Of Performance...			66.125
Expected Result At Optimum Condition...			50.75

Figure 9.13   Optimum condition and performance expected.

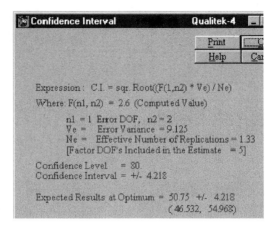

Figure 9.14 Confidence interval (C.I.) for performance expected.

bining columns into higher-level columns, such as combining 7 two-level columns into 1 eight-level column are not covered in this book.

Method of Downgrading Column Levels.

Just as columns can be combined to form a new column of higher number of levels, it can also be modified to have lower number of levels when the situation demands. In this process, a three-level column can be downgraded to a two-level column, or a four-level column can be reduced to a three-level column. Suppose that you have 1 two-level factor (A) and 3 three-level factors (B, C, and D). There is no standard array available to design this experiment economically. You could use an L-18 array, which has 1 two- and 7 three-level columns. But it will not be cost-effective to run 18 experimental conditions to study four factors. Instead, an L-9, which has four 3-level columns, would be a good selection for modification. The four factors in this case require 7 ($2 \times 3 + 1$) DOFs. An L-8 has 7 DOFs; it does not have any three-level column. Thus L-9, which has 8 DOFs, will be a good candidate for the design.

The 3 three-level factors (B, C, and D) could be assigned to three of the 4 three-level columns of the L-9 array. The remaining factor needs to go in the fourth three-level column after it is modified to a two-level column. The process of downgrading a column involves taking the standard column and simply changing the highest level to one of the other levels. This method of downgrading is commonly known also as a *dummy treatment* (Table 9.4).

To modify the column of an L-9 array to accommodate the two-level factor and to assign the other 3 three-level factors, follow these three steps:

1. Select any of the 4 three-level columns.

2. Replace level 3 with 1 and mark it as 1′ (identifying the changed level).

3. Assign the other three-level factors at random to the columns available.

TABLE 9.4   L-9 Array with Downgraded Column[a]

	Factor (column)			
	---	---	---	---
	B	C	A	D
Experiment	(1)	(2)	(3)	(4)
1	1	1	1	1
2	1	2	2	2
3	1	3	1′	3
4	2	1	2	3
5	2	2	1′	1
6	2	3	1	2
7	3	1	1′	2
8	3	2	1	3
9	3	3	2	1

[a]A prime indicates new modified levels 1′ instead of 3. Any column can be selected for dummy treatment.

In Example 9.1, the level 3 of the downgraded column is replaced by level 1. As a result, there are six 1's and three 2's in the column. The downgraded column is therefore no longer balanced for experimental purposes. It is however, functional and an acceptable practice. Note that you are not actually modifying the L-9 array by replacing level 3 with 1. Instead, you are simply changing the meaning of the level notation such that experiments can be described in terms of available factor levels.

Another question that arises: Should you always replace level 3 by 1? Could you consider replacing level 3 with 2? Generally, this decision is arbitrary (remember that we imposed no restriction as to what to call level 1 or 2). On special occasions when factor behavior is known, you may use the variability the factor is expected to exhibit as a criterion for your decision. Of course, you want more tests in the levels where response is expected to vary more. If the factor you would assign to the downgraded column is known to display higher variability at level 2 than that at level 1 (Figure 9.15),

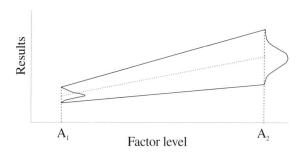

Figure 9.15   Response variability at factor levels.

TABLE 9.5   Factors and Levels for Example 9.2

	Level		
Factor	1	2	3
A: pressure	Low	Medium	High
B: cell gaps	Narrow	Current gap	Wider
C: temperature	15° lower	Present temp.	15° higher
D: LC fluids	Type 1	Type 2	

you should prefer to replace level 3 with level 2. In this way six of nine experiments include level 2 of the factor.

You could also downgrade a four-level column to make it into a three-level column just as you have done with a three-level column. The method here is to replace level 4 of the column by level 1 (also 2 or 3). QT4 can downgrade columns automatically, as you will see in Example 9.2.

Example 9.2: Color Uniformity of Liquid-Crystal Displays   Process   specialists of an electronic firm planned to carry out a designed experiment to study the process variables of one of their liquid-crystal display (LCD) panels. After team brainstorming, 1 two-level factor and 3 three-level factors were identified (Table 9.5). Design an experiment and describe the conditions of the individual trials.

Solution: The experiment design in this case will require an L-9 array with one of its columns downgraded to two-level. In this case, too, the *Automatic design* option of QT4 downgrades the columns automatically and assigns the factors to the appropriate column. To proceed with experiment design, go to the main screen of QT4 and select the *Automatic design* option from the *Design* menu item. Indicate the design input and the array selected for the design shown in Figures 9.16 and 9.17.

QT4 always wants you to describe the lower-level factors first. Describe the 1 two-level factor as shown in Figure 9.18. Describe the 3 three-level factors next (Figure 9.19). As you proceed, QT4 shows you the complete list of factors, their descriptions, and the columns to which they are assigned (Figure 9.20). QT4 shows you the modified L-9 array next. It selected the last column of the array to downgrade into a two-level column and assigned factor *D* to it (Figure 9.21).

When QT4 prompts you for a file name, *click cancel* and return to the main screen. Recall that the version of the program you have disallows saving any file other than those that utilize an L-8 array. This experiment has already been designed and saved as *BKEX-92A.Q4W*. To review the nine experimental conditions, select *Trial condition* from the *conditions* menu item. (Results of these experiments are also provided for optional analysis and exercises.)

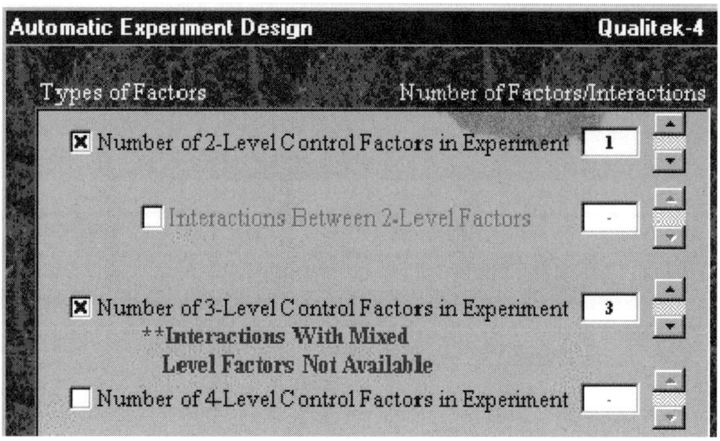

Automatic Experiment Design   **Qualitek-4**

Types of Factors   Number of Factors/Interactions

☒ Number of 2-Level Control Factors in Experiment   `1`

☐ Interactions Between 2-Level Factors   `·`

☒ Number of 3-Level Control Factors in Experiment   `3`
    ****Interactions With Mixed
    Level Factors Not Available**

☐ Number of 4-Level Control Factors in Experiment   `·`

Figure 9.16   Automatic design input.

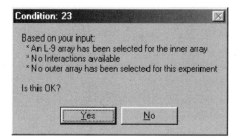

Condition: 23   ☒

Based on your input:
 * An L-9 array has been selected for the inner array
 * No Interactions available
 * No outer array has been selected for this experiment

Is this OK?

           [Yes]   [No]

Figure 9.17   Design array selection.

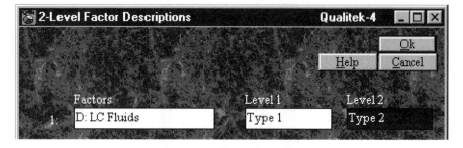

2-Level Factor Descriptions   **Qualitek-4**   _ ☐ ☒

                                                    [Ok]
                                        [Help]   [Cancel]

    Factors   Level 1   Level 2
1: D: LC Fluids   Type 1   Type 2

Figure 9.18   Description of three-level factor.

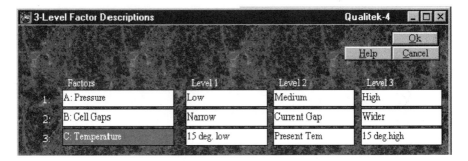

Figure 9.19 Descriptions of 3 three-level factors.

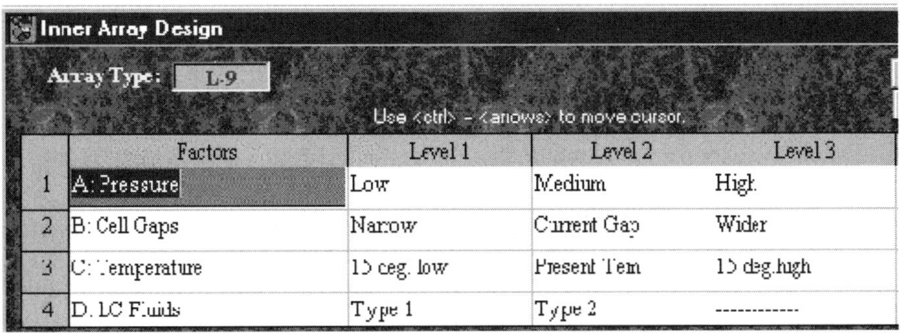

Figure 9.20 Factor descriptions and column assignments.

Figure 9.21 L-9 array with a downgraded column.

REVIEW OF ARRAY MODIFICATION TECHNIQUES

From the discussions so far, you know two methods of modifying the columns of an array:

1. *Upgrading columns.* In this method, 3 two-level columns are combined to form a four-level column. Following this method, 3 two-level columns of an L-8, L-16, or L-32 array can be upgraded to a four-level column. You will not apply this technique to upgrade columns of an L-12 array; it has no interacting group of columns.

2. *Downgrading a column.* This technique allows you to reduce the number of levels of one or more columns of an array. You can apply this to modify columns of any array that has three or more levels. Commonly, you will downgrade a three-level column to a two-level, or reduce the levels of a four-level column to a three-level.

Using the downgrade and upgrade procedures, it is now possible for you to create a three-level column within a two-level array. To do so, you will first combine 3 two-level columns into a four-level column, then downgrade this column to a three-level column.

Example 9.3: Lost Foam Casting Process Engineering and quality specialists of an automotive foundry division proposed to investigate the variables involved in the lost foam casting (LFC) process for engine blocks. In the LFC process, prototype parts made with Styrofoam materials are used to form molds by packing a mixture of sand around it. When liquid metal (i.e., aluminum) is poured into the mold, the Styrofoam evaporates, leaving the cast part. Upon brainstorming, nine factors at mixed levels were identified (Table 9.6). There were personnel from four different plants involved in this project. When it came to factors such as gating, sand compaction, and coating,

TABLE 9.6 Factors and Levels for Example 9.3

Factor	Level			
	1	2	3	4
A: metal head	Low	High		
B: sand supplier	Supplier 1	Supplier 2		
C: sand perm.	200 perm.	300 perm.		
D: metal temperature (°F)	1430	1460		
E: quench type	450 F	725 F		
F: gas level	None	High		
G: coating type	Type 1	Type 2	Type 3	
H: sand compaction	Plant X	Plant Y	Plant Z	
I: gating type	Plant X	Plant Y	Plant Z	Plant W

each plant had their different ways of doing things. To study different practices in these areas, the factors had to have different levels. Describe the strategy for designing an experiment for the group.

Solution   This study includes a total of nine mixed-level factors, which cannot be fitted into any of the standard arrays. We must therefore select an array that has enough DOFs and can be modified such that it ends up with at least nine columns to accommodate all the factors. The minimum number of DOFs required by all the factors considered for the study is 13. An L-16 (2^{15}) array, which has 15 DOFs (15 two-level columns) is a potential design solution.

$$\text{DOFs for 6 two-level factors } 6 \times (2-1) = \quad 6$$

$$\text{DOFs for 2 three-level factors } 2 \times (3-1) = \quad 4$$

$$\text{DOFs for 1 four-level factors } 1 \times (4-1) = \quad 3$$

$$\text{Total DOFs} = \quad 13$$

Strategy for array modification.   An L-16 array has 15 two-level columns. Six of the nine factors in this study could be assigned to the six columns of the array. That leaves three factors that are not at two levels, and there are 9 two-level factors available for modification. Remember that the way to prepare a three-level column from two-level columns is first to make it into a four-level column, then downgrade it into a three-level one. Therefore, for the 2 three-level factors and 1 four-level factor in this experiment, the strategy will be to create 3 four-level columns first, then downgrade 2 of the 3 four-level columns to 2 three-level columns.

Three two-level columns, which are part of an independent interacting group of columns, are needed to prepare a four-level column. When it comes to an L-16 array, there are many interacting groups. Five of these groups are independent (review the triangular table in Step 8). The three such independent groups shown here are selected for upgrading into 3 four-level columns.

$$1\ 2\ 3 \qquad 4\ 8\ 12 \qquad 7\ 9\ 14$$

(These sets are only one selection among many other possible interacting groups of columns.)

In the column modifications shown in Table 9.7, the third columns (3, 12, and 14) in each group are discarded. The remaining two columns are then combined following the rules established earlier (see Example 9.1). Two of the three new four-level columns are then degraded to a three-level column. To accomplish the design using QT4, go to the main screen and select the *Automatic design* option from the *Design* menu items. The input you need to provide for QT4 to select the array and upgrade its columns automatically is shown in Figure 9.22.

TABLE 9.7 Upgrading of Columns of an L-16 Array[a]

1	2	3	New 1	4	8	12	New 4	7	9	14	New 7
1	1	1	1	1	1	1	1	1	1	1	1
1	1	1	1	1	2	2	2	1	2	2	2
1	1	1	1	2	1	2	3	2	1	2	3
1	1	1	1	2	2	1	4 = 1'	2	2	1	4 = 1'
1	2	2	2	1	1	1	1	2	1	2	3
1	2	2	2	1	2	1	2	2	2	1	4 = 1'
1	2	2	2	2	1	1	3	1	1	1	1
1	2	2	2	2	2	1	4 = 1'	1	2	2	2
2	1	2	3	1	1	1	1	2	2	1	4 = 1'
2	1	2	3	1	2	1	2	2	1	2	3
2	1	2	3	2	1	1	3	1	2	2	2
2	1	2	3	2	2	1	4 = 1'	1	1	1	1
2	2	2	4	1	1	1	1	1	2	2	2
2	2	2	4	1	2	1	2	1	1	1	1
2	2	2	4	2	1	1	3	2	2	1	4 = 1'
2	2	2	4	2	2	1	4 = 1'	2	1	2	3

[a] Column modification rules: 1 1 > 1   1 2 > 2   2 1 > 3   2 2 > 4.

After you finish describing factors and their levels, your design screen (factor and column assignments) should look like Figure 9.23. QT4 does not design interaction studies when there are three-level or four-level factors included in the study. However, you can edit the design to replace factors and/or include interactions, depending on column availability. The array modified by QT4 and used for the design is shown in Figure 9.24. Notice that columns 1, 4, and 7 are the same as those shown in Table 9.4 labeled as new columns.

As you proceed with the design process, QT4 asks you enter a file name to save the experiment. Click *Cancel* at this prompt and return to the main screen. The design of this experiment has already saved an experiment file named *BKEX-93A.Q4W*. You can now review the 16 experimental conditions by loading this file and selecting *Trial conditions* from the *Conditions* menu item.

In each of the following examples, you can use the QT4 *Automatic design* option to design the experiment. You may also review the experiment file already created (*BKEX-94A.Q4W, BKEX-94B*, etc.) and examine the rationale for array selection and steps used to modify the columns.

Example 9.4a: Display Screen Filling Process Study This study intended to minimize "Voids," which is suspected to be influenced by 1 three- and 3 two-level factors (Table 9.8).

Solution: This experiment file has been saved as *BKEX-94A.Q4W*. You may either create your own experiment following the *Automatic design* option or review this file.

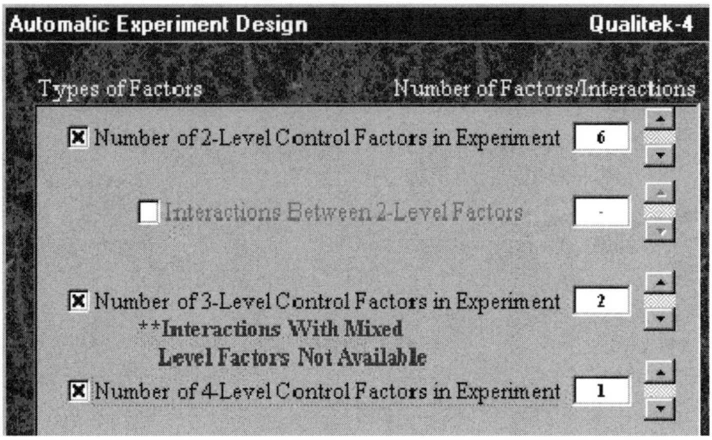

Figure 9.22 Input for automatic design.

This experiment requires a total of 5 DOFs (2 for 1 three-level factor and 3 for 3 two-level factors). Therefore, an L-8 orthogonal is suitable for the design. The interacting groups of columns, 1, 2, and 3 are modified first to form a new four-level column, then downgraded to a three-level column. QT4 places the new column in column 1 and changes the status of columns 2 and 3 as unused. Since only three of the four columns of the L-8 array are needed for the 3 two-level factors, column 7 is also set as unused.

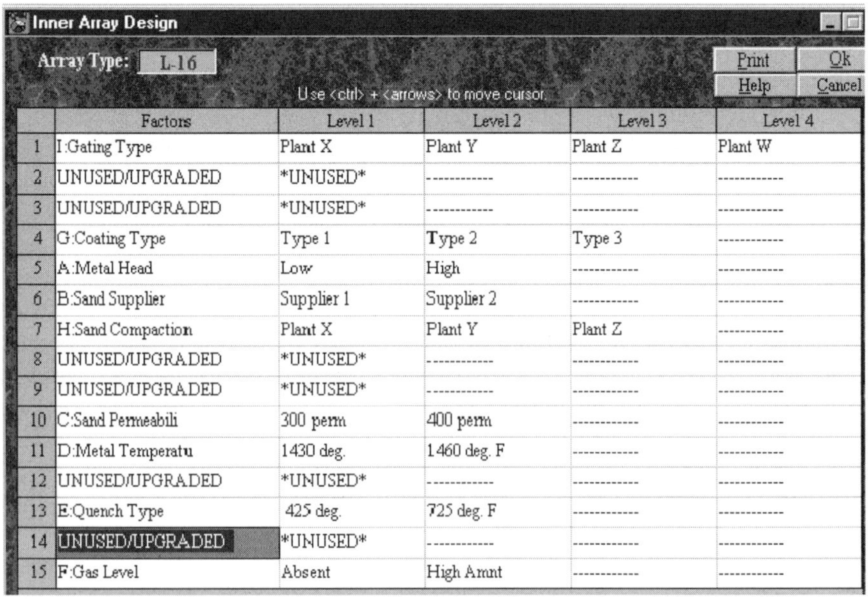

	Factors	Level 1	Level 2	Level 3	Level 4
1	I:Gating Type	Plant X	Plant Y	Plant Z	Plant W
2	UNUSED/UPGRADED	*UNUSED*	-----------	-----------	-----------
3	UNUSED/UPGRADED	*UNUSED*	-----------	-----------	-----------
4	G:Coating Type	Type 1	Type 2	Type 3	-----------
5	A:Metal Head	Low	High	-----------	-----------
6	B:Sand Supplier	Supplier 1	Supplier 2	-----------	-----------
7	H:Sand Compaction	Plant X	Plant Y	Plant Z	-----------
8	UNUSED/UPGRADED	*UNUSED*	-----------	-----------	-----------
9	UNUSED/UPGRADED	*UNUSED*	-----------	-----------	-----------
10	C:Sand Permeabili	300 perm	400 perm	-----------	-----------
11	D:Metal Temperatu	1430 deg.	1460 deg. F	-----------	-----------
12	UNUSED/UPGRADED	*UNUSED*	-----------	-----------	-----------
13	E:Quench Type	425 deg.	725 deg. F	-----------	-----------
14	UNUSED/UPGRADED	*UNUSED*	-----------	-----------	-----------
15	F:Gas Level	Absent	High Amnt	-----------	-----------

Figure 9.23 Factor descriptions and column assignments.

Edit Inner Array

Array Type: L - 16

	1	2	3	4	5	6	7	8	9	10	11	12	13	14	15
1	1	0	0	1	1	1	1	0	0	1	1	0	1	0	1
2	1	0	0	2	1	1	2	0	0	2	2	0	2	0	2
3	1	0	0	3	2	2	3	0	0	1	1	0	2	0	2
4	1	0	0	1	2	2	1	0	0	2	2	0	1	0	1
5	2	0	0	1	1	2	3	0	0	2	2	0	1	0	2
6	2	0	0	2	1	2	1	0	0	1	1	0	2	0	1
7	2	0	0	3	2	1	1	0	0	2	2	0	2	0	1
8	2	0	0	1	2	1	2	0	0	1	1	0	1	0	2
9	3	0	0	1	2	1	1	0	0	1	2	0	2	0	2
10	3	0	0	2	2	1	3	0	0	2	1	0	1	0	1
11	3	0	0	3	1	2	2	0	0	1	2	0	1	0	1
12	3	0	0	1	1	2	1	0	0	2	1	0	2	0	2
13	4	0	0	1	2	2	2	0	0	2	1	0	2	0	1
14	4	0	0	2	2	2	1	0	0	1	2	0	1	0	2
15	4	0	0	3	1	1	1	0	0	2	1	0	1	0	2
16	4	0	0	1	1	1	3	0	0	1	2	0	2	0	1

Figure 9.24 Modified L-16 array.

The modified array is as shown in Figure 9.25. Factor *D*, being the three-level factor, is assigned to the newly created column 1. The 3 two-level factors *A*, *B*, and *C* are assigned to the original columns 4, 5, and 6 in the order shown in Figure 9.26.

Example 9.4b: Study Handle Twist Assembly Problem This study was intended to minimize rejects, which are suspected to be caused by 2 three- and 3 two-level factors (Table 9.9).

Solution: This experiment file has been saved as *BKEX-94B.Q4W*. You may either create your own experiment following the QT4 *Automatic design* or *Manual design*

TABLE 9.8 Factors and Levels for Example 9.4a

	Level		
Factor	1	2	3
A: temperature (°C)	Ambient	30	
B: vacuum level (millitorr)	110	80	
C: fill port opening (inches)	0.075	0.100	
D: vacuum saturation (minutes)	10	30	60

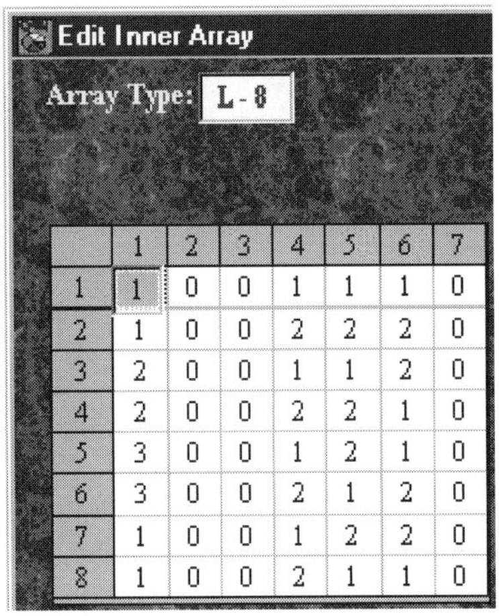

Edit Inner Array

Array Type: L-8

	1	2	3	4	5	6	7
1	1	0	0	1	1	1	0
2	1	0	0	2	2	2	0
3	2	0	0	1	1	2	0
4	2	0	0	2	2	1	0
5	3	0	0	1	2	1	0
6	3	0	0	2	1	2	0
7	1	0	0	1	2	2	0
8	1	0	0	2	1	1	0

Figure 9.25 Modified L-18 array.

option, or review this file. The number of DOFs needed for this experiment is 7 (4 for 2 three-level factors and 3 for 3 two-level factors). An L-8 array has 7 DOFs, but it will not be suitable for the design since only 1 three-level column can be upgraded in an L-8 array. As explained earlier, the L-8 array has only one independent interacting group of columns available for upgrading purposes. The next higher arrays are L-9, L-12, L-16, and so on. An L-9 array does not have enough columns for the five factors included in the study. Similarly, an L-12 array does not have an interacting group of columns and cannot be upgraded. In this case, your choice of the array for the design will be an L-16.

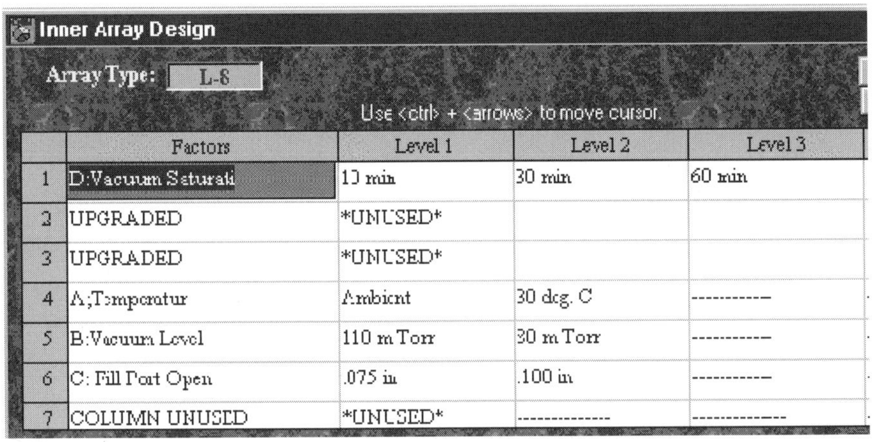

Inner Array Design

Array Type: L-8

Use <ctrl> + <arrows> to move cursor.

	Factors	Level 1	Level 2	Level 3
1	D:Vacuum Saturati	10 min	30 min	60 min
2	UPGRADED	*UNUSED*		
3	UPGRADED	*UNUSED*		
4	A;Temperatur	Ambient	30 deg. C	-----------
5	B:Vacuum Level	110 m Torr	30 m Torr	-----------
6	C: Fill Port Open	.075 in	.100 in	-----------
7	COLUMN UNUSED	*UNUSED*	-------------	-------------

Figure 9.26 Factor descriptions and column assignments.

TABLE 9.9   Factors and Levels for Example 9.4b

	Level		
Factor	1	2	3
A: twister setup	Current position	Opened °	
B: twister deflection	On	Off	
C: gauging skill	Unskilled	Skilled	
D: fixture type	Supplier 1	Supplier 2	Supplier 3
E: torque	10% under	Specified	10% over

To modify two columns, QT4 works with two interacting group of columns, [1 2 3] and [4 8 12]. These groups are first made into a four-level column, then downgraded to three-level. The new three-level columns are placed in columns 1 and 4. The other columns used to upgrade the two (i.e., columns 2, 3, 8, 12) are set as unused. The modified array is as shown in Figure 9.27.

Six of the 15 columns of the array are used for upgrading the columns. This leaves nine of the original columns for the two-level factors. The 3 two-level factors (A, B, and C) are assigned to columns 5, 6, and 7, and the remaining 6 two-level columns are set as unused (Figure 9.28).

Edit Inner Array

Array Type: **L-16**

	1	2	3	4	5	6	7	8	9	10	11	12	13	14	15
1	1	0	0	1	1	1	1	0	0	0	0	0	0	0	0
2	1	0	0	2	1	1	1	0	0	0	0	0	0	0	0
3	1	0	0	3	2	2	2	0	0	0	0	0	0	0	0
4	1	0	0	1	2	2	2	0	0	0	0	0	0	0	0
5	2	0	0	1	1	2	2	0	0	0	0	0	0	0	0
6	2	0	0	2	1	2	2	0	0	0	0	0	0	0	0
7	2	0	0	3	2	1	1	0	0	0	0	0	0	0	0
8	2	0	0	1	2	1	1	0	0	0	0	0	0	0	0
9	3	0	0	1	2	1	2	0	0	0	0	0	0	0	0
10	3	0	0	2	2	1	2	0	0	0	0	0	0	0	0
11	3	0	0	3	1	2	1	0	0	0	0	0	0	0	0
12	3	0	0	1	1	2	1	0	0	0	0	0	0	0	0
13	1	0	0	1	2	2	1	0	0	0	0	0	0	0	0
14	1	0	0	2	2	2	1	0	0	0	0	0	0	0	0
15	1	0	0	3	1	1	2	0	0	0	0	0	0	0	0
16	1	0	0	1	1	1	2	0	0	0	0	0	0	0	0

Figure 9.27   Modified L-16 array.

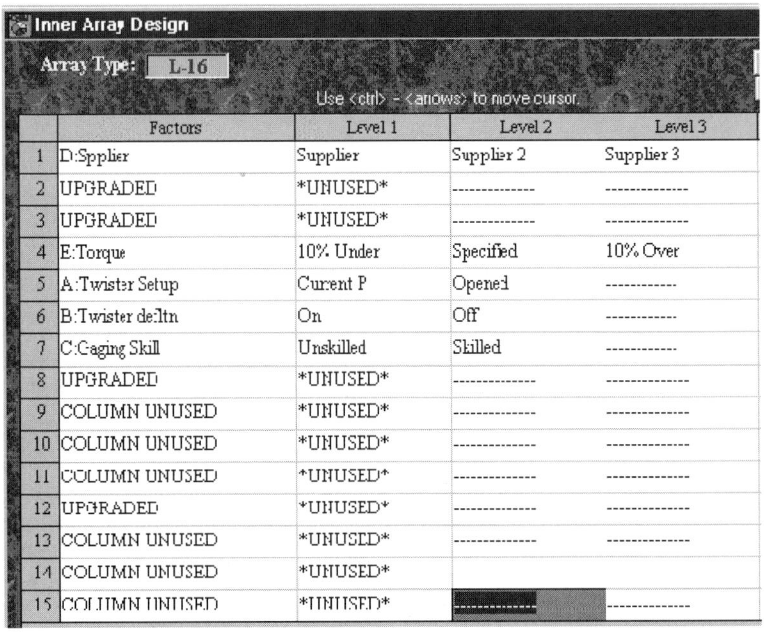
Inner Array Design

Array Type: L-16

Use <ctrl> - <arrows> to move cursor.

	Factors	Level 1	Level 2	Level 3
1	D:Spplier	Supplier	Supplier 2	Supplier 3
2	UPGRADED	*UNUSED*	-------------	-------------
3	UPGRADED	*UNUSED*	-------------	-------------
4	E:Torque	10% Under	Specified	10% Over
5	A:Twister Setup	Current P	Opened	-----------
6	B:Twister deltn	On	Off	-----------
7	C:Gaging Skill	Unskilled	Skilled	-----------
8	UPGRADED	*UNUSED*	-------------	-------------
9	COLUMN UNUSED	*UNUSED*	-------------	-------------
10	COLUMN UNUSED	*UNUSED*	-------------	-------------
11	COLUMN UNUSED	*UNUSED*	-------------	-------------
12	UPGRADED	*UNUSED*	-------------	-------------
13	COLUMN UNUSED	*UNUSED*	-------------	-------------
14	COLUMN UNUSED	*UNUSED*		
15	COLUMN UNUSED	*UNUSED*	-------------	-------------

Figure 9.28 Factor descriptions and column assignments.

This is an example of an inefficient use of an array. In this setup it is possible for you to include an additional 6 two-level factor with your current list of factors and not increase the total number of experiments you need to carry out. On the other hand, if you intend to keep the experiment small, either drop 1 two-level factor (this allows you to use an L-9 array with two columns downgraded) or reduce the level of 1 three-level factor from three to two (allows use of an L-8 array).

Example 9.4c: Discoloration Variation in Medical Cart Handles This study was undertaken to minimize color variation, which is known to be influenced by 2 three- and 2 two-level factors (Table 9.10).

Solution: This experiment file has been saved as *BKEX-94C.Q4W*. You may either create your own experiment following the QT4 *Automatic design* or *Manual design* option or review this file. The number of DOFs needed for this experiment is 6 (4 for 2 three-level factors and 2 for 2 two-level factors). An L-8 array has 7 DOFs, but it will not be suitable for the design, as only 1 three-level column can be upgraded in an L-8 array. The next bigger array, L-9, which has 4 three-level columns, will be your choice for the design. The QT4 *Automatic design* option prefers to keep the first two columns unchanged and downgrade the last two columns (arbitrary selection). The modified column is shown in Figure 9.29. Accordingly, the 2 three-level factors (C and D) are assigned to columns 1 and 2, and the 2 two-level factors (A and B) are assigned to downgraded columns 3 and 4 as shown in Figure 9.30.

TABLE 9.10 Factors and Levels for Example 9.4c

	Level		
Factor	1	2	3
A: mixing time (minutes)	2	4	
B: temperature (°F)	Ambient	120	
C: dye type	Supplier 1	Supplier 2	Supplier 3
D: Spray nozzle	Fine	Medium	Misty

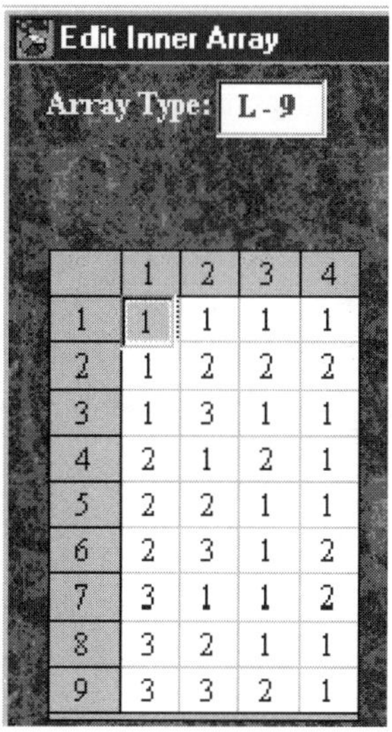
Edit Inner Array

Array Type: L-9

	1	2	3	4
1	1	1	1	1
2	1	2	2	2
3	1	3	1	1
4	2	1	2	1
5	2	2	1	1
6	2	3	1	2
7	3	1	1	2
8	3	2	1	1
9	3	3	2	1

Figure 9.29 Modified L-9 array.

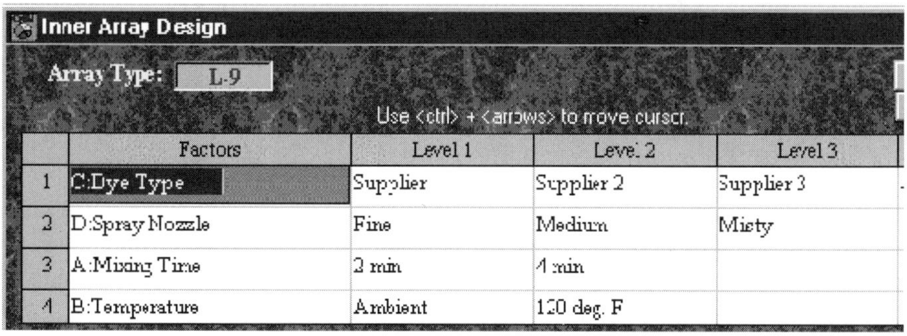
Inner Array Design

Array Type: L-9

Use <ctrl> + <arrows> to move cursor

	Factors	Level 1	Level 2	Level 3
1	C:Dye Type	Supplier	Supplier 2	Supplier 3
2	D:Spray Nozzle	Fine	Medium	Misty
3	A:Mixing Time	2 min	4 min	
4	B:Temperature	Ambient	120 deg. F	

Figure 9.30 Factor descriptions and column assignments.

TABLE 9.11 Factors and Levels for Example 9.4d

	Level		
Factor	1	2	3
A: machine type	Type 1	Type 2	
B: tool type	Carbide	Ceramic	
C: steel grade	SAE 1010	SAE 1020	SAE 1035
D: speed (rev/min)	1000	1200	1500
E: design variation	Plant 1	Plant 2	Plant 3
F: shift of manufac.	Shift 1	Shift 2	Shift 3
G: coolant rate	Low	Medium	High

Example 9.4d: Handlebar Design of a Racing Mountain Bike This study is intended to reduce rejects and warranty cost, which is affected by several assembly and manufacturing factors (5 three- and 2 two-level factor; Table 9.11).

Solution: This experiment file has been saved as *BKEX-94D.Q4W*. You may either create your own experiment following the QT4 *Automatic design* or *Manual*

	1	2	3	4	5	6	7	8
1	1	1	1	1	1	1	1	0
2	1	1	2	2	2	2	2	0
3	1	1	3	3	3	3	3	0
4	1	2	1	1	2	2	3	0
5	1	2	2	2	3	3	1	0
6	1	2	3	3	1	1	2	0
7	1	1	1	2	1	3	2	0
8	1	1	2	3	2	1	3	0
9	1	1	3	1	3	2	1	0
10	2	1	1	3	3	2	2	0
11	2	1	2	1	1	3	3	0
12	2	1	3	2	2	1	1	0
13	2	2	1	2	3	1	3	0
14	2	2	2	3	1	2	1	0
15	2	2	3	1	2	3	2	0
16	2	1	1	3	2	3	1	0
17	2	1	2	1	3	1	2	0
18	2	1	3	2	1	2	3	0

Figure 9.31 Modified L-18 array.

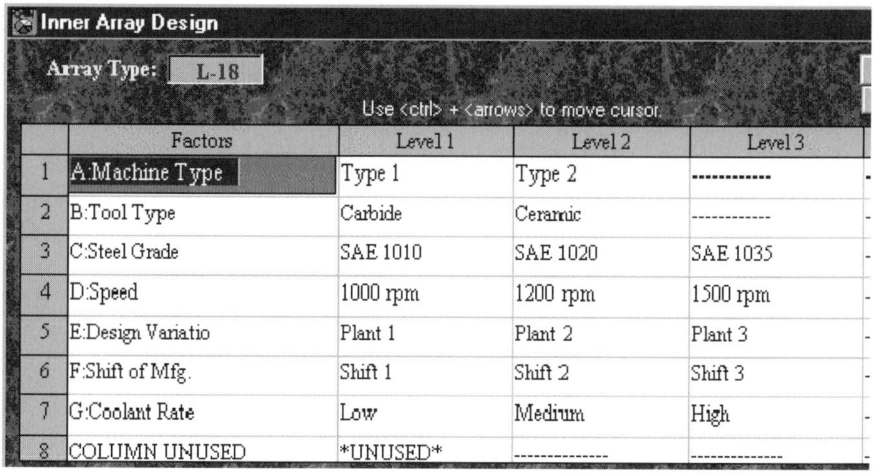
Inner Array Design

Array Type: L-18

Use <ctrl> + <arrows> to move cursor.

	Factors	Level 1	Level 2	Level 3
1	A:Machine Type	Type 1	Type 2	------------
2	B:Tool Type	Carbide	Ceramic	------------
3	C:Steel Grade	SAE 1010	SAE 1020	SAE 1035
4	D:Speed	1000 rpm	1200 rpm	1500 rpm
5	E:Design Variatio	Plant 1	Plant 2	Plant 3
6	F:Shift of Mfg.	Shift 1	Shift 2	Shift 3
7	G:Coolant Rate	Low	Medium	High
8	COLUMN UNUSED	*UNUSED*	--------------	--------------

Figure 9.32 Factor descriptions and column assignments.

design option or review this file. For the 5 three- and 2 two-level factors, the total number of DOFs needed for the experiment is 12. Because of the presence of a majority of three-level factors and adequate DOFs, an L-18 array will be the choice for the design. An L-18 array has seven 3-level columns and one 2-level column (column 1). To accommodate the 2 two-level factors, an additional two-level column is made by downgrading one of the 7 three-level column (the column selected for downgrading is arbitrary). Five of the remaining 6 three-level columns are used for the 5 three-level factors, which leave 1 three-level column as unused. The modified array for the experiment is shown in Figure 9.31. The experimental design is completed by assigning the 2 two-level factors (A and B) to columns 1 and 2, and by assigning the 5 three-level factors to columns 3 to 7, as shown in Figure 9.32.

SUMMARY

In this step you have learned how to design experiments when factors at mixed levels are included in the study. You are now aware that for such designs you will need to select the array and modify its columns before assigning the factors to the columns. In terms of modification of the array, you have learned (1) how to create four-level columns from groups of 3 two-level columns and (2) how to reduce the number of levels in a column from four to three or from three to two levels.

REVIEW QUESTIONS

Q. What is a mixed-level-factor situation?

A. When factors you are studying are not all at the same level (i.e., all at two-levels), you have a mixed-level-factor situation.

Q. Do you always have to modify arrays whenever you have a mixed-level factor?

A. Usually, except for a few special situations. For example, if you have 1 two-level factor and several three-level factors, you will be able to use an L-18 array as is.

Q. How do you select the 3 two-level columns needed to combine into a four-level column?

A. These are any three columns that form an interacting group of columns. An interacting group of columns are the 3 two-level columns that constitute a valid reading from the triangular table.

Q. Can you upgrade columns of any two-level array?

A. All except L-12 and L-4. L-12 does not have an interacting group of columns, and L-4 has too few columns to combine.

Q. Is it possible to make 2 four-level columns in an L-8 array?

A. No. It has only one interacting group.

Q. How do you make two-level columns into three-level columns?

A. By first making them four-level columns, then downgrading to reduce the levels to three.

Q. What happens to the degrees of freedom (DOFs) of the array when its columns are downgraded?

A. DOFs are lost and cannot be utilized. For example, when 3 two-level columns of an L-8 array are made into a three-level column, the new column will have 2 DOFs and the array will have 6 DOFs (2 + 4).

EXERCISES

9.1 Design experiments to study the following situations with factors at mixed levels. Select the appropriate orthogonal array and indicate the column assignments.

(a) 3 two-level factors and 1 four-level factor

(b) 4 two-level factors and 1 three-level factor

(c) 2 three-level factors and 1 two-level factor

(d) 7 three-level factors and 1 two-level factor

(e) 4 four-level factors and 3 two-level factors

(f) 3 four-level factors, 2 three-level factors, and 5 two-level factors.

9.2 Consider an experiment with factor descriptions and arrays as shown in Tables 9.12 and 9.13.

(a) Determine the trial number for the condition described below.

TABLE 9.12   Factors and Levels for Exercise 9.2

Column	Factor	Level			
		1	2	3	4
1	Casement structure	Present	Textured	Bolted	Present fit
2		(column used for upgrading)			
3		(column used for upgrading)			
4	Air gap	Present	Increase		
5	Impregnation	Present	Harder		
6	Contact brush	Type 1	Type 2		
7	Stator structure	Present	Epoxy		

       Casement structure = bolted

       Air gap = increased gap

       Impregnation = present type

       Contact brush = type 2

       Stator structure = present design

  (b)   Describe trial condition 4 by writing out the factor levels for casement structure, air gap, impregnation, contact brush, and stator structure.

9.3 To accommodate a four-level factor in an L-8 array, which three columns will you work with? Select all appropriate answers.

  (a) Any three interacting group of columns

  (b) Columns 1, 2, and 3

  (c) Columns 2, 6, and 4

  (d) Columns 1, 5, and 4

  (e) Columns 1, 6, and 7

TABLE 9.13   Array for Exercise 9.2

Trial	Column						
	1	2	3	4	5	6	7
1	1	0	0	1	1	1	1
2	1	0	0	2	2	2	2
3	2	0	0	1	1	2	2
4	2	0	0	2	2	1	1
5	3	0	0	1	2	1	2
6	3	0	0	2	1	2	1
7	4	0	0	1	2	2	1
8	4	0	0	2	1	1	2

9.4   Which array will you use to design an experiment with 2 four-level factors and 1 two-level factor? L8, L12, or L16?

9.5   To investigate a process with 2 factors at four-levels, 2 factors at three-levels, and 3 factors at two-levels, an experimenter proposed using an L-16 in the following way:

Use columns 1, 2, and 3 for the first four-level factor.

Use columns 4, 8, and 12 for the second four-level factor.

Use columns 5, 15, and 10 for the first three-level factor (with dummy treatment).

Use columns 7, 14, and 9 for the second three-level factor (with dummy treatment).

Use columns 6, 11, and 13 for the 3 two-level factors.

Is this design correct?

9.6   [*Degrees of Freedom*] Consider an experiment designed to study 4 three-level factors (Table 9.14). Three samples per trial condition were tested.

(a)   Determine the degrees of freedom (DOFs) of factor A.

(b)   Determine the total DOFs of all the factors.

(c)   Determine the DOFs of the fourth column of the array.

(d)   Determine the DOFs of the array used for the experiment.

(e)   Determine the total DOFs of the experiment.

(f)   Determine the extra DOFs of the experiment (over that which the factors need).

(g)   Identify the orthogonal array that has enough DOFs to satisfy the needs of an experiment planned to study 3 three-level factors, 1 four-level factor, and 2 two-level factors.

TABLE 9.14   Array for Exercise 9.6

Trial	Factor				Results			Average
	A	D	B	C				
1	1	1	1	1	3	4	5	4
2	1	2	2	2	7	6	5	6
3	1	3	3	3	4	5	3	4
4	2	1	2	3	6	7	5	
5	2	2	3	1	8	9	7	
6	2	3	1	2	9	10	8	
7	3	1	3	2	6	8	7	
8	3	2	1	3	5	7	6	
9	3	3	2	1	5	6	7	

TABLE 9.15   L-8 Array for Exercise 9.7

				Factor (column)				
	A		B	D	C		E	
Trial	(1)	(2)	(3)	(4)	(5)	(6)	(7)	Result
1	1	1	1	1	1	1	1	6
2	1	1	1	2	2	2	2	4
3	1	2	2	1	1	2	2	8
4	1	2	2	2	2	1	1	7
5	2	1	2	1	2	1	2	9
6	2	1	2	2	1	2	1	4
7	2	2	1	1	2	2	1	12
8	2	2	1	2	1	1	2	15

9.7 [*Array Modifications*] In the experiment shown in Table 9.15, the three-level factor D is assigned to column 4. All other factors in the experiment are two-level. Modify the orthogonal array to design your experiment correctly.

   (**a**)   Which columns would you combine to form a three-level column?

   (**b**)   Indicate the level numbers for modified column 4.

   (**c**)   Is it possible to determine if interaction exists between A and B?

   (**d**)   For the *Smaller is better* quality characteristic, plot the main effects and identify the level of factor D that is desirable for the optimum condition.

9.8 [*Validation of Results*] An experiment was designed and tested to study 5 two-level factors and two interactions between factors as follows:

   col. 1 A: chemical ratio (all factors at two levels)

   col. 2 H: flow rate

   col. 3 Interaction $A \times H$

   col. 4 Interaction $A \times D$

   col. 5 D: pour weight

   col. 6 B: mold temperature

   col. 7 C: lid close time

   Three samples were tested for each trial condition and the results were analyzed. To confirm the estimate of performance, five additional samples were tested at the optimum condition (Figure 9.33).
   The results of three samples tested in each trial: QC = B.
   Calculated values are (Figure 9.34)

$$\text{Grand average} = 167.29$$

$$\text{Levels: 2 1 2 1 1}$$

TRIAL#/	R(1)	R(2)	R(3)	Average
1	145	152	156	151.0000
2	134	142	146	140.6667
3	119	122	128	123.0000
4	234	211	216	220.3333
5	254	231	221	235.3333
6	198	191	176	188.3333
7	142	143	156	147.0000
8	123	132	143	132.6667

Figure 9.33 Test results for Exercise 9.8.

$$Y_{opt} = 232.83 \quad \text{(without interaction)}$$

$$= 258 \quad \text{(with one interaction)}$$

$$\text{C.I. from ANOVA: } \pm 9.36$$

Confirmation tests with five samples produced:

$$255 \quad 248 \quad 264 \quad 268 \quad 258 \Rightarrow \frac{1293}{5} = 258.6 \text{ (average)}$$

Perform standard analysis.

(a) Determine the optimum condition without interaction consideration and Y_{opt}.

(b) Which of the two interactions is more significant?

(c) Determine the optimum condition and Y_{opt} with interaction effect of II.

(d) Compare the confirmation test results with the optimum performance calculated and the C.I. What course of action will you follow, and why?

(*Hint*: Ignore interaction and accept results, refine the prediction of optimum with interaction, repeat the experiment with new interactions, brainstorm for more factors, go for robust design, and so on)

Expression:	C.I.	=	Sqr ((F(1,n2) × Ve)/Ne)
Where:	F(n1,n2)	=	Computed Value of F = 2.599999
	n1	=	1 n2 = error DOF = 16
	Ve	=	Error Variance = 101.2266
	Ne	=	Effective Number of Replications = 3
Confidence Level		=	90%
CONFIDENCE INTERVAL		=	+/– 9.366412
EXPECTED RESULT AT OPTIMUM		=	260.4999 +/– 9.366412

Figure 9.34 Calculated C.I. for Exercise 9.8.

9.9 An L-8 array was used to study 1 four-level factor (E, column 1) and 4 two-level factors (A, B, C, and D in columns 4, 5, 6, and 7). One sample tested at each trial condition produced the following results: 71, 68, 70, 75, 68, 65, 65, and 74. If a smaller value is desired, determine:

(**a**) The optimum condition

(**b**) Performance at the optimum condition

(**c**) The main effect of factor D

EXERCISE ANSWERS

9.1 (**a**) L-8 array. Columns 1, 2, and 3 upgraded to a four-level column for the four-level factor. Assign the tree two-level factors to columns 4, 5, and 6. Leave column 7 unused.
(**b**) L-8 array. All columns used.
(**c**) L-16. Use interacting groups [1 2 3] and [4 8 12] to form three-level columns. Assign two-level factors to any remaining column.
(**d**) L-18. Assign factors to the column matching the levels.
(**e**) L-16. Use interacting groups [1 2 3], [4 8 12], [7 9 14], and [5 10 15] to create upgraded columns for the 4 four-level factors. Assign the 3 two-level factors to columns 6, 11, and 13.
(**f**) L-32.

9.2 (**a**) Trial 6; (**b**) levels 2 2 2 1 1 (i.e., textured, increased, harder, type 1, and present).

9.3 All answers are correct. Generally, answer (a).

9.4 L-16.

9.5 Yes.

9.6 (**a**) 2; (**b**) 8; (**c**) 2; (**d**) 8; (**e**) 2; (**f**) 18; (**g**) L-16.

9.7 (**a**)–(**c**) Combine columns 2, 4, and 6 (discard column 6 and combine columns 2 and 4) to form a four-level column, then downgrade it to a three-level column. Review the experiment file *BKEX-97A.Q4W* (created using the *Manual design* option) and perform the standard analysis.
(**d**) Main effect of factor D: 9.25, 4, and 10.0. D_2 is the best level.

9.8 (**a**) 217.496; (**b**) $A \times B$; (**c**) 260.495; (**d**) results confirm. (*Hint*: Design experiment using the *Manual design* option. To calculate expected performance without interaction, click *Estimate* when in the *Optimum* screen. Type 0 for level of interaction to exclude them from the estimate of performance. Review experiment file *BKEX-98A.Q4W* and analyze results using standard analysis option.)

9.9 (a) $E_2 A_2 B_1 C_1 D_2$; (b) 77, (c) (70–69). Create your own design or review experiment file *BKEX-99A.Q4W*.

Combination Designs

What You Will Learn in This Step

For some applications, the factors and levels are such that standard use of the orthogonal array does not produce economical experimental strategy. In such situations a special experiment design technique such as combination design may offer a significant savings in number of experiments. This step deals with how to lay out experiments using a combination design technique and the necessary assumptions that must be made.

Thought for the day

Experience shows that success is due less to ability than to zeal. The winner is he who gives himself to his work, body and soul.

—Charles Buxton

FACTOR-LEVEL COMPATIBILITIES

Before we discuss principles of combination design, it is important that you understand the need for factor-level compatibility in experiments designed using orthogonal arrays. All experiments designed using orthogonal arrays require that all levels of all factors be compatible (coexist) with all levels of the other factors included in the experiment. Suppose that in one of your experiments there are 2 two-level factors, A and B, along with several other factors, as shown in Figure 10.1.

In experimental design using an L-8 array as shown, the numbers in the rows (representing the levels of the factor assigned to the columns) describe the eight trial conditions in each level of the array. The level numbers in the rows are arranged such that any level of a factor, say A_1, must exist with all levels of other factors, such as B_1, B_2, C_1, C_2, and so on. Considering the two factors A and B, since they have two levels each, there could be four possible ways (A_1B_1, A_2B_1, A_1B_2, and A_2B_2) for them to be present in the rows of the array. Observe that these conditions are satisfied by the trail conditions. As a matter of fact, arrays are designed such that all combinations not only exist but are present in equal numbers (A_1B_1 is present in trials 1 and 2, A_1B_2 is in trials 3 and 4, etc.). You can verify that this is true between any other pair of factors.

TABLE 10.1 Experiment Design with L-8 Array

	Factor (column)						
Trial	A (1)	B (2)	C (3)	D (4)	E (5)	F (6)	G (7)
1	1	1	1	1	1	1	1
2	1	1	1	2	2	2	2
3	1	2	2	1	1	2	2
4	1	2	2	2	2	1	1
5	2	1	2	1	2	1	2
6	2	1	2	2	1	2	1
7	2	2	1	1	2	2	1
8	2	2	1	2	1	1	2

What if, for some reason, it is not feasible to experiment with all combinations of a pair of factors? This is a likely scenario. Hopefully, you would discover this situation before you completed the experiment design, because if this condition were to occur, you would not be able to complete all experiments that your design calls for. Such an application exercise would not be a good use of your team's time and effort.

Suppose that the two factors *engine* (A: 3.0 liters and 3.2 liters) and *transmission* (B: T320 and T445) are among the factors involved in the study of an automobile front structure design study. Business plans were such that the combination of a 3.2-liter engine and a T445 transmission (A_2B_2) were not planned for production. If you were not aware of this situation beforehand and you proceeded with the experiment treating engine and transmission as two separate two-level factors, you will not be able to carry out the experiment. How can you pursue your study? What can you do about these two factors?

Two two-level factors create four conditions (e.g., A_1B_1, A_2B_1, A_1B_2, and A_2B_2). For convenience, these two factors can also be studied by treating them as one combined factor (AB, not A and B) with four levels. In other words, in the case above, instead of treating engine and transmission separately, it is possible to call it a new factor, power plant (PP), with four levels (PP$_1 = A_1B_1$, PP$_2 = A_2B_1$, PP$_3 = A_1B_2$, and PP$_4 = A_2B_2$). By studying PP at four levels (by assigning it to a four-level column), you can obtain the same information as you would have obtained by studying the factors separately. Of course, if you are able to prepare experimental samples in all conditions, there is no need to study them together. But when all combinations are not available, this option of being able to combine two or more factors into a single factor becomes an effective solution. When the combination A_2B_2 is not available, for reasons mentioned earlier, it is possible for you to treat the combined factor as a three-level factor (all combinations except A_2B_2), perform the experiment and collect useful information about the two factors.

COMBINATION DESIGN TECHNIQUE

This is a special technique of fitting 2 two-level factors into a three-level column. You will prefer this technique when it results in larger savings in number of experiments compared to the conventional approach.

Consider a situation where you want to study 3 three-level factors (A, B, and C) and 2 two-level factors (X and Y). By intuition, or by calculating the degrees of freedom (DOFs, $3 \times 2 + 2 \times 1 = 8$), you will find that an L-9 array (has 4 three-level columns) has the DOFs needed, but it does not have the five columns needed for the factors. Following the normal process of upgrading and downgrading columns, an L-16 will be necessary for the design. If there were a way to make use of the L-9 to design the experiment, it will mean a savings of 7 (16–9) experiments.

Combination design is a clever way to make an array do more than what it is normally used for. In this method 2 two-level factors can be assigned to a single three-level column. For 2 two-level factors and 3 three-level factors, an L-9 array can indeed be used to design the experiment. An L-9 array has 4 three-level columns. Three of these three-level columns can be used to assign the 3 three-level factors (A, B, and C in columns 1, 2, and 3). The remaining three-level column (column 4) is available for the 2 two-level factors. In combination design technique, the 2 two-level factors, X and Y, are combined into a new single fictitious factor (XY) with four levels. Subsequently, any one of the four levels is discarded and the remaining three levels are assigned to the available three-level column. In the new notation, the four levels of the factor (XY) will have the following meanings.

$$(XY)_1 = X_1 Y_1 \qquad (XY)_2 = X_2 Y_1$$

$$(XY)_3 = X_1 Y_2 \qquad (XY)_4 = X_2 Y_2$$

TABLE 10.2 Combination Design Using L-9 Array

	Factor (column)			
Experiment	A (1)	B (2)	C (3)	(XY) (4)
1	1	1	1	1
2	1	2	2	2
3	1	3	3	3
4	2	1	2	3
5	2	2	3	1
6	2	3	1	2
7	3	1	3	2
8	3	2	1	3
9	3	3	2	1

Now, since the available three-level column can accommodate only a three-level factor, the number of levels of the factor (XY) is reduced by discarding any of the four levels of (XY), say the fourth level [i.e., $(XY)_4$]. The factor is now assigned to the column. Understand that while (XY) is a fictitious factor, its levels expressed in terms of the levels of the original factors X and Y do have physical meanings. Thus after assignments of the factors to the column as shown in Table 10.2, trail condition 1, for example, will mean $(A_1B_1C_1X_1Y_1)$, as level 1 of the factor in column 4, $(XY)_1$, is X_1Y_1.

MAIN EFFECTS OF COMBINED FACTORS

While combining the 2 two-level factors and dropping one level to make it into a three-level factor, stands to reason, are we still able to determine the individual factor effects? In other words, is it still possible to determine the main effects of the 2 two-level factors X and Y? Are there any restrictions on the behavior of the two factors being combined?

Once the experiments are carried out and the results collected, the average effects of factors $(A, B, C,$ and $XY)$ assigned to various columns can be calculated. Thus the three average effects of the fictitious factor (XY) [i.e., $(XY)_1$, $(XY)_2$, and $(XY)_3$ are obtained. The main effects of individual factors X and Y are computed from the average effects of (XY).

To establish the main effects of factors, we need two average effects. For example, the main effect of factor X can be expressed (or plotted) when the average effects at $(XY)_1$ and $(XY)_2$ are available. From the results of the combination design, the difference between the two average effects of XY as shown below yields the difference between the effects of one factor (X), while the other (Y) is held fixed. It turns out that the average effect of only three levels of XY can produce the main effects of both the combining factors. (You may verify that three levels will always produce the individual main effects no matter which of the four levels of XY you discard. You do not always need to discard the fourth level.)

$$\text{Main effect of } X = \overline{(XY)}_2 - \overline{(XY)}_1$$

$$= \overline{X_2Y_1} - \overline{X_1Y_1} \qquad \text{at fixed } Y = Y_1$$

$$\text{Main effect of } Y = \overline{(XY)}_3 - \overline{(XY)}_y$$

$$= \overline{X_1Y_2} - \overline{X_1Y_1} \qquad \text{at fixed } X = X_1$$

Notice in the calculation above for the main effect; say for X, factor Y is assumed a constant at level 1 (Y_1). Similarly, while expressing the main effects for Y, factor X is assumed held at level 1 (X_1). This conclusion of the trend of influences of the factors (main effects of X and Y) would be true only if there is no change in the behavior of X when Y changes, and vice versa. This means that the main effects calculated from the combined factor experiments would hold true only if there is no interaction between X and Y. This is a necessary assumption about the behavior of factors being combined. (General practice is to assume that there is no interaction, and go for the savings and confirm the prediction of results.)

Example 10.1: H-Series Crawler Tractor Noise Study To study the dynamic sound power of a redesigned tractor, engineers of an earthmoving equipment manufacturer planned an experiment to study several mixed-level factors (Table 10.3). The objective of the experiment was to determine insulation and sound-dampening parameters that meet the operating sound level within the ISO 6395 requirements (105 dBA). The sound level was recorded and pressure level calculated according to standard practice (QC = S).

Solution: This project utilized an L-9 array to study the 2 two-level factors and 3 three-level factors. Combine the 2 two-level factors (X and Y) into a single three-level factor, (XY), described below.

$(XY)_1 = X_1 Y_1$ [i.e., metal (fan blade) and venturi (shroud design)]

$(XY)_2 = X_2 Y_1$ [i.e., nylon (fan blade) and venturi (shroud design)]

$(XY)_3 = X_1 Y_2$ [i.e., metal (fan blade) and box (shroud design)]

$(XY)_4 = X_2 Y_2$ [i.e., nylon (fan blade) and box (shroud design)–level discarded]

The experimental design was accomplished by assigning the combined factors in column 1 and the remaining 3 three-level factors to the other three columns of the array. The experiment file is saved for your review as *BKEX-101.Q4W*. The procedure for designing the experiment using the *Manual design* option of QT4 is described here. You may either follow this or the *Automatic design* option, which you are already familiar with. If you follow *Automatic design* option, be sure to indicate 4 three-level factors (X and Y combined as one factor, XY).

TABLE 10.3   Factors and Levels for Example 10.1

	Level		
Factor	1	2	3
X: fan blade material	Metal	Nylon	
Y: shroud design	Venturi	Box	
A: fan pulley ratios	0.50	0.77	0.83
B: undercarriage	1 guide	2 guides	3 guides
C: engine insulation	Absent	Foam	Fiberglass

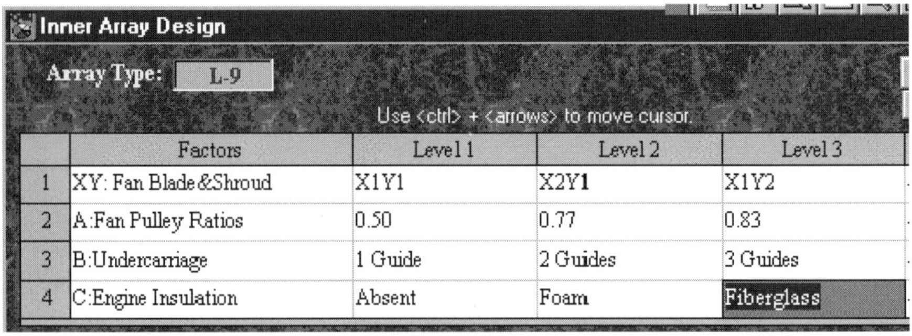

Figure 10.1   Factor description and column assignments.

Once the factors are combined, the combination design can be obtained in the same manner as any other experimental designs. To design this experiment, select the *Manual design* option from the main screen and check the box for an L-9 array. After you describe the factors and their levels, your *Inner array design* screen should look like the one shown in Figure 10.1. Remember that in the *Manual design* option, you must select the array for the design and also determine the column for the factor description. Notice that the combined factor and its three levels are assigned to the column. It is immaterial whether the levels of the combined factor are described in detail or notation form as long as you are able to interpret the results (*main effects, optimum condition*, etc.) after analysis is completed.

Upon completion of the design, you will not be able to save your experiment file, as only L-8 array experiments can be saved using the QT4 version you have. Open the file *BKEX-101.Q4W* and review the experimental conditions from the *Condition* menu item in the main screen. The nine experiments were carried out by testing three samples in each. The results recorded for the sound pressure level are shown in Figure 10.2.

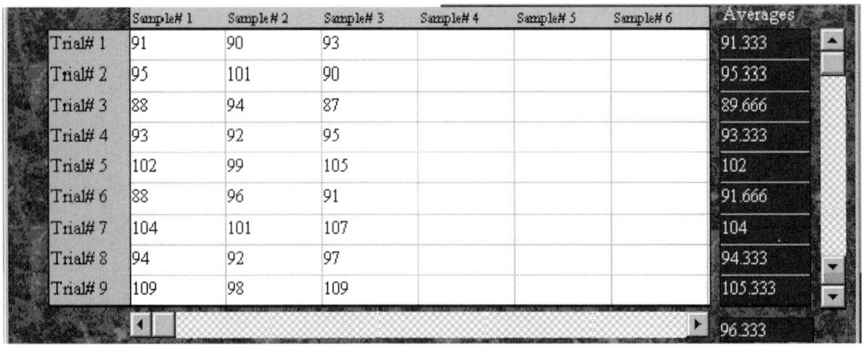

	Sample# 1	Sample# 2	Sample# 3	Sample# 4	Sample# 5	Sample# 6	Averages
Trial# 1	91	90	93				91.333
Trial# 2	95	101	90				95.333
Trial# 3	88	94	87				89.666
Trial# 4	93	92	95				93.333
Trial# 5	102	99	105				102
Trial# 6	88	96	91				91.666
Trial# 7	104	101	107				104
Trial# 8	94	92	97				94.333
Trial# 9	109	98	109				105.333
							96.333

Figure 10.2   Experimental results and the trial averages.

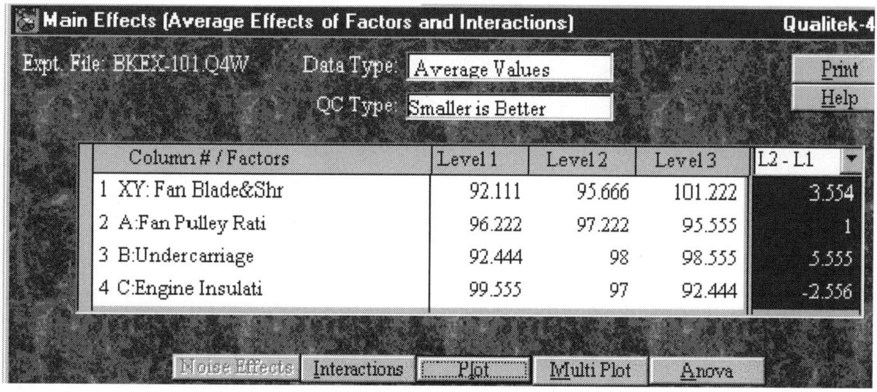

Main Effects (Average Effects of Factors and Interactions)				Qualitek-4

Expt. File: BKEX-101.Q4W   Data Type: Average Values   Print

QC Type: Smaller is Better   Help

Column # / Factors	Level 1	Level 2	Level 3	L2 - L1 ▼
1 XY: Fan Blade&Shr	92.111	95.666	101.222	3.554
2 A:Fan Pulley Rati	96.222	97.222	95.555	1
3 B:Undercarriage	92.444	98	98.555	5.555
4 C:Engine Insulati	99.555	97	92.444	-2.556

Noise Effects   Interactions   Plot   Multi Plot   Anova

Figure 10.3   Main effects of factors.

To perform an analysis, select the *Standard analysis* from the *Analysis* menu item in the main screen. The main effects of the 3 three-level factors (*A*, *B*, and *C*) and the three-level combined factor (*XY*) are as shown in Figure 10.3. In this case, our interest is to obtain the main effects of individual factors (*X* and *Y*) which were combined as *XY* for the study. You can obtain the plot of the main effects of the combined factor (*XY*) by clicking the *Multiple plots* and *Plot* buttons (Figures 10.4 and 10.5). Since

$(XY)_1$ or $X_1Y_1 = 92.111$ (from experiment, first row of Figure 10.4)

$(XY)_2$ or $X_2Y_1 = 95.666$

the main effects of factor *X* can be extracted as $(XY)_2 - (XY)_1$ (the difference between the second-level average effect and the first-level value), or

$X_2Y_1 - X_1Y_1 = 95.666 - 92.111$ (shows the effect of *X* as *Y* is fixed)

In graphical form, the main effects of *X* (the first segment of the straight line shown in Figure 10.5) can be plotted as shown in Figure 10.6.

Similarly, the main effect of *Y* is obtained as $(XY)_3 - (XY)_1$, or

$X_1Y_2 - X_1Y_1 = 101.222 - 92.111$ (shows the effect of *Y* as *X* is fixed)

which is the difference between the average effects at the third and first levels of the combined factor *XY* (see Figure 10.6). Note that although QT4 displays all the necessary data for the main effects of the combining factors (*X* and *Y*), it currently does not have the option to extract and plot the main effects for them. Should you be interested in the individual effects of these factors, you must create the plot manually, as shown here.

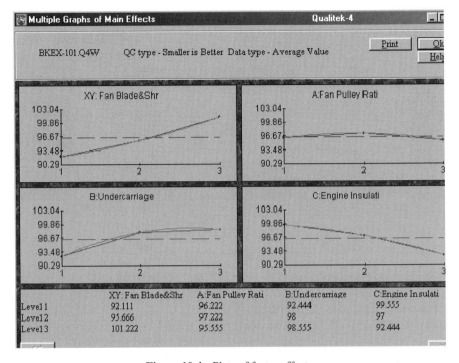

Figure 10.4   Plots of factor effects.

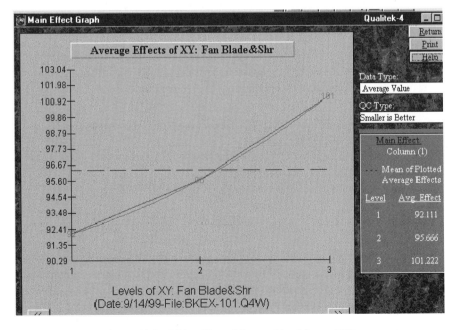

Figure 10.5   Main effect of the combined factor (*XY*).

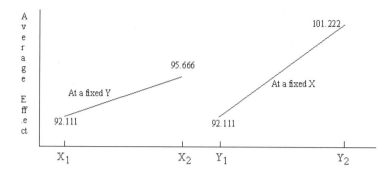

Figure 10.6 Main effects of factors X and Y.

The rest of the analysis for these experimental results follows the same procedure as experiments discussed earlier. The ANOVA for the experimental results is shown in Figure 10.7. Note that the percent influence 32.256 represents the influence of the combined factor (XY) to the variation of the results, and not that of the individual factors (X and Y). There is no easy way to extract the influence of the individual factors (X and Y) from the ANOVA data. In the ANOVA table shown, factor A, which has minimal influence and considered insignificant, has been pooled.

As a general practice, QT4 computes the optimum condition and the performance at the optimum condition with only the significant factors, as shown in Figure 10.8. QT4 views each column of the array occupied by a single factor or an interaction. Therefore, in your combination design the contribution of XY represents that of the combined factors (X and Y). The effects of the individual factors (X or Y) are not obtainable from the program directly. As far as the optimum condition is concerned, the combined level (in this case level 1 of XY) is always selected from the main effect of the combined factor (XY, 92.111 in Figure 10.4) and on the basis of the applicable

ANOVA Table

Expt. File: BKEX-101.Q4W   Data Type: Average Value   Qualitek-4

QC Type: Smaller is Better

Print | Ok
Help | Cancel

Col # / Factor	DOF (f)	Sum of Sqrs. (S)	Variance (V)	F - Ratio (F)	Pure Sum (S')	Percent P(%)
1 XY: Fan Blade&Shr	2	379.564	189.782	13.888	352.234	32.256
2 A:Fan Pulley Rati	(2)	(12.675)		POOLED	(CL= *NC*)	
3 B:Undercarriage	2	205.55	102.775	7.52	178.22	16.32
4 C:Engine Insulati	2	233.577	116.788	8.546	206.247	18.887
Other/Error	20	273.302	13.665			32.537
Total:	26	1091.995				100.00%

Figure 10.7 ANOVA of the experimental results.

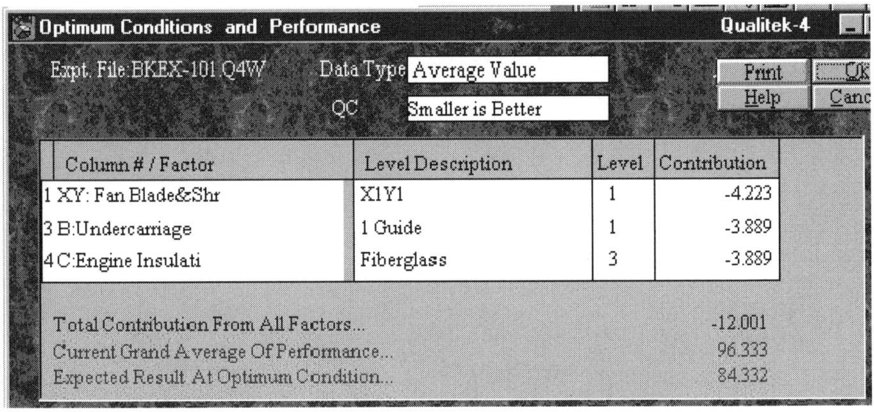

Figure 10.8 Optimum condition and performance.

quality characteristic (*smaller is better* in this case). Here the level of the combined and individual factors selected for optimum condition is $(XY)_1$, which is X_1Y_1 or metal (fan blade) and venturi (shroud design). As a result of this study, the sound pressure level was reduced 12 decibels.

Example 10.2: Axle Iron Generation Study This study (Table 10.4) was initiated by a group of test engineers involved in investigating the causes of oil breakdown in the brake pack due to increased brake wear, which has been the primary item of customer complaint. The purpose of the investigation was to determine design parameters for brake function with the least amount of heat. The temperature measured at a critical

TABLE 10.4   Factors and Levels for Example 10.2

	Level		
Factor	1	2	3
U: ventilation	Present	Absent	
V: plate design	Thick	Thin	
X: plate hardness	R_c 30	R_c 50	
Y: pad materials	Composite	Metal-based	
Z: disk surface finish	Rough	Smooth	
A: starting iron content (ppm)	25	100	150
B: input speed (rev/min)	200	400	700
C: service brake pressure (lb/in^2)	100 psi	150	200
D: case pressure	None	10	15
E: initial particle count	ISO 20/18	ISO 27/25	ISO 32/35

structure was assumed to be proportional to the heat generated due to friction. The design was selected based on the lower temperature (QC = *smaller is better*).

Solution: This example presents a situation in which to apply the technique of combining factors. The number of degrees of freedom (DOFs) for all the factors is 15. An L-16 array has 15 DOFs, but does not allow upgrading of enough columns to make rooms for 5 three-level factors (which will take 15 two-level columns) and five two-level factors. An L-18 array (1 two-level and 7 three-level columns) has enough DOFs but can be used to study only 8 factors. Under normal circumstances, you will need to use an L-27, which has 13 three-level columns. By applying the combination design technique, this experiment can be designed using an L-18 array, which will result in savings of 9 experiments (18 experiments instead of 27).

Design strategy.   Assign one (Z) of the 5 two-level factors to column 1, which comes as a standard two-level column. Combine factors U and V and factors X and Y as new factors UV and XY. Make the combined factors UV and XY as three-level factors by discarding the last combinations of the combining factors. Assign these factors to the three-level columns 2 and 3. Assign the 5 three-level factors (A, B, C, D, and E) to the three-level columns 4 through 8, as shown in Figure 10.9.

Practice designing this experiment using *Manual design* from the *Design* menu item of the QT4 main screen. Review experiment file *BKEX-102.Q4W*, which contains the results (Figure 10.10), and use it for analysis purposes.

You can perform a standard analysis and determine the main effects of all the original and combined factors (eight columns) shown in Figure 10.11. You can now use the main effects of the combined factors to extract the main effects of the constituting factors shown here. From the main effects (third row in Figure 10.11) of the combined factor (UV),

	Factors	Level 1	Level 2	Level 3
	Array Type: L-18			
	Use <ctrl> + <arrows> to move cursor.			
1	Z:Disk Surface Fi	Rough	Smooth	------------
2	UV:	U1V1	U2V1	U1V2
3	XY:	X1Y1	X2Y1	X1Y2
4	A:Satrting Iron C	25 ppm	10 ppm	150 ppm
5	B:Input Speed	200 RPM	400 RPM	700 RPM
6	C:Service Brake P	100 psi	150 psi	200 psi
7	D:Case Pressure	None	10 psi	15 psi
8	E:Initial Particl	ISO 20/18	ISO 27/25	ISO 32/35

Inner Array Design

Figure 10.9   Factor descriptions and column assignments.

	Sample# 1		Sample# 1		Sample# 1
Trial# 7	341	Trial# 1	380	Trial# 13	255
Trial# 8	295	Trial# 2	342	Trial# 14	316
Trial# 9	275	Trial# 3	394	Trial# 15	397
Trial# 10	230	Trial# 4	420	Trial# 16	281
Trial# 11	311	Trial# 5	460	Trial# 17	250
Trial# 12	335	Trial# 6	280	Trial# 18	271

Average of all trial results

= 324.055

Figure 10.10   Results of L-18 experiments.

$$(UV)_1 \text{ or } U_1V_1 = 332$$

$$(UV)_2 \text{ or } U_2V_1 = 354.666$$

$$(UV)_3 \text{ or } U_1V_2 = 285.5$$

which yields

main effect of factor $U = U_2V_1 - U_1V_1 = 354.666 - 332$ (at a fixed V)

main effect of factor $V = U_1V_2 - U_1V_1 = 285 - 332$ (at a fixed U).

This would indicate that U_1V_2 are the suitable levels of factors U and V for a lower temperature rise.

Similarly, from the main effects (third row in Figure 10.11) of the combined factor (XY),

n Effects (Average Effects of Factors and Interactions)				Qualitek-
File: BKEX-102.Q4W Data Type: Average Values				Print
QC Type: Smaller is Better				Help
Column # / Factors	Level 1	Level 2	Level 3	L2 - L1 ▼
1 Z:Disk Surface Fi	354.111	294		-60.111
2 UV:	332	354.666	285.5	22.665
3 XY:	317.833	329	325.333	11.166
4 A:Satrting Iron C	338.833	334	299.333	-4.834
5 B:Input Speed	316.5	345	310.666	28.5
6 C:Service Brake P	299.166	309	364	9.834
7 D:Case Pressure	341.166	306.666	324.333	-34.5
8 E:Initial Particl	338.833	290.666	342.666	-48.168

Figure 10.11   Main effects of factors.

Col # / Factor	DOF (f)	Sum of Sqrs. (S)	Variance (V)	F - Ratio (F)	Pure Sum (S')	Percent P(%)
1 Z:Disk Surface Fi	1	16260.077	16260.077	35.1	15796.831	22.263
2 UV:	2	14920.067	7460.033	16.103	13993.576	19.722
3 XY:	(2)	(388.856)		P O O L E D	(CL= *NC*)	
4 A:Satrting Iron C	2	5570.855	2785.427	6.012	4644.364	6.545
5 B:Input Speed	2	4050.073	2025.036	4.371	3123.581	4.402
6 C:Service Brake P	2	14650.074	7325.037	15.812	13723.583	19.341
7 D:Case Pressure	2	3571.404	1785.702	3.854	2644.913	3.727
8 E:Initial Particl	2	10077.408	5038.704	10.876	9150.917	12.897
Other/Error	4	1852.982	463.245			11.103
Total:	17	70952.944				100.00%

ANOVA Table — Expt. File: BKEX-102.Q4W — Data Type: Average Value — QC Type: Smaller is Better — Print — Ok — Help — Cancel

Figure 10.12 ANOVA of the experimental results.

$$(XY)_1 \text{ or } X_1Y_1 = 317.833$$

$$(XY)_2 \text{ or } X_2Y_1 = 329$$

$$(XY)_3 \text{ or } X_1Y_2 = 325.33,$$

which yields

main effect of factor $X = X_2Y_1 - X_1Y_1 = 329 - 317.833$ (at a fixed Y)

main effect of factor $Y = X_1Y_2 - X_1Y_1 = 325.33 - 317.833$ (at a fixed X).

This indicates that X_1Y_1 are the suitable levels of factors X and Y for a lower temperature rise.

Proceeding with the analysis, you will find that the combined factor XY has a minimum effect on the variation of results and may be pooled as shown in Figure 10.12. Since all of the combined factor XY is found to be insignificant (and pooled), regardless of their main effects, any of its three levels can be used for the optimum condition. However, if the main effects indicated that the fourth combination (X_2Y_2, which is not tested in the experiment) is the suitable factor level, it must be selected as the optimum condition. The factor level suitable for the best design condition and the performance expected is shown in Figure 10.13. The levels of the combined factors U and V are indicated by U_1V_2. This experiment resulted in lowering the temperature from an average of 324° to 141° (Figure 10.13).

Optimum Conditions and Performance			Qualite

Expt. File:BKEX-102.Q4W	Data Type	Average Value	Print
	QC	Smaller is Better	Help

Column # / Factor	Level Description	Level	Contribution
1 Z:Disk Surface Fi	Smooth	2	-30.056
2 UV:	U1V2	3	-38.556
4 A:Satrting Iron C	150 ppm	3	-24.723
5 B:Input Speed	700 RPM	3	-13.389
6 C:Service Brake P	100 psi	1	-24.889
7 D:Case Pressure	10 psi	2	-17.389
8 E:Initial Particl	ISO 27/25	2	-33.389

Total Contribution From All Factors...	-182.392
Current Grand Average Of Performance...	324.055
Expected Result At Optimum Condition...	141.664

Figure 10.13 Optimum condition and performance.

SUMMARY

In this step you have learned how to economize your experiment design by combining factors and making an array (e.g., L-9 or L-18) do more than what it is generally used for. This technique will allow you to make practical compromises, which may result in a reduced number of experiments for the project.

REVIEW QUESTIONS

Q. Is combination design generally done with two-level factors?

A. Yes. Two two-level factors combined creates four situations. The combined factor can then be studied by assigning it to three- or four-level columns. Combining other mixed factors (two- with three-level) with higher number of levels is beyond the scope of this book.

Q. What is the primary benefit of using combination design?

A. A saving in the number of experiments, as a smaller array is generally needed to do the experiment.

Q. What is the risk in designing experiments using the combination design technique?

A. It is assumed that the factors being combined do not interact with each other. If strong interaction is present, the performance predicted may not be accurate. It is a good idea to run tests at the optimum condition to determine if the results confirm the predicted value.

Q. Which arrays are generally used for a combination design?

A. You will find that L-9 and L-18 arrays are more suitable for common projects.

Q. Are there common characteristics that one should look for in projects before proceeding to apply this technique?

A. Generally when you have many three-level factors and one or more two-level factors, you should consider if a combination design could save you some experiments.

EXERCISES

10.1 Design an experiment to study economically 3 three-level factors (A, B, and C) and 2 two-level factors (D and E). Assume that interaction between factors D and E is negligible.

10.2 An experimenter used an L-9 array to study 3 three-level factors (A, B, and C) and 2 two-level factors (X and Y). Using the principles of combination design, three combinations of X and Y (i.e., X_1Y_1, X_2Y_1, and X_1Y_2) were assigned to the three levels of column 1. Using the results shown in Table 10.5, determine:

(a) Average effect $\overline{X_1Y_1}$, $\overline{X_2Y_1}$, $\overline{X_1Y_2}$

(b) Main effect of X

(c) Main effect of Y

10.3 An experiment to study 3 three-level factors (A, B, and C) and 2 two-level factors (X and Y) was designed using an L-9 array (Table 10.6). The factors X and

TABLE 10.5 L-9 Array for Exercise 10.2

	Factor (column)				
Trial	XY (1)	A (2)	B (3)	C (4)	Result
1	1	1	1	1	60
2	1	2	2	2	50
3	1	3	3	3	40
4	2	1	2	3	55
5	2	2	3	1	45
6	2	3	1	2	35
7	3	1	3	2	65
8	3	2	1	3	60
9	3	3	2	1	55

TABLE 10.6 L-9 Array for Exercise 10.3

Trial	A	(XY)	B	C	Results	Average
1	1	1	1	1	3 4 5	4
2	1	2	2	2	7 6 5	6
3	1	3	3	3	4 5 3	4
4	2	1	2	3	6 7 5	
5	2	2	3	1	8 9 7	
6	2	3	1	2	9 10 8	
7	3	1	3	2	6 8 7	
8	3	2	1	3	5 7 6	
9	3	3	2	1	5 6 9	

The heading spans "Factor" over A, (XY), B, C.

Y were combined to form a four-level factor and assigned to column 2. Assuming that the first three levels $(X_1Y_1, X_2Y_1,$ and $X_1Y_2)$ of XY were used for the experiment, determine:

(a) Main effect of X (assume that there is no interaction between X and Y)

(b) Optimum condition when the quality characteristic is *Smaller is better*

$$\overline{A}_1 = \qquad\qquad \overline{B}_1 =$$

$$\overline{A}_2 = \qquad\qquad \overline{B}_2 =$$

$$\overline{A}_3 = \qquad\qquad \overline{B}_3 =$$

$$\overline{C}_1 =$$

$$\overline{C}_2 =$$

$$\overline{C}_3 = \qquad\qquad \text{Optimum condition:}$$

10.4 An experimenter used an L-9 array to study 3 three-level factors (A, B, and C in columns 1, 2, and 3) and 2 two-level factors (X and Y in column 4 as X_1Y_1, X_2Y_1, and X_1Y_2). The results of the nine trials are as follows: 42, 45, 38, 56, 36, 52, 64, 68, and 54. If a larger value of the result is desired, determine:

(a) The optimum condition

(b) The main effects of factor X

(c) Performance under the optimum condition

EXERCISE ANSWERS

Use QT4 capability to solve exercises when appropriate.

10.1 Combine factors D and E to form a new combined factor DE. Select three of the four combinations of D and E. Assign this factor to column 1 of an L-9 array. Assign the 3 three-level factors (A, B, and C) to columns 2, 3, and 4.

10.2 Refer to experiment file *BKEX-106.Q4W*. (**a**) 50 45 60; (**b**) 45–50; (**c**) 60–50.

10.3 Review experiment file *BKEX-107.Q4W*. Use the Standard analysis option.
(**a**) 5.666 6.666 6.555
(**b**) A: 4.666 7.666 6.555; B: 6.333 6.222 6.333; C: 6.222 7.333 5.333.
(**c**) Optimum condition: $A_1B_2C_3X_1Y_1$. (Any levels of B may be selected, as B is found to be insignificant.)

10.4 Review experiment file *BKEX-108.Q4W*. (**a**) $A_3B_1C_1X_2Y_1$; (**b**) 53.66–44.00; (**c**) 72.331.

Strategies for Robust Design

What You Will Learn in This Step

Variations among parts manufactured to the same specifications are common even when all factors are properly controlled. Reduction of variation is our ultimate goal for it will favorably affect one aspect of quality by bringing consistency in performance. The most common variations are considered to be caused by factors that are not controllable or are too expensive to control (noise factors). In robust design strategy, the approach is not to go about controlling the noise factors, but to minimize their influence by adjusting the controllable factors that are included in the study.

Thought for the Day

You cannot teach a man anything; you can only help him find it within himself.

—Galileo

The word *robust* in engineering design terms means insensitive or immune. The effort in robust design strategy is to make a product or process insensitive to the probable causes of performance variation. The objective is to determine the combination of the product/process parameters that reduces variation. Variation is a law of nature. All things found in nature vary. Two apples from a tree are not of equal weight. We are not all created equal; we come in different heights and weights. Can humans and machines make things alike? From a distance they may look alike, but they are probably not alike when you take a closer look. Two automobiles of the same model, two 9-volt batteries, two wristwatches of one kind, or two ink cartridges for the same printer do not last the same amount of time. Two light bulbs of the same type do not burn out at the same time, and two components made for the same job do not fail at the same time. Parts we manufacture or components we make using the same specifications also vary. Thus things that we buy, components that our suppliers provide for assembly, things we manufacture will exhibit variation in one characteristic or another. When we conduct experiments with fabricated hardware, performances will vary.

What causes variation? The sources of variation are mainly uncontrollable factors. As a design matures, most known causes of variation that it is possible to control, adjust, or fix are taken care of. When all that can be done to reduce variation is done, when we feel that the design of the product/process is at its best, there will still be variation. In the long run, the primary sources of variation remain to be those factors that we cannot control. No matter the form, variation is undesirable. Obviously, if there were an easy and economical way to control these factors, someone would have done it. In the past, statistical scientists called these factors *nuisance factors*; Dr. Taguchi called them *noise factors*.

Since the early days of experimental engineering, people knew about the sources of variation being those factors that they couldn't do much about. Attempts to identify the major causes and to try to fix them have generally been unsuccessful. In the robust design methodology, Taguchi offers a totally new way to address the concern for variation. His approach is different in what he does with the uncontrollable factors. Contrary to the conventional approach of attempting to control the uncontrollable factors, Taguchi proposed to build insensitivity into the design such that the influence of the uncontrollable factors is minimized.

Robust design helps achieve greater customer satisfaction and increased sales by making a product or process deliver the desired performance consistently. Consider the process of baking a pound cake by the manufacturer of baked goods. The ingredients for a cake, such as sugar, flour, butter, and milk are factors specified in the recipe for users to follow. These are factors that all who wish to bake their own cake following the recipe can set and adjust. These are controllable factors. What other factors influence how the cake looks and tastes? Experience has shown that oven type, kitchen temperature, and kitchen humidity also affect the process. It has been shown that using a gas oven rather than an electric oven makes a difference, as does low or high temperature and humidity. Should the manufacturer specify the need for a specific type of oven for the baking process and a special temperature-controlled kitchen? What will that do to the market share of sales? Is that a prudent business decision?

Perhaps it all depends on the application. If the cake recipe were intended for high-end users such as costly hotels and restaurants, perhaps for the sake of higher quality it would be all right to require that a costly electric oven or an air-conditioned kitchen be required for the baking process. But if it were a recipe for general consumption, such equipment needs will make the recipe unattractive to the majority of customers. These factors are definitely the noise factors in the process. Will they vary while a customer is baking the cake? What should be done about them? Obviously, customers should not have to do anything, if the recipe is robust, that is immune to the variation caused by the uncontrollable factors.

DRIVING PHILOSOPHY

For the manufacturer of baked goods described above, what sort of recipe should it hope to develop? Business goals will dictate that the recipe be such that any user, no matter the oven used and kitchen environment, can bake good cakes.

Consider a design engineer responsible for developing the suspension system for a luxury vehicle. His or her goal is to design a suspension system that gives a good ride no matter who uses it, where it is used, and under what weather and road conditions. Experience and analyses have shown that ride quality depends on many controllable factors, such as springs, shocks, body structure, and seat cushions. The ride is also found to depend on the type of road (e.g., gravel, concrete) and the tire *inflation pressure*. How do you counter the influences of these uncontrollable factors? A progressive design definitely ought not be affected naturally by road conditions or the inflation pressure, which may or may not be maintained at a specified level by all drivers.

What should the design objectives be for the manufacturer of a wristwatch? What makes a watch robust? A robust watch must be designed such that it shows the correct time all the time; no matter who wears it and under what conditions it is worn. It should withstand the most extreme environment and the most abusive applications.

So, how can designs be made robust? Certainly, noise factors cannot be controlled for most products after they leave the manufacturing plant. The only thing possible is to adjust the controllable factors so as to minimize the influence of the uncontrollable factors. The idea of robust design is to reduce variability without actually removing the cause of variation. Thus for the robust recipe, selection of the controllable factors (sugar, butter, milk, etc.) is made by considering the results of many cakes baked under extreme or all possible combinations of noise conditions.

COLLECTING INFORMATION ABOUT VARIATION

When things vary we need to collect sample performance data to get an indication of the population. When designed experiments are carried out, the results of each experiment (trial condition) will vary when tests are repeated. This is because factor levels will vary and the same setup done repeatedly will influence the results. In addition, the influence of uncontrollable factors such as tool wear, ambient temperature, humidity, dust in the wind, operator fatigue, time of day, and so on, will influence the results. Since the results will surely vary when experiments are repeated, we must repeat experiments and collect multiple sample data to get a reading of the variability present. About the only time the results will not be expected to vary and you need to repeat experiments is when you are experimenting with a simulations (analytical/computer) model where the effects of noise factors are not incorporated. So when experimenting with hardware samples, we must repeat. Questions such as how many to repeat—in other words, how many samples should be tested in each experimental condition— often is determined based on cost and expected variability. Generally, the more variation expected, the greater the number of samples. You should consider running more samples in each experiment when the cost of repeating samples is lower.

When reducing variability is of interest, you must run multiple samples per trial condition. How would you know how many is too many or how few is good enough? No definite guidelines are available. Running three samples in each condition is common in the low end. There is no need to exceed 10 samples in each trial conditions in any situation (take 20 samples if time and cost is not an issue). There is a diminishing benefit from the increased number of samples at the higher end. When you follow the principles of robust design and include noise factors in your experiments, there may be some guidelines about the number of samples desired.

EXPERIMENTAL STRATEGY FOR ROBUST DESIGN

The experiments you need to design to develop robust products and processes will necessarily require you to test more samples and may also require more resources. Most often, though, the return on investment realized from robust design makes it all worth it. Here are the steps that apply to most common experiments for robustness studies. (Robust experiments designs apply to static and dynamic systems. Depending on the objectives of the product/process it may be static or dynamic. All experiments in this book are on static systems. Detailed discussion of dynamic systems is beyond the scope of this book.)

1. Identify noise factors that are controllable in the laboratory (uncontrollable in the user environment).
2. Combine the noise factors using an orthogonal array to produce a number (at least two) of extreme noise conditions.
3. Repeat samples in each trial condition, exposing each to the noise conditions
4. Analyze results and select the levels of the controllable factors that produce the least variation in results.

FORMALITIES OF COMBINING NOISE FACTORS

When there are too many noise factors, testing samples under all possible combinations could become prohibitive. For example, if there are three noise factors and it is possible to hold them at two conditions (say, high and low), there can be eight possible combinations, requiring eight samples in each trial condition. Instead, if the three noise factors are combined by using an L-4 orthogonal array in the same manner that three control factors are combined in an L-4 experiment, the number of samples needed for the experiment can be reduced. Use of an orthogonal array in this manner to combine the noise factors and create conditions to which the samples are exposed under different trial conditions creates an *outer array* (the same L-4 array rotated 90° anticlockwise) in the experiment. Table 11.1 shows an experiment design with 7 two-level control factors (*A, B, C*, etc.) and 3 two-level noise factors (*X, Y,* and *Z*).

To distinguish the array used to combine control factors, it is called the *inner array*. Once you use an outer array in your experiment, its size dictates the number of samples you need to repeat each trial condition. The combination of the various noise conditions created by the outer array and the trial conditions create a number of unique test conditions, generally called a *cell*. One or more samples can be tested in each of these unique conditions (cells). In the design shown (Table 11.1), there are 32 (8×4) unique conditions in which one sample is tested. For example, result R11 is obtained by testing trial 1 while the noise condition is held at 1 ($X_1 Y_1 Z_1$).

The size of the *outer array*, which dictates the number of samples, required to complete an experiment, is independent of the inner array and depends strictly on the number of noise factors and their levels. It is also independent of the size of the inner array

TABLE 11.1 Experiment Design with Noise Factors Using an Outer Array

								Outer Array (L-4)			
						Z		1	2	2	1
						Y		1	2	1	2
						X		1	1	2	2
			Control Factors						Results		
Trial	A	B	C	D	E	F	G	1	2	3	4
1	1	1	1	1	1	1	1	R11	R12	R13	R14
2	1	1	1	2	2	2	2	R21	...	...	...
3	1	2	2	1	1	2	2	:			
4	1	2	2	2	2	1	1				
5	2	1	2	1	2	1	2	R51	R52	R53	R54
6	2	1	2	2	1	2	1	:			
7	2	2	1	1	2	2	1				
8	2	2	1	2	1	1	2	R81	R82	R83	R84
			Inner Array (L-8)								

and the control factors included in the study. The purpose of an outer array in the experiment is to formally combine the noise factors and run multiple samples in the same trial condition exposed to the conditions. Although it is possible to combine the noise factors by means other than an orthogonal array, when so done, results of the experiments can produce valuable information about the effects of the noise factors and the interaction between them and the control factors. In the design shown in Table 11.1, the 8 by 4 matrix of numbers (R11, R12, R84) represents that of an L-8 experiment with four samples in each of the eight trial conditions. With respect to the outer array design, it is an L-4 experiment with eight results in each of the four trial conditions, consisting of the three noise factors. Thus the effects of noise factors are also contained in the same matrix of results (8 × 4 matrix), which contains information about the control factors. Extraction of the effects of noise factors along with the steps in outer array design are described in detail in Example 11.1. Designing experiments with different combinations of numbers of noise and control factors is demonstrated in subsequent examples in this step.

Example 11.1: Nylon Grip Molding Process Study The nylon grip is one of the assembly components of a consumer appliance and has been found to suffer from wall thickness variations. The thickness variation is attributed to the effects of shrinkage after the injection molding process. The objective of the study was to determine the combination of process parameters that is less sensitive to the influences of the noise factors. For the control and noise factors identified in Tables 11.2 and 11.3 and the results obtained (file *BKEX-111.Q4W*):

(a) Design the experiment for robustness.

TABLE 11.2 Control Factors for Example 11.1

Factor	Level			
	1	2	3	4
A: cure time (min)	Present	+ 5		
B: cooling rate	Slower	Rated		
C: cooling airflow	Natural	Fan		
D: additive	Added	Absent		
E: mold temperature (°F)	Lower	Current	+10	+30

(**b**) Identify the nature of influence of the noise factors.

(**c**) Determine the process description that is least sensitive to the influence (robust) of the noise factors.

Solution: This experiment has been saved as *BKEX-111.Q4W* and the shrinkage (results) measured for samples tested for the trial conditions are included. Design the experiment following the procedure shown here and use the experiment file to determine the optimum condition and the effects of the noise factors. You can use QT4 to design your experiment with both an inner and an outer array by selecting the *Automatic design* or *Manual design* option. The *Automatic design* option will be followed in this example. You will have opportunity to learn the *Manual design* option for outer array design in later examples.

(**a**) *Experiment design*

1. Run QT4 and go to the main screen.
2. From the main screen, select the *Automatic design* option from the *Design* menu item.
3. At the *Automatic experiment design* screen, check the appropriate factor boxes for 4 two- and 1 four-level control factors and 3 two-level noise factors (Figure 11.1). (For designs with noise factors, you must check the two-level factor box on top of the screen even when all control factors are at the two-level)

TABLE 11.3 Noise Factors for Example 11.1

Factor	Level	
	1	2
X: nylon condition	Air-dried	Vacuum-dried
Y: ambient temperature	Low	High
Z: humidity(%)	45	95

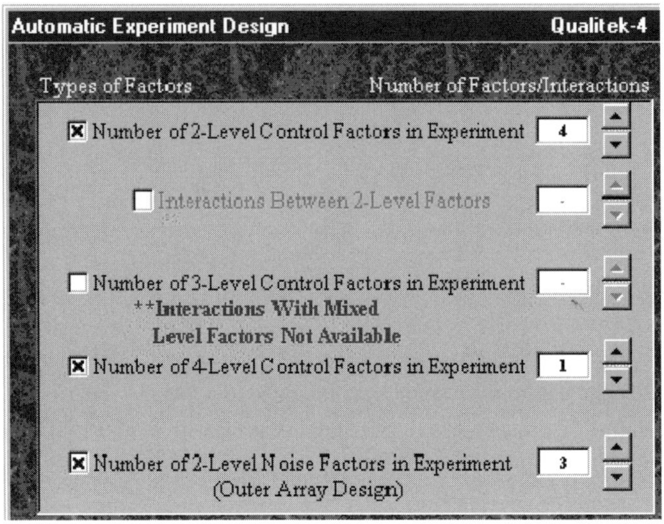

Figure 11.1 Input for automatic experiment design.

4. Proceed and describe the 4 two-level factors first, the 1 four-level factor next, and the 3 two-level noise factors last. As you proceed after describing the factors, QT4 will show you the factors and the column assignment as in Figure 11.2. As there is 1 four-level factor, QT4 automatically modifies the L-8 array it selected for the design by combining columns 1, 2, and 3 and making column 1 into a four-level column. The modified array is shown in Figure 11.3. The 3 two-level noise factors require an L-4 array as the outer array. The noise factor descriptions, column assignments, and the outer array are shown in Figures 11.4 and 11.5. It is a good idea always to make sure that the array that QT4 has selected is appropriate and that the columns are modified correctly (in the

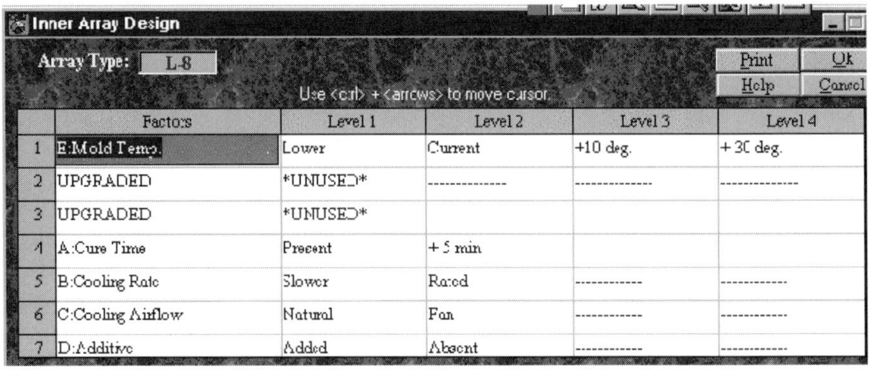

	Factors	Level 1	Level 2	Level 3	Level 4
1	E:Mold Temp.	Lower	Current	+10 deg.	+30 deg.
2	UPGRADED	*UNUSED*	-------------	-------------	-------------
3	UPGRADED	*UNUSED*			
4	A:Cure Time	Present	+5 min		
5	B:Cooling Rate	Slower	Raised	-----------	-----------
6	C:Cooling Airflow	Natural	Fan	-----------	-----------
7	D:Additive	Added	Absent	-----------	-----------

Figure 11.2 Control factor descriptions and column assignment.

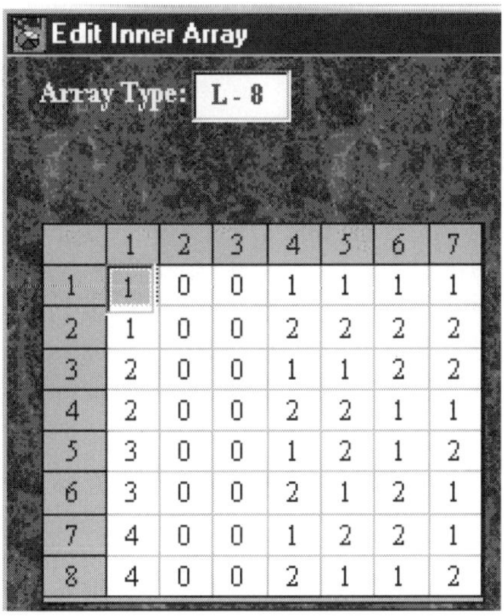

	1	2	3	4	5	6	7
1	1	0	0	1	1	1	1
2	1	0	0	2	2	2	2
3	2	0	0	1	1	2	2
4	2	0	0	2	2	1	1
5	3	0	0	1	2	1	2
6	3	0	0	2	1	2	1
7	4	0	0	1	2	2	1
8	4	0	0	2	1	1	2

Figure 11.3 Modified L-8 array for the control factors.

Automatic design option). If for some reason the array does not look right, click on the *Reset* button to reset the array and modify manually if necessary.

5. As before, for experiments like this example that utilize an L-8 array, you will be able to save your experiment under a file name of your choice. After you are done reviewing the arrays, you may name your file or open file *BKEX-111.Q4W* once you return (click *Cancel* at the *Save screen*) to the main screen. Your experiment file should now display both the inner and outer arrays as shown in Figure 11.6.

6. Because this experiment utilizes an L-8 and an L-4 array, there are 32 unique cells, which requires a minimum of one sample per cell. The results of one sample tested in each cell are also shown in Figure 11.6. The experimental conditions for each cell are obtained by combining the eight trial conditions with

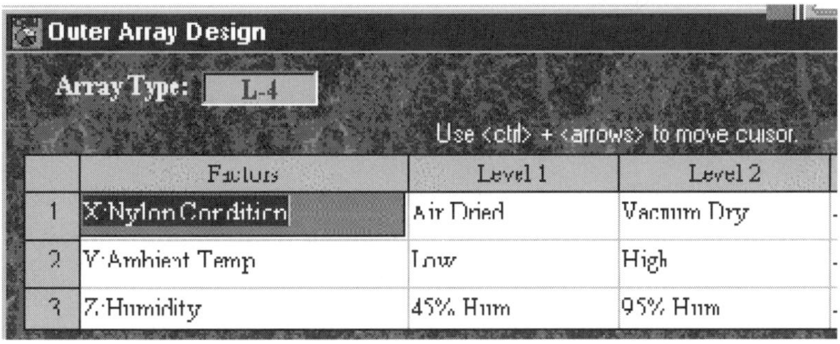

	Factors	Level 1	Level 2
1	X:Nylon Condition	Air Dried	Vacuum Dry
2	Y:Ambient Temp	Low	High
3	Z:Humidity	45% Hum	95% Hum

Figure 11.4 Noise factors and column assignment.

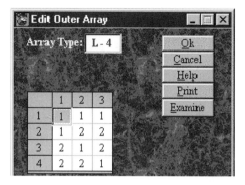

Figure 11.5 L-4 used as outer array.

the four noise conditions. To obtain the trial conditions and the noise conditions, select these options one at a time from the *Condition* menu item in the main screen. Sample trial and noise conditions are shown in Figure 11.7 (only one condition shown) and 11.8 (all four noise conditions shown). The combinations of noise factors shown in Figure 11.8 are considered as the main causes of variability in the results. When four samples at trial condition 1 (Figure 11.7) are tested by setting the four noise conditions, the results shown in the first row of Figure 11.6 are obtained. Results for samples tested at other trial conditions with

<table>
<tr><td colspan="12">Experiment Configuration</td></tr>
<tr><td colspan="12">File Edit Design Conditions Results Analysis Report Loss/Savings Options Practice Tips Help</td></tr>
</table>

Expt File: BKEX-111.Q4W   Qualitek-4

Review Noise Factors

Review Control Factors

								Outer Array			
							3	1	2	2	1
							2	1	2	1	2
							1	1	1	2	2

	1	2	3	4	5	6	7		1	2	3	4
1	1	0	0	:	1	1	1		:.04	1.47	1.39	1.65
2	1	0	0	2	2	2	2		:.68	2.43	1.77	1.73
3	2	0	0	:	1	2	2		:.13	.81	.36	.54
4	2	0	0	2	2	1	1		2.21	2.06	1.17	2.13
5	3	0	0	.	2	1	2		.42	.65	.36	1.25
6	3	0	0	2	1	2	1		.65	.63	1.23	.98
7	4	0	0	:	2	2	1		:.19	1.33	1.18	1.12
8	4	0	0	2	1	1	2		:.2	2.16	2.16	1.05

Inner Array   Results

Figure 11.6 Experiment configuration with outer array.

Descriptions Trial Conditions		Quali
Trial Condition 1 (Random order for running this Trial is 3)		
Factors	Level Description	Level #
E:Mold Temp.	Lower	1
A:Cure Time	Present	1
B:Cooling Rate	Slower	1
C:Cooling Airflow	Natural	1
D:Additive	Added	1

Figure 11.7 First of the eight trial conditions.

Descriptions Trial Conditions		Qua
Noise Condition 1 (Randomly selected order # 1)		
Factors	Level Description	Level #
X:Nylon Condition	Air Dried	1
Y:Ambient Temp.	Low	1
Z:Humidity	45% Hum.	1
Noise Condition 2 (Randomly selected order # 4)		
Factors	Level Description	Level #
X:Nylon Condition	Air Dried	1
Y:Ambient Temp.	High	2
Z:Humidity	95% Hum.	2
Noise Condition 3 (Randomly selected order # 3)		
Factors	Level Description	Level #
X:Nylon Condition	Vacuum Dr	2
Y:Ambient Temp.	Low	1
Z:Humidity	95% Hum.	2
Noise Condition 4 (Randomly selected order # 2)		
Factors	Level Description	Level #
X:Nylon Condition	Vacuum Dr	2
Y:Ambient Temp.	High	2
Z:Humidity	45% Hum.	1

Figure 11.8 The four noise conditions.

the same four noise conditions are obtained. Similarly, if you are using the experiment file you just created, you need to enter the results shown in your file.

Analysis of results. For the most part, the analysis of results of an experiment designed for robustness is similar to that for any other experiment you are familiar with. Opportunities for additional analysis and information are presented in two areas. First, if the results from multiple samples are obtained by testing them under noise conditions designed using an outer array, information about the effects of the noise factors can be obtained easily by a few additional calculations. Second, regardless of the presence of an outer array, whenever multiple samples are tested in each trial condition, there exist better ways to analyze the results than the standard practice of comparing average results. The better method of analyzing the results of multiple samples is to compare the variation data (*S/N* or signal-to-noise ratio) for each trial result. Analysis using *S/N* ratios is covered in Step 12. You will learn how influences (main effects) of noise factors are extracted from the result matrix (multiple sample results per trail condition) of the example (*BKEX-111.Q4W*) under discussion.

1. From the main screen of QT4, select the *Standard analysis* option in the *Analysis* menu item. The preferred (recommended) practice for analysis of multiple sample results is *S/N* analysis. QT4 warns you about this but will allow you to proceed as you click the *Yes* button in the prompt screen shown in Figure 11.9. (The word *run* refers to the number of samples tested in each trial condition.)

2. Select the quality characteristic appropriate for the project (QC = S in this case) and proceed to the result screen shown in Figure 11.10. Notice that this is the normal arrangement (eight rows and four columns) of the results from the point of view of experiments with control factors. This screen shows the results of all samples tested in this experiment and the averages of results in each trial condition. This screen allows you to review the accuracy of the results and to study variability within the data. You can examine variability with and between trial results by clicking the *Graph* button (Figure 11.10). In Figure 11.11, the total lengths of the horizontal bars indicates the highest values of the results in each trial, whereas bar length represents the ranges (variability) within the

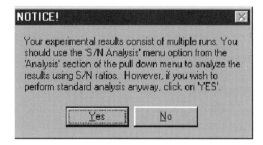

Figure 11.9 Reminder regarding *S/N* analysis for multiple sample results.

Experimental Results
Qualitek-4
Expt. File: BKEX-111.Q4W
Data Type: Average Value
QC Type: Smaller is Better
Ok
Cancel
Help
Print

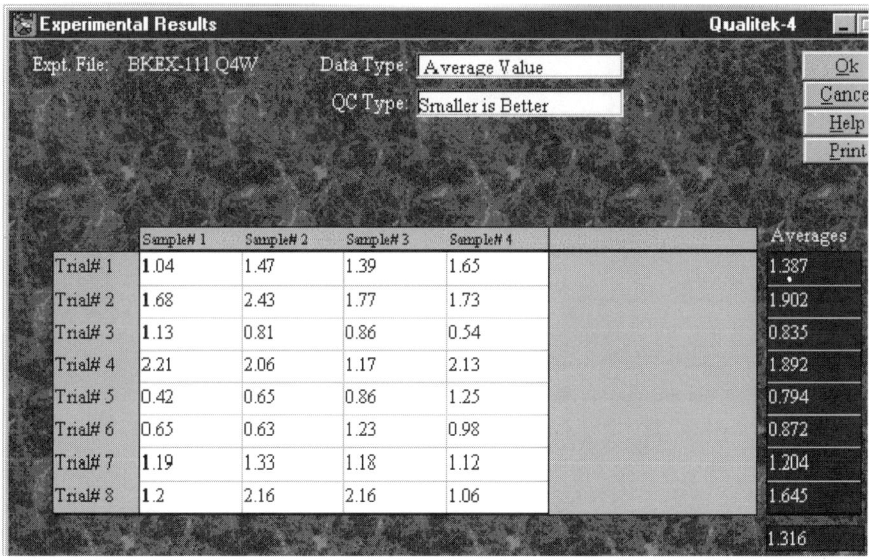

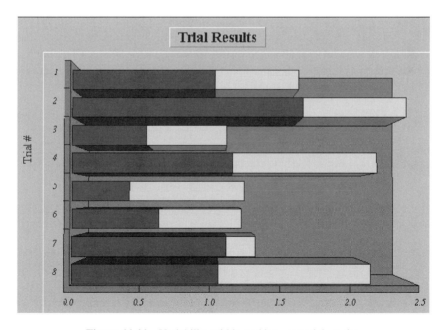

	Sample# 1	Sample# 2	Sample# 3	Sample# 4	Averages
Trial# 1	1.04	1.47	1.39	1.65	1.387
Trial# 2	1.68	2.43	1.77	1.73	1.902
Trial# 3	1.13	0.81	0.86	0.54	0.835
Trial# 4	2.21	2.06	1.17	2.13	1.892
Trial# 5	0.42	0.65	0.86	1.25	0.794
Trial# 6	0.65	0.63	1.23	0.98	0.872
Trial# 7	1.19	1.33	1.18	1.12	1.204
Trial# 8	1.2	2.16	2.16	1.06	1.645
					1.316

Figure 11.10 Results of all trials under all noise conditions.

Trial Results
Trial #
1
2
3
4
5
6
7
8
0.0
0.5
1.0
1.5
2.0
2.5

Figure 11.11 Variability within and between trial results.

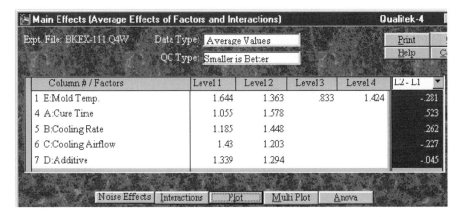

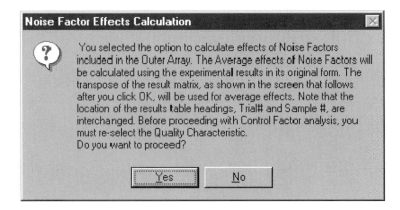

Figure 11.12   Average effects of control factors.

results in each trial condition. Click *Return* in this screen and proceed with the analysis when done reviewing results.

3. Using the average of results of each trial condition, the factor averages are calculated and displayed in the *Main effects* screen (Figure 11.12). As with analysis of any other experiments, plots of individual factor influence (plot of main effects) and interaction between factors can be studied by clicking the appropriate buttons from this screen.

   (**b**) *Noise factor effects.*   Notice that the *Noise effects* button in this screen is now visible. This button is available only when an outer array is present in the experiment. The button allows you to analyze the results for noise factor effects. Click the *Noise factor* button, and click *Yes* when prompted about the strategy for calculating the noise factor effects as indicated in Figure 11.13.

Figure 11.13   Prompt about noise factor effects.

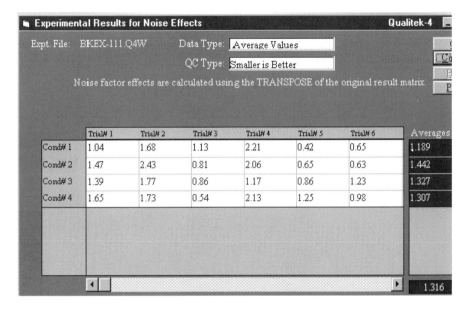

Figure 11.14   Result matrix for analysis of noise factors.

4.   The result matrix, which now has four rows and eight columns from the viewpoint of an L-4 experiment with the 3 two-level noise factors, is shown in Figure 11.14. Note that QT4 shows only six columns of results at a time, but allows you the option to scroll the entire results. No matter how many columns are displayed in the screen at one time, the average column always includes all results (eight in this case). Understand that this result matrix is the same original (eight rows and four columns, see Figure 11.6) viewed vertically from the point of view of the L-4 array, which combines the 3 two-level noise factors. Review the results and click *OK* when done.

5.   Examine the noise factor effects shown in Figure 11.15. This knowledge about the influence of noise factors may be useful in future studies with noise effects. In robust design strategy, we need to be able to control the noise factors, which we would not need to control in an actual application environment. So from a robust design point of view, effects of noise factors are of little immediate benefit. But since this information is available without running additional experiments or incurring any cost, it is a common practice to learn about them. You can view the trend of influence of noise factors by selecting options available in this screen.

6.   Click the *Plot multiple noise factor* effects button and view the plot of the noise factor effects shown in Figure 11.16.

7.   Click *Return* in this screen and click *OK* in the *Noise effect* screen (Figure 11.15) to finish the analysis of noise factors. This action returns you to the beginning of analysis, where you select the quality characteristic. If you wish to continue

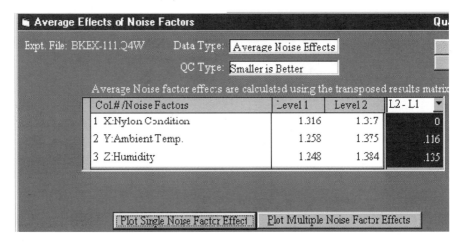

Figure 11.15 Noise factor effects.

with the analysis of results of the experiments, proceed (click *OK*) with the analysis and be in the *Main effects* screen displayed earlier (see Figure 11.11). In this screen, click *OK* to obtain *ANOVA* and *Optimum* screens as done for examples in previous steps.

8. In the *ANOVA* screen shown in Figure 11.17, pool the factor *Additive*, as its influence to the variation was zero, by double-clicking on its description and clicking *Yes* at the prompt. Attempt to pool any other factor and find that it has over a 90% confidence level. If 90% confidence is a satisfactory level, there is

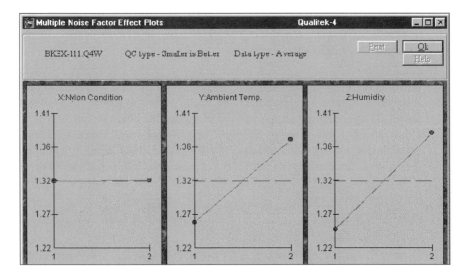

Figure 11.16 Plot of noise factor effects.

ANOVA Table						qualitek-4	

Expt. File: BKEX-111.Q4W   Data Type: Average Value   Print   Ok

QC Type: Smaller is Better   Help   Cancel

Col. # / Factor	DOF (f)	Sum of Sqrs. (S)	Variance (V)	F - Ratio (F)	Pure Sum (S)	Percent P(%)
1 E:Mold Temp.	3	2.839	.946	7.429	2.457	26.784
4 A:Cure Time	1	2.184	2.184	17.141	2.056	22.415
5 B:Cooling Rate	1	.556	.556	4.367	.429	4.676
6 C:Cooling Airflow	1	.409	.409	3.214	.282	3.074
7 D Additive	(1)	(.016)		POOLED	(CL= *NC*)	
Other/Error	25	3.185	.127			43.051
Total.	31	9.175				100.00%

Figure 11.17 ANOVA of experimental results.

no need to pool any other factor. You may view from this screen plots of factor influences (right column) on the variation of results in the form of a bar graph or pie chart. Click *OK* to proceed to the optimum screen shown in Figure 11.18.

(**c**) The *Optimum screen* (Figure 11.18) shows that the shrinkage can be reduced to 0.324 at the optimum condition shown. At this condition, performance is expected to be affected least by the influence of the noise factors within the range tested in the experiment. Notice that factor *D*, which was pooled, is not included in the estimate. This makes the expected performance conservative and you will have the option to set a less costly level. If necessary, you can obtain the confidence interval (C.I.) on this estimated (expected) performance from this screen also.

Optimum Conditions and Performance				Qualitek

Expt. File:BKEX-111.Q4W   Data Type Average Value   Print

QC   Smaller is Better   Help

Column # / Factor	Level Description	Level	Contribution
1 E:Mold Temp.	+10 deg.	3	-.484
4 A:Cure Time	Present	1	-.262
5 B:Cooling Rate	Slower	1	-.132
6 C:Cooling Airflow	Fan	2	-.114

Total Contribution From All Factors...	-.992
Current Grand Average Of Performance...	1.316
Expected Result At Optimum Condition...	.324

Figure 11.18 Optimum condition and performance.

TABLE 11.4 Control Factors for Example 11.2

Factor	Level			
	1	2	3	4
A: type of seal	EPDM	SBR		
B: type of grease	Brycoat 1728	418 disc brake		
C: temperature (°F)	Ambient	150		
D: amount of grease (milligrams)	2000	1600		
E: master cylinder design	Design 1	Design 2		

Example 11.2: Prefilled Clutch Hydraulics Design Optimization Engineers in-volved in developing a more efficient clutch system for automotive transmission per-formed an experiment to study the process under the influence of several noise conditions and determine the design parameters that exhibit the least fluid leakage. During the planning session, the project team identified five control factors (Table 11.4) and four noise factors (Table 11.5). Two interactions ($A \times B$ and $B \times C$) among the control factors and one noise factor at three levels was included in the study.

For the control and noise factors identified and the results obtained (review file *BKEX-112.Q4W*):

(a) Design an experiment using inner and outer arrays (use QT4 *Manual design*).

(b) Robust design conditions for the least fluid leakage (measured in milliliters).

Solution: This example has been designed and the experimental results saved as ex-periment file *BKEX-112.Q4W*. You may either review this experiment file or follow the steps shown here on experiment design using the QT4 *Manual design* option and design your own experiment. As should be intuitively obvious, the inner array for the

TABLE 11.5 Noise Factors for Example 11.2

Factor	Level		
	1	2	3
X: grease aging	Fresh	1 week old	1 month old
Y: seasonal temperature	Summer	Winter	
X: humidity	Low	High	
W: pollutants	Dusty air	Clean air	

design is an L-8 (for two-level factors and two dependent interactions). Since there are 3 two-level and 1 three-level noise factors, an L-8 outer array will have to be modified to accomplish the design. The outer array will be modified by combining its columns 1, 2, and 3 into a four-level column first, then *dummy treating* it to a three-level column. Because the design utilizes an L-8 inner array and an L-8 outer array, a total of 64 (8 × 8) samples were tested, each at the unique combination of control and noise factors prescribed by the design.

   (a) Experiment Design:

1. Run QT4 and go to the main screen. Select the *Inner array* in the *Manual design* option from the DESIGN menu item. From the available inner array shown in Figure 11.19, select the L-8 array and click *OK* to proceed.

2. Describe factors A, B, C, D, and E in rows (which are columns of the array) 1, 2, 4, 5, and 7 of the experiment design screen, respectively. Reserve column 3 for interaction A × B and column 6 for interaction B × C as shown in Figure 11.20. Use either of the *Col inter* or *Inter table* buttons from the bottom of the screen. Remember that in the *Manual design* option of QT4, you must decide about the column assignments and array modification, if any were applicable. If you are not sure about which column you should reserve, say for interaction A × B, you should use the *Inter table* button to let QT4 set the interaction column. After you are done describing factors, QT4 will ask for a project description. This input is

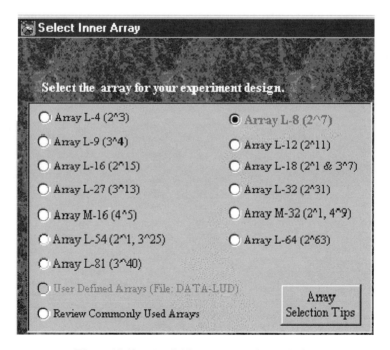

Figure 11.19.   Available arrays experiment design

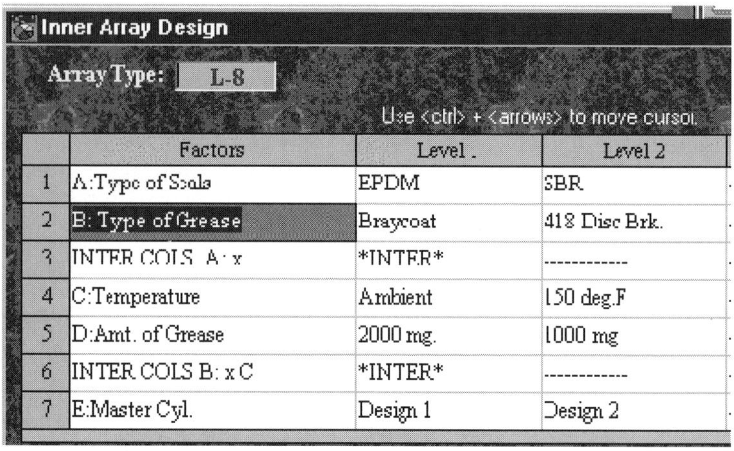

	Factors	Level 1	Level 2
1	A:Type of Seals	EPDM	SBR
2	B: Type of Grease	Braycoat	418 Disc Brk.
3	INTER COLS A : x	*INTER*	------------
4	C:Temperature	Ambient	150 deg.F
5	D:Amt. of Grease	2000 mg.	1000 mg
6	INTER COLS B : x C	*INTER*	------------
7	E:Master Cyl.	Design 1	Design 2

Figure 11.20 Control factors and column assignments.

optional. You need not enter anything if you do not want to. QT4 next shows you the inner array you selected for the design. In the *Manual design* option, QT4 changes a column to zero if it is set as unused, but it does not upgrade or downgrade any column. Complete the inner array design by saving the experiment file using a name of your choice and return to the main screen.

To add the outer array to your experiment design, select *Outer array* in the *Manual design* option from the *Design* menu item. Select the L-8 as the outer array from the arrays available from a screen similar to the one shown in Figure 11.19 and proceed to describe the noise factors shown in Figure 11.21. Describe the three-level factor, *X*,

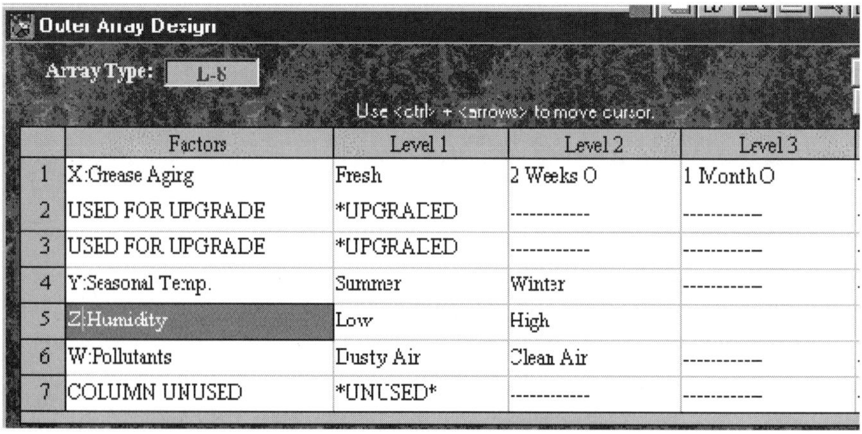

	Factors	Level 1	Level 2	Level 3
1	X:Grease Aging	Fresh	2 Weeks O	1 Month O
2	USED FOR UPGRADE	*UPGRADED	------------	------------
3	USED FOR UPGRADE	*UPGRADED	------------	------------
4	Y:Seasonal Temp.	Summer	Winter	------------
5	Z:Humidity	Low	High	
6	W:Pollutants	Dusty Air	Clean Air	------------
7	COLUMN UNUSED	*UNUSED*	------------	------------

Figure 11.21 Noise factors and column assignments.

and its levels in column 1. Recall that columns 1, 2, and 3 are used to upgrade column 1 as a four-level column. Therefore, these two columns must be set as unused. (Click on this button when the cursor is in the proper row.) Assign and describe the 3 two-level factors to columns 4, 5, and 6, respectively. Set column 7 as unused. Click *OK* to proceed to review and edit the outer array you selected for your experiment.

4.   For columns set as unused, QT4 automatically sets the levels to zero, as shown in Figure 11.22 (columns 2, 3, and 7). Since the factor (*X*) assigned to column 1 has three levels, you must modify the column to a three-level column, as shown (Figure 11.22). To modify the column, click on the level location and type in the number. Click *OK* to finish the design and return to the main screen, which should look as shown in Figure 11.23. (Since only four columns of results are displayed at one time, your results may look different until displayed columns are the same.)

Analysis of results:

1. To perform the analysis, select the *Standard analysis* option from the *Analysis* menu item in the main screen (Figure 11.23). Proceed with the analysis by checking the box for the *Smaller is better* quality characteristic.

2. Review the factor influences and effects of noise factors from the *Main effects* screen (Figure 11.24). To review noise factor effects, follow the steps used in Example 11.1. When finished reviewing the factor effects, click *OK* to view the *ANOVA* screen shown in Figure 11.25.

3. Revise the ANOVA results by pooling interaction columns 3 and 6 (less than 80% confidence level; CL = NC indicates not calculated by QT4). Click *OK* to view the optimum condition (Figure 11.26).

Edit Outer Array

Array Type: **L - 8**

	1	2	3	4	5	6	7
1	1	0	0	1	1	1	0
2	1	0	0	2	2	2	0
3	2	0	0	1	1	2	0
4	2	0	0	2	2	1	0
5	3	0	0	1	2	1	0
6	3	0	0	2	1	2	0
7	1	0	0	1	2	2	0
8	1	0	0	2	1	1	0

Figure 11.22   Modified outer array.

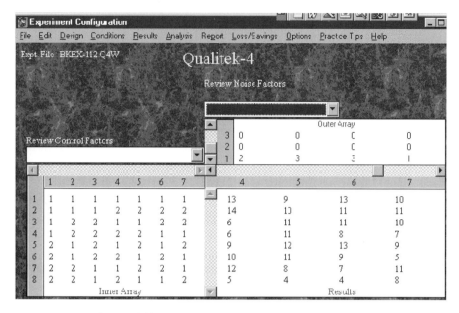

Figure 11.23 Experiment configuration with outer array.

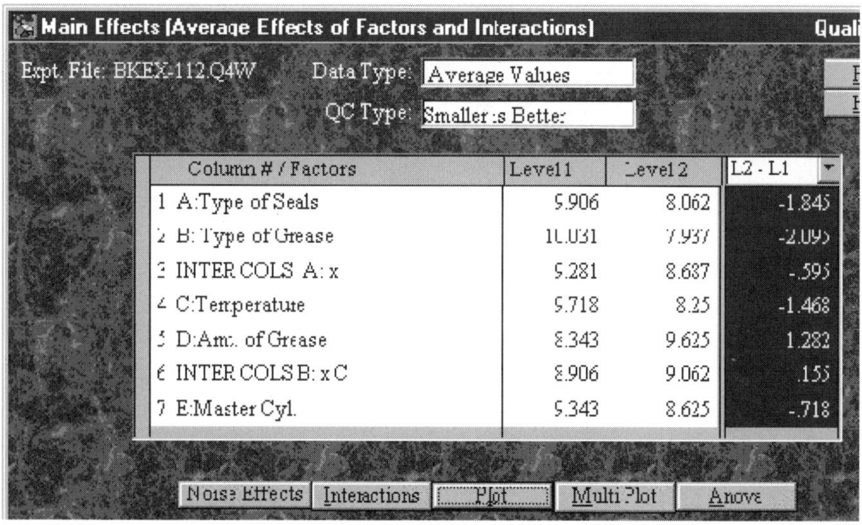

Column # / Factors	Level 1	Level 2	L2 - L1
1 A:Type of Seals	5.906	8.062	-1.843
2 B: Type of Grease	10.031	7.937	-2.093
3 INTER COLS A: x	5.281	8.687	-.593
4 C:Temperature	5.718	8.25	-1.468
5 D:Amt. of Grease	8.343	9.625	1.282
6 INTER COLS B: x C	8.906	9.062	.153
7 E:Master Cyl.	9.343	8.625	-.718

Figure 11.24 Average effects of control factors.

ANOVA Table — Qualitek-4

Expt. File: BKEX-112.Q4W Data Type: Average Value
QC Type: Smaller is Better

Col # / Factor	DOF (f)	Sum of Sqrs. (S)	Variance (V)	F - Ratio (F)	Pure Sum (S)	Percent P(%)
1 A:Type of Seals	1	54.39	54.39	12.96	50.153	11.486
2 B: Type of Grease	1	70.14	70.14	16.713	65.943	15.09
3 INTER CCLS A: x	(1)	(5.64)		POOLED	(CL=78.61%)	
4 C:Temperature	1	34.515	34.515	8.224	30.318	6.938
5 D:Amt. of Grease	1	26.265	26.265	6.258	22.068	5.05
6 INTER CCLS B: x C	(1)	(.39)		POOLED	(CL= *NC*)	
7 E:Master Cyl.	1	8.265	8.265	1.969	4.068	.93
Other/Error	58	243.405	4.196			60.505
Total	63	436.984				100.00%

Figure 11.25 ANOVA of experimental results.

(**b**) From the optimum condition, the robust design parameters are found to be: SBR seal, 418 disk brake grease, 150° F temperature, 2000 milligrams of grease, and design 2 master cylinder. This design condition is expected to reduce the leakage to 5.279 milliliters (see Figure 11.26) even when exposed to the range of noise conditions tested.

Optimum Conditions and Performance — Qualitek

Expt. File: BKEX-112.Q4W Data Type: Average Value
QC: Smaller is Better

Column # / Factor	Level Description	Level	Contribution
1 A:Type of Seals	SBR	2	-.922
2 B: Type of Grease	418 Disc Brk.	2	-1.047
4 C:Temperature	150 deg.F	2	-.735
5 D:Amt. of Grease	2000 mg.	1	-.641
7 E:Master Cyl.	Design 2	2	-.36

Total Contribution From All Factors... -3.706
Current Grand Average Of Performance... 8.984
Expected Result At Optimum Condition... 5.279

Figure 11.26 Optimum condition and performance expected.

Inner Array Design

Array Type: L-8

Use <ctrl> + <arrows> to move cursor.

	Factors	Level 1	Level 2
1	Collar Rotor	.020 in.	.070 in.
2	Shaft to Bearing	.015	.040
3	INTERACT 1 X 2	*INTER*	------------
4	Fringer to drive	.0150	.20
5	INTERACT 1 X 4	*INTER*	------------
6	INTERACT 2 X 4	*INTER*	------------
7	Rotor Chuchk	.025	.040

Figure 11.27 Control factor descriptions and column assignment.

Example 11.3: Robust Bearing Design Study Design an experiment to study the design parameters for a journal bearing with control and noise factors as in Figures 11.27 and 11.28.

Control factors: 4 two-level factors (A, B, C, and D; notations not shown)
Interactions: three interactions among control factors ($A \times B$, $B \times C$, and $C \times A$)

The arrays used for the design, factor descriptions, and column assignments are as shown in Figure 11.27 and 11.28. Determine:
(**a**) The number of samples needed to complete the experiments
(**b**) The description of the test condition for the third sample in the fifth trial condition

Outer Array Design

Array Type: L-4

Use <ctrl> + <arrows> to move cursor.

	Factors	Level 1	Level 2
1	Temperature	70 Deg. F	150 Deg.F
2	Pressure	200	350
3	Fuel Type	Type A	Type B

Figure 11.28 Noise factor descriptions and column assignments.

Descriptions Trial Conditions		Qual
Trial Condition 3 (Random order for running this Trial is 7)		
Factors	Level Description	Level #
Collar Rotor	.070 in.	2
Shaft to Bearing	.015	1
Fringer to drive	.0150	1
Rotor Chuchk	.040	2

Figure 11.29   Fifth trial condition.

Solution: This example experiment has been designed and saved under the file name *BKEX-113.Q4W.* You can practice designing this experiment using the *Manual design* option from the *Design* menu items of QT4, or review this file. In either case, the experiment designs should look as shown in Figures 11.27 and 11.28.

(**a**) Since there are four control factors and three dependent interactions, the inner array used for the experiment is an L-8. To combine the noise factors, you will need an L-4 outer array. The number of samples (with only one sample per cell) you will need to complete the experiments is 32 (8×4).

(**b**) Each of the 32 samples tested will have unique combinations of control and noise factors. The main interest will be the fifth trial condition. This condition is obtained from the *Conditions* menu item of the QT4 main screen. Select the *Trial condition* option and display the fifth trial condition as shown in Figure 11.29. Four samples are tested in this trial condition. Each sample is exposed to a separate noise condition (four noise conditions). The third samples in all trial conditions are tested by exposing the test samples to the third noise condition, as shown in Figure 11.30. To obtain the noise condition, select the *Noise condition* option in the *Conditions* menu item from the QT4 main screen.

Descriptions Trial Conditions		Qualit
Noise Condition 3 (Randomly selected order # 1)		
Factors	Level Description	Level #
Temperature	150 Deg.F	2
Pressure	200	1
Fuel Type	Type B	2

Figure 11.30   Third noise condition.

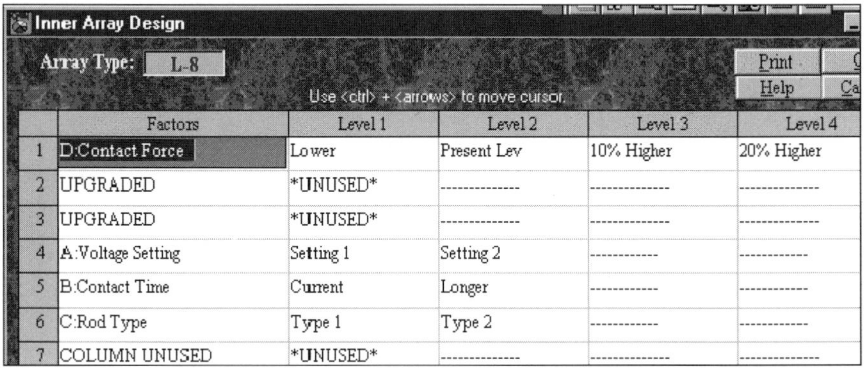

Figure 11.31 Control factors and column assignments.

Example 11.4: Spot Weld Optimization Study Process engineers responsible for the spot welding operations in a welding facility wished to carry out an experiment to make the process insensitive to the thermal expansion of the part. In the fully auto-mated production setup, the environmental temperature to which the parts are exposed has been found to make a difference in the weld strength. In the planning session, the project team identified the control factors listed in Table 11.6. The ambient tempera-ture of the part storage was considered as a noise factor, as it varied widely throughout the year with seasonal temperature variations. The effects of other uncontrollable fac-tors were considered negligible. Three samples in each trail conditions were tested and the breakaway strength of the parts was measured in terms of force in pounds (160 to 350 pounds). Discuss the experiment design strategy.

Solution: Since there is a four-level control factor (E) along with 3 two-level factors, the experiment is designed by modifying an L-8 array. Columns 1, 2, and 3 of the array can be combined to form a four-level column. The four-level factor can be assigned to this modified column and the other two-level factors can be assigned to columns 4, 5, and 6. You can use either the *Manual design* or *Automatic design* option of QT4 to create your design. Remember, if you use the *Manual design* option, you must set col-umns 2 and 3 as an upgraded status (use this button while describing the factor), and

TABLE 11.6 Factors and Levels for Example 11.4

Factor	Level			
	1	2	3	4
A: voltage setting	Setting 1	Setting 2		
B: contact time	Current	Longer		
C: rod type	Type 1	Type 2		
D: contact force	Lower	Present level	10% higher	20% higher

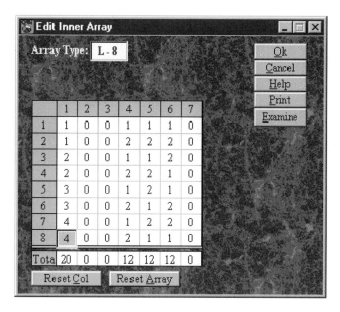

Figure 11.32 Modified L-8 array.

modify the first column to a four-level column. This experiment has been designed and results saved for your review under the file name *BKEX114.Q4W*.

Since there is only one noise factor (temperature) in this application, there is no need to utilize an array to combine the noise factors or design an outer array. The noise conditions were created by setting the part temperature at different levels for the experiment (not described by QT4, as there is no outer array). Samples in each trial condition were tested at three levels of part temperature (parts were maintained at a fixed temperature for test purposes). Six samples in each trial condition were tested by using two samples in each of the three part temperatures. In all, 48 (8 × 6) samples were needed to complete the experiments.

Your completed experiment should indicate the factor assignments shown in Figure 11.31. The 3 two-level factors could be placed in any three of columns 4 to 7. (The QT4 *Automatic design* option leaves column 7 unused.) The L-8 array modified to accommodate the four-level factors should be as shown in Figure 11.32. If for some reason, the array does not match the columns shown, use the *Reset array* button and then modify column 1 for the four-level factor. QT4 automatically turns the column to zero when its status is set to unused or upgraded, but it does not upgrade or downgrade the column automatically. You must remember to do so before proceeding to analyze the results.

The results of this experiment, shown in Figure 11.33, are for your review. You may perform analysis in the manner used for other examples by selecting the *Standard analysis* option. Note that since the outer array is absent in this experiment, QT4 cannot give you the effects of the noise factor. Should you be interested in the effects of

	Sample# 1	Sample# 2	Sample# 3	Sample# 4	Sample# 5	Sample# 6
Trial# 1	287	248	239	237	281	264
Trial# 2	315	256	343	335	341	289
Trial# 3	217	206	321	303	338	245
Trial# 4	240	233	315	216	234	267
Trial# 5	198	166	223	245	209	258
Trial# 6	201	268	226	283	290	222
Trial# 7	220	245	234	231	244	264
Trial# 8	244	265	236	217	210	233

Figure 11.33 Results of six samples tested in each trial condition.

the noise factors, you will need to hand calculate the information from the results (the average of two columns of results).

PREFERRED WAYS TO TREAT NOISE FACTORS

There are several ways to include noise factors in your design. You may want to subject your experiments to the full impact of noise factors by combining them using an orthogonal array, or you may choose to ignore them. Usually, dealing with noise factors requires a higher-order discipline and generally cost more time and money. But the degree of insensitivity (robustness) you can build into your product will depend on the extent to which you were able to identify the applicable noise factors and include them in the experiment. Based on how you wish to handle the noise factors, you may use one of the following strategies for designing your experiment.

1. *Formal outer array design.* Here your design includes orthogonal array as the outer array. The noise conditions and the number of unique (called *cell*) test conditions are defined by the outer array, and it is possible for you to control the noise level under laboratory test conditions. Such an experiment may cost more and take longer to complete, but it can be expected to produce the highest return on investment.

2. *Extreme noise conditions.* In this setup you will not have all possible noise conditions. To keep the number of samples to a minimum, you will identify two

or more extreme noise conditions, then run one or more samples under these conditions.

3. *Random noise conditions.* You may prefer this strategy when noise factors are identified but are difficult to control even in the laboratory conditions. Here you can decide the number of samples arbitrarily and let the noise factors vary at random. The random condition in this instance would mean that no effort is made either to control the noise factor while running the experiment or to complete all the experiments under a particular noise condition.

4. *Simply repeat.* In this design you prefer to run multiple samples in each trial condition to capture normal variability in results. You would prefer this when the scope of the project does not allow you to identify or control noise factors. The number of samples for such experiments generally depends on the cost of repeating samples and the variability that is expected. Between three and nine samples are generally recommended.

5. *Poor person's experiment.* Here you run only one sample in each trial condition. This strategy lets you complete the experiment with the lowest number of samples. However, when there is only one data point from each trial condition, it is hard to tell which side of the population distribution it is from. Such experiments serve well as practice runs or to screen (sort out) factors and determine that levels are all right. But not when reduction of variability is sought from the new design.

SUMMARY

In this step you have learned how to identify and incorporate uncontrollable (noise) factors in the experiments. You have learned that the causes of variation in performance are the influences of the noise factors and that an economical way to make the product and process robust (insensitive) is by adjusting the controllable factors such that the influence of the uncontrollable (noise) factors are minimized. You have also learned how to lay out the conditions of noise factors by using an outer array, which minimizes the number of sample requirements.

REVIEW QUESTIONS

Q. What is a noise factor?

A. It is a factor (variable or parameter) that has a definite influence on the performance (result), but it is impossible, too expensive, or difficult to control. Examples: *tool wear* and *skill levels of operators* are two noise factors for a machining process.

Q. How do you determine noise factors?

A. Noise factors are identified along with the control factors during an experiment planning session. Separate the noise factor from the control factor by considering the condition of the factor in actual applications. If this factor cannot be controlled, specified, or otherwise maintained at a desirable level, it is a noise factor.

Q. Are environmental factors such as temperature, humidity, and air pollution considered noise factors for most applications?

A. No such general guidelines apply. A factor is only a noise factor when it is determined to be uncontrollable for the project or application. For a manufacturer of computer chips, room temperature and humidity are not necessarily noise factors. For most they are control factors, and consequently, such manufacturing processes require that the facility be controlled for air, humidity, and dust.

Q. Do we need an outer array whenever we have noise factors?

A. The purpose of an outer array in an experiment design is to create the noise factor conditions to which the trial conditions are exposed while completing the experiments for robustness.

Q. What is an outer array?

A. Any structure or format that combines noise factor levels to create different combinations is called an outer array. Often, standard orthogonal arrays are used to reduce the number of combinations of noise factor used to combine the control factors.

Q. What is the purpose of running test samples exposed to noise conditions?

A. The results from tests exposed to noise conditions captures the variability expected. The optimum condition determined from analysis of such results is expected to be even better under similar noise conditions in real-life applications.

Q. Is it possible to determine the optimum design for reduced variability by analyzing averages of trial results?

A. An average does not utilize the information about variation captured in the test results. A better yardstick is to compare the variances (σ^2) of the trial results rather than the averages. Taguchi strongly recommends use of the signal-to-noise (S/N) ratio, which is a logarithmic transformation of the ratio between the average and the variance (see Step 12).

EXERCISES

11.1 Design experiments to suit the following experimental study objectives. Determine the appropriate inner and outer arrays.

(a) 4 two-level factors and 3 two-level noise factors

(b) Five two-level factors and 3 three-level noise factors

(c) 10 two-level factors and 5 two-level noise factors

11.2 Answer the following questions as they relate to projects in which you are involved.

(a) What are noise factors? Give an example of noise factors related to a project with which you are familiar. What does robust design mean as applied to your project?

(b) Is analysis using trial result averages satisfactory? Discuss how the analysis can be improved.

11.3 What are noise factors? Select all appropriate answers.

(a) Factors that influence performance but cannot be controlled in field applications or production

(b) Factors that are difficult to control

(c) Control factors that are not included in the experiment

(d) Factors that have a strong influence on the outcome

(e) Environmental factors such as temperature and humidity

11.4 A group of process engineers involved in a soldering process optimization study selected the following factors for investigation:

3 four-level control factors

1 two-level control factor

3 two-level noise factors

For evaluation of performance, each soldered sample was to be evaluated by four separate criteria (solder overflow, flexibility, resistance, and shape). Determine the following for the experiment:

(a) The orthogonal array for the inner array

(b) The orthogonal array for the outer array

(c) The total number of samples required

(d) The number of trial conditions

(e) The number of repetitions

(f) The total number of observations/evaluations

11.5 In experiments with noise factors:

(a) What is the smallest outer array?

(b) Which orthogonal array will you use to formally include 5 two-level noise factors?

(c) For an experiment designed using an L-16 inner array and an L-9 outer array, how many samples will you need to complete the experiment?

(d) An experiment was designed to study 7 two-level factors and 3 two-level noise factors as shown in Table 11.7. Using descriptions of the control and noise factors,

(1) Describe the conditions (levels) of the noise for the second sample in trial 1.

(2) Calculate the average effect of tool holder type A (Table 11.8).

TABLE 11.7   Outer Array Design (L-8) with Three Noise Factors

								Outer Array			
					Noise	X	1	2	2	1	
					Factor	Y	1	2	1	2	
						Z	1	1	2	2	
			Control Factor (column)					Results			
	A	B	C	D	E	F	G				
Trial	(1)	(2)	(3)	(4)	(5)	(6)	(7)	1	2	3	4
1	1	1	1	1	1	1	1	4	5	4	3
2	1	1	1	2	2	2	2	6	6	7	4
3	1	2	2	1	1	2	2	8	5	6	7
4	1	2	2	2	2	1	1	6	5	5	4
5	2	1	2	1	2	1	2	3	4	4	5
6	2	1	2	2	1	2	1	5	4	5	4
7	2	2	1	1	2	2	1	4	3	3	5
8	2	2	1	2	1	1	2	8	6	7	6
				Inner Array							

11.6   Process specialists for a blow molding process ran an experiment incorporating the noise factors to reduce the process variations. The control and noise factor descriptions, and the experiment design using an L-8 inner array and an L-4 outer array, are shown in Figures 11.34 and 11.35.

Experiments were carried out and the shoulder strengths of the bottle samples were measured. The results of 32 (8×4) tests are shown in Figure 11.36. Determine the robust design condition (optimum condition) and the performance expected.

TABLE 11.8   Factors and Levels for Exercise 11.5

	Level	
Noise Factor	1	2
X: operator	Average	Above average
Y: coolant	Oil-base	Water-base
Z: toolholder	Type A	Type B

	Factors	Level 1	Level 2	Level 3
1	E:Blow Pressure	10% Lower	Specified	10% Higher
2	UPGRADED	*UNUSED*	--------------	--------------
3	UPGRADED	*UNUSED*	--------------	--------------
4	A:Melt temp.	Current	Higher	------------
5	B:Hydraulic Temp.	Ambient	Cooled	------------
6	C:Cycle Time	Rated	Short	------------
7	COLUMN UNUSED	*UNUSED*	--------------	--------------

Figure 11.34 Control factors and column assignments.

	Factors	Level 1	Level 2
1	X: Water Temp.	Summer	Winter
2	Y:Humidity	45%	95%
3	COLUMN UNUSED	*UNUSED*	------------

Figure 11.35 Noise factors and column assignments.

Experimental Results

Expt. File: BKEX-116.Q4W

	Sample# 1	Sample# 2	Sample# 3	Sample# 4
Trial# 1	16	14	17	12
Trial# 2	26	28	19	24
Trial# 3	16	15	12	17
Trial# 4	16	19	23	20
Trial# 5	19	25	18	23
Trial# 6	21	26	22	26
Trial# 7	23	19	16	24
Trial# 8	20	20	28	27

Figure 11.36 Results of experiments exposed to noise conditions.

EXERCISE ANSWERS

Use QT4 capability to solve exercises when applicable.

11.1 (a) L-8 inner array and L-4 outer array; (b) L-8 inner array and L-9 outer array; (c) L-12 inner array and L-8 outer array.

11.2 Answers will vary.

11.3 Answers (a) and (b) are correct.

11.4 (a) L16; (b) L-4; (c) 64 samples; (d) 16; (e) 4; (f) 64 × 4.

11.5 (a) L-4; (b) L-8; (c) 144
(d) See example file *BKEX-115.Q4W*. (1) Toolholder, Type A; coolant, oil; and operator, average; (2) average effect of toolholder Type A = 5.125 (experiment *BKEX-115.Q4W*; Figure 11.37).

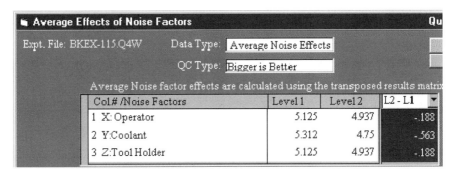

Figure 11.37   Noise factor effects.

11.6 Robust design condition: 10% higher blow pressure, higher melt temperature, cooled hydraulic temperature, and short cycle time. Performance: 26.529. Use experiment file *BKEX-116.Q4W*.

Analysis Using Signal-to-Noise Ratios

What You Will Learn in This Step

The traditional method of calculating the average effects of factors and thereby determining the desirable factor levels (optimum condition) is to look at the simple averages of the results (performance). Although average calculation is simpler, it does not capture the variability of data within the group. A better way to compare the population behavior is to use the mean-squared deviation (MSD) of the results. For convenience of linearity and to accommodate wide-ranging data, a log transformation of MSD (called *S/N ratio*) is recommended for analysis of experimental results. In this step you learn how the MSD is calculated for various quality characteristics and how analysis using *S/N* ratios differs from standard practice.

Thought for the Day

Judge a man by his questions rather than his answers.

—Voltaire

The traditional method of handling the results of multiple samples per trial condition has been to use the average (mean) of the trial results to calculate the factor effects. So far you have followed this in your analyses when you performed by selecting the *Manual analysis* option in QT4. But when averages of the trial results are used to determine the optimum design condition, it can only predict the mean of the results from the new design, not the variation around the mean. Reduction in variation, however, is the purpose of robust design (Step 11) and is the reason why you would test multiple samples in the trial condition. For us to utilize the variation information captured in the multiple sample test results, we must have a means of analyzing results that recognizes both the mean and the variations.

Although the use of average results is a common practice for comparing populations of results, it is quite inadequate when consistency is desired over mean. Consider a situation where you are comparing voltages produced from two separate processes and you are to decide the better of the two.

Process 1: 7, 9, and 11 volts (three data points); average = 9 volts

Process 2: 8, 9.5, 9, 8.5, and 10 (five data points); average = 9 volts

If we go strictly by the average values, the two processes would look alike. It is only when we look at the original data and compare the spread (range) or standard deviations that we can get information about their variations. Obviously, from the variation point of view, process 2 is desirable.

It is clear from the example above that the average alone is never an adequate measure when we want to compare two populations of data. If variation is of concern, the use of average alone will be misleading. What should we use to compare performances when variation is of major concern? There are several ways to measure the variation within a set of data. No matter the form, such a new measure must include average and standard deviation. The mean-squared deviation of the results seems to satisfy these characteristics.

MEAN-SQUARED DEVIATION

Taguchi has adopted *consistency of performance* as the generalized definition of quality. To measure the performance of a population's test samples in terms of its consistency, a quantity called *mean-squared deviation* (MSD) has been defined. Since consistent performance is obtained when performances are nearer the target more often, the MSD is calculated by adding squares (to avoid the effect of the sign) of the deviations of all the sample results from the target (when present).

If $y_1, y_2, y_3, \ldots, y_n$ are n data points (results), its MSD can be calculated as (see Figure 12.1)

$$\text{MSD} = \frac{\Sigma(Y_i - Y_0)^2}{n}$$

$$= \frac{(Y_1 - Y_0)^2 + (Y_2 - Y_0)^2 + (Y_3 - Y_0)^2 + \cdots}{n}$$

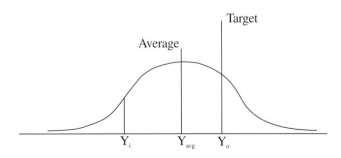

Figure 12.1   MSD formulation for distribution of any type.

Further, it can be shown that

$$MSD = \sigma^2 + (Y_{avg} - Y_0)^2$$

where Y_0 is the target value and σ is the standard deviation (population, using n as the divisor).

The definition of MSD for nominal quality characteristic above can be minimized either by lowering standard deviation, reducing the distance of the average to the target, or by doing both. In other words, MSD is reduced by an equal amount for equal reduction in σ^2 or $(Y_{avg} - Y_0)^2$. This single measure of MSD takes account of changes in the mean as well as variation (σ^2 is called the *variance*).

Definition of MSD for the Three Quality Characteristics

The concept of target does not apply in cases where the original evaluation of the results hold smaller or bigger quality characteristics (QC). In these cases MSD can be viewed as deviations with respect to the origin (target at zero). Since, like deviations, MSD is always desired to be a smaller quantity in all three quality characteristics, in case of bigger QC, the inverse of the deviation squares are used as shown below. In this way, regardless of the quality characteristics of the original results, the desirability of MSD is always retained as *smaller is better*.

$$Nominal: \quad MSD = \frac{(Y_1 - Y_0)^2 + (Y_2 - Y_0)^2 + (Y_3 - Y_0)^2 + \cdots}{n}$$

$$Smaller: MSD = \frac{Y_1^2 + Y_2^2 + Y_3^2 + \cdots}{n}$$

$$Bigger: MSD = \frac{1/Y_1^2 + 1/Y_2^2 + Y_3^2 + \cdots}{n}$$

RECOMMENDED YARDSTICK FOR ANALYSIS

Transformation of results into a *logarithmic* scale before analysis is a common practice in experimental engineering. Many scientific data are plotted in log versus log or log versus natural scale to make a wide-ranging data fit into a graph and to make the plot appear linear. The linear plot is, of course, highly desired, for it offers the ability to form conclusions by extrapolation and interpolation of results. The effect of plotting results on log paper is the same as transforming the results by taking the logarithm first, then plotting it in a natural scale. This is commonly done with sound pressure [measured in pounds per square inch, (lb/in^2)] expressed in decibels (dB), which is nothing but a constant multiplier times $\log_{10}$ of the sound pressure measured in lb/in^2 (log to the base 10 is preferred over the natural log).

For all the rationale above and much more, Taguchi strongly recommends (see the note below) use of *signal-to-noise (S/N) ratio*, which is expressed as a log transformation of the MSD, as the yardstick for analysis of experimental results. A simple form of *S/N* is defined by multiplying the log of MSD by -10:

$$S/N = -10 \log_{10} \text{MSD}$$

The multiplier 10, as used above, is a scale factor. As such, its magnitude (10, 20, 100, etc.) is arbitrary and has no effect on the conclusions derived from the result. The negative sign is applied purposefully, however, to assure that *S/N* increases for decreasing MSD. So for all QCs of the original results, lower MSD and higher *S/N* values will be desirable.

NOTE: In his latest publications, Taguchi urges viewing engineering systems as a means of energy transformer whose response characteristics change with input signal. Under such a definition, *S/N*, which we always want to maximize, can be expressed as a ratio of response due to signal factor to that due to the noise factor. For analysis of results, he recommends a two-step optimization process for adjusting factor levels for robust design. This methodology is beyond the scope of this book.

For the *nominal is best* quality characteristic, there are a few alternative expressions available for *S/N* ratio. A popular one (available in QT4) is

$$\frac{S}{N} = +10 \log_{10} \left(\frac{S_m - V_e}{nV_e} \right)$$

which is approximately same as

$$\frac{S}{N} = +10 \log_{10} \left(\frac{Y_{avg}^2}{\sigma^2} \right)$$

where $V_e = \sigma^2$ (σ for sample calculated using $n - 1$) and $S_m = n \, (Y_{avg})^2$.

Benefits and Complexities of Analyses Using *S/N* Ratios

- It uses a single measure (MSD) which incorporates the effects of changes in mean as well as the variation (standard deviation) in equal priority.
- The results behave more linearly when expressed in terms of *S/N* ratios. The linear behavior of results is an assumption necessary to express performance (Y_{opt}) in the optimum condition. Therefore, when the results are analyzed using *S/N* ratios, the prediction of optimum performance is more likely to come true.

- When results are converted into *S/N* ratios (log scale), a much wider-ranging data (0.01 to 10,000, on multicycle log paper) can easily be included in the calculation.
- For comparison purposes, a larger value is always desired regardless of the QC of the original results.
- The *S/N* ratio can be used as a yardstick for comparison of any population performance. The use in analysis of results of designed experiments is just one application.
- The only disadvantage of *S/N* analysis is that some numbers (which would be positive or negative) may look strange. Some results may also need to be transformed back into units of the original results for interpretation purposes.

Overall, the benefits of *S/N* analysis outweigh the minor complexities it creates. As a general guideline, use of *S/N* analysis is recommended for all experiments with multiple samples per trial condition.

Example 12.1: Practice Computation of MSD and *S/N* Ratios   The output voltages from two different designs of automobile generators were:

Design 1: 11, 12.5, 10.5, 14.3, 12.25 and 11.6; sample size $n = 6$
Design 2: 12.2, 11.5, 10.2, 13.8, and 14.6; sample size $n = 5$

(a) If the target value is 12 volts, determine the design that is likely to produce more consistent voltage output (QC = *nominal is best*).
(b) Which design should you prefer for higher consistent voltages (QC = *bigger is better*)?

Solution: (a) *Design 1*: target = 12, $n = 6$

$$\text{MSD} =$$

$$\frac{(11 - 12)^2 + (12.5 - 12)^2 + (10.5 - 12)^2 + (14.3 - 12)^2 + (12.25 - 12)^2 + (11.6 - 12)^2}{6}$$

$$= 1.502$$

$$S/N = -10 \log_{10} \text{MSD}$$

$$= -10 \times 0.1767$$

$$= -1.767$$

Design 2: target $= 12$, $n = 5$

$$\text{MSD} = \frac{(12.2 - 12)^2 + (11.5 - 12)^2 + (10.2 - 12)^2 + (13.8 - 12)^2 + (14.6 - 12)^2}{5}$$

$$= 2.706$$

$$S/N = -10 \log_{10} 2.706$$

$$= -10 \times 0.4323$$

$$= -4.323$$

Since -1.767 (*S/N* for design 1) is greater than -4.323 (*S/N* for design 2), design 1 is expected to perform more consistently close to the target. (*Note*: The *S/N* is negative or positive depending on the QC of the data and its magnitude. Regardless of the sign of *S/N*, a bigger value is always preferred.)

You can use calculation capabilities of QT4 to compute the MSD and *S/N* for any arbitrary set of data. To calculate MSD and *S/N* ratio for design 1 data, select *Population statistics* from the *Loss/savings* menu item from the main screen. Click on the *User data* field and enter one datum at a time as shown in Figure 12.2. Be sure to check the nominal quality characteristic and enter 12 for the target value. Click the *Calculate* button when ready to compute MSD and *S/N* ratio.

(*b*) *Design 1*: QC = B, $n = 6$

$$\text{MSD} = \frac{(1/11)^2 + (1/12.5)^2 + (1/10.5)^2 + (1/14.3)^2 + (1/12.25)^2 + (1/11.6)^2}{6}$$

$$= 0.007$$

$$S/N = -10 \log_{10} 0.007$$

$$= -10 \times (-2.154)$$

$$= +21.54$$

Design 2: QC = B, $n = 5$

$$\text{MSD} = \frac{(1/12.2)^2 + (1/11.5)^2 + (1/10.2)^2 + (1/13.8)^2 + (1/14.6)^2}{5}$$

$$= 0.006$$

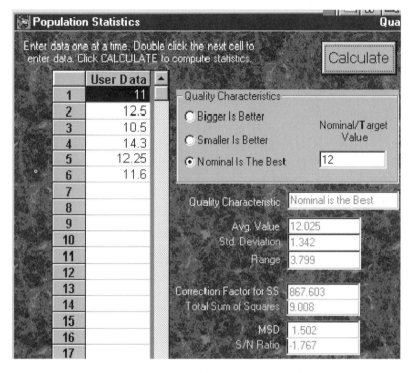

Figure 12.2   MSD and S/N calculation screen in QT4.

$$S/N = -10 \log_{10} 0.006$$

$$= -10 \times (-2.2218)$$

$$= +22.2$$

Based on higher consistent values, design 2 is desirable (as $22.2 > 21.54$).

You can use QT4 to perform all of the calculations above. Be aware that QT4 results may vary slightly due to rounding of numbers.

Example 12.2: Cam Lifter Design Study   An experiment conducted with 3 two-level factors [A: spring rate (current/new), B: cam profile (type 1/type 2), and C: weight of push rod (lighter/heavier)] and three samples per trial condition yielded results shown in Table 12.1. For the least amount of wear (QC = S), perform analysis using S/N ratios.

(a) Determine the desirable design condition.

(b) Express the performance in terms of the original units of measurement.

TABLE 12.1   Results of Cam Lifter Study

| Experiment | Factor | | | Result | | | | |
	A	B	C	R1	R2	R3	S/N	Average
1	1	1	1	2	3	4	−9.86	3
2	1	2	2	4	5	3	−12.2	4
3	2	1	2	4	5	6	−14.1	5
4	2	2	1	3	5	7	−14.4	5

(c) Compare and comment on the conclusions drawn with those obtained from standard analysis.

Solution: This example experiment has been designed and saved under the file name *BKEX-122.Q4W*. To perform analysis, you will need to load (select the *Open* option from the *FILE* menu item) this file and follow the steps prescribed. The calculations below are shown for your understanding only. Sample calculation of *S/N*:

First row: results 2, 3, and 4; QC = smaller

$$\text{MSD} = \frac{2^2 + 3^2 + 4^2}{3} = 9.666$$

$$S/N = -10 \log_{10} \text{MSD} = -9.86$$

Second row: results 4, 5, and 3; QC = smaller

$$\text{MSD} = \frac{4^2 + 5^2 + 3^2}{3} = 16.666$$

$$S/N = -10 \log_{10} \text{MSD} = -12.218$$

and so on.

The *S/N* numbers are shown in the column immediately to the right of the result columns (Table 12.1). The averages of the trial results are also shown in the rightmost column. Depending on the type of analysis (standard or *S/N*) you wish to perform, you will use one or the other column (labeled as *S/N* or average). When performing analysis using QT4, it will show the column appropriate for type of analysis only.

Unrelated to analysis of the total results of the experiment, if we were to compare results of trials 3 and 4 to determine which one is better, we can now easily do that by comparing the *S/N* ratios. Based on averages, condition results of trials 3 and 4 are

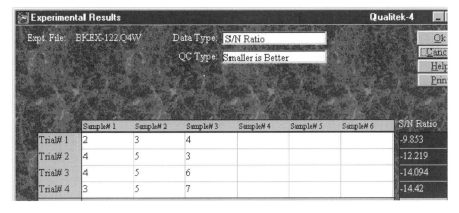

Figure 12.3   Trial results and *S/N* ratios.

equal. But comparing the *S/N* ratios, condition 3 is better as −14.1 > −14.4 (−14.1 is greater than −14.4).

To analyze the results of the experiments using *S/N* ratios, follow these steps:

1. Be in the main screen (file *BKEX-122.Q4W*) and select the *S/N analysis* option from the *Analysis* menu item.

2. Check the *Smaller is better* quality characteristic and click *OK* to proceed.

3. QT4 shows you the results and the corresponding *S/N* for the trail results as in Figure 12.3. These numbers should be same as the hand-calculated numbers shown earlier (except rounding error). Click *OK* to proceed.

4. Based on the *S/N* ratios, the factor average effects are calculated in the same manner as trial averages are used in case of standard analysis. The average effects indicated in the screen shown in Figure 12.4 are all in *S/N* ratios. This is because all calculations under this analysis are performed using *S/N* ratios of the results originally collected. You may now explore all options available in this

Main Effects (Average Effects of Factors and Interactions)   Qua

Expt. File: BKEX-122.Q4W   Data Type: S/N Ratio

QC Type: Smaller is Better

Column # / Factors	Level 1	Level 2	L2 - L1
1 A:Spring Rate	-11.036	-14.257	-3.222
2 B:Cam Profile	-11.974	-13.32	-1.347
3 C:Wt.of Push Rod	-12.137	-13.157	-1.02

Figure 12.4   Factor average effects in *S/N*.

screen as usual, keeping in mind that the numbers are all in *S/N*. One key point of difference is that when examining the factor effects, it is always the higher value that will indicate which level of the factor is preferable, no matter what the quality characteristic of the original result. For example, consider the desirable level for the factor *spring rate* (*A*). Since -11.036 is greater than -14.257, level 1 for this factor is desirable. Such observations will be reflected in the determination of the optimum condition. Click *OK* to proceed.

5. In the *ANOVA* screen shown in Figure 12.5, notice that the error degrees of freedom (DOFs) is zero. This is because for the sake of ANOVA calculation, the number of results now refers to the number of *S/N* values, which is 4 in this case. The error DOF is calculated as total DOF (number of results -1) minus the factor DOF ($0 = 3 - 3$). Interesting to note also is that even though all calculations are done using *S/N* ratios, the percent numbers in the last column are all free of units, as it is always a ratio. As before, the percent numbers indicate the relative influence of the factors to the variation of results (*S/N* in this case). Click *OK* when done reviewing ANOVA.

6. The optimum condition and the results expected at the optimum condition in terms of *S/N* ratio are as shown in Figure 12.6. Recall that the factor levels are identified by comparing the factor average effects and selecting the larger *S/N* values shown earlier.

(a) Optimum condition (see Figure 12.6): all factors at level 1, as shown.

(b) The performance at optimum condition is -9.856 (last line in Figure 12.6).

When optimum performance is expressed in terms of the *S/N* ratio, as in answer (b), it bears any meaning to a person who is not familiar with the transformation of the results to *S/N* ratios. So you must interpret its meaning in simpler terms. First, what does the expected *S/N* = -9.856 indicate? This means that if a large number of samples were tested at the optimum condition, the *S/N* ratio of all results is expected to be close to -9.856. But how close is close enough? You can get the answer to this question by

ANOVA Table

Expt. File: BKEX-122.Q4W   Data Type: S/N Ratio   QC Type: Smaller is Better

Col # / Factor	DOF (f)	Sum of Sqrs. (S)	Variance (V)	F - Ratio (F)	Pure Sum (S')	Percent P(%)
1 A:Spring Rate	1	10.374	10.374	-----	10.374	78.44
2 B:Cam Profile	1	1.811	1.811	-----	1.811	13.693
3 C:Wt. of Push Rod	1	1.04	1.04	-----	1.04	7.864
Other/Error	0					
Total:	3	13.226				100.00%

Figure 12.5 ANOVA with *S/N* ratios of results.

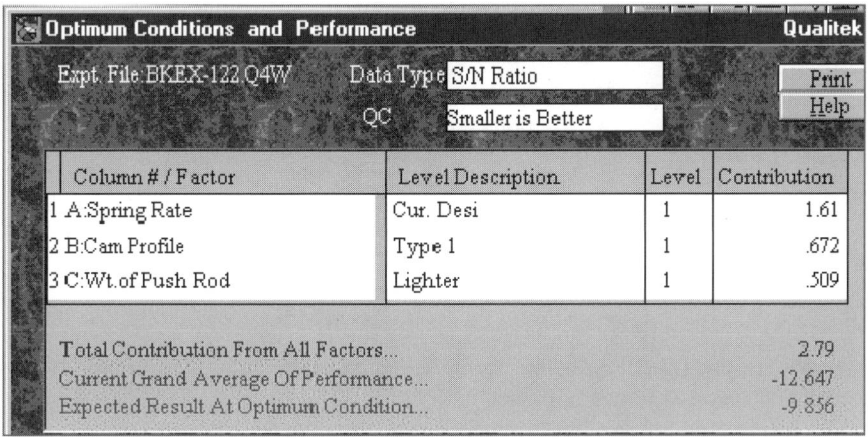
Optimum Conditions and Performance
Qualitek
Expt. File:BKEX-122.Q4W
Data Type S/N Ratio
QC Smaller is Better
Print
Help
Column # / Factor
Level Description
Level
Contribution
1 A:Spring Rate
Cur. Desi
1
1.61
2 B:Cam Profile
Type 1
1
.672
3 C:Wt.of Push Rod
Lighter
1
.509
Total Contribution From All Factors...
2.79
Current Grand Average Of Performance...
-12.647
Expected Result At Optimum Condition...
-9.856

Figure 12.6 Optimum condition and performance in terms of *S/N* ratios.

calculating the confidence interval (C.I.) on the expected performance when the error DOF is nonzero. Calculation of C.I. is not possible in this case as the error DOF.

Second, what does $S/N = -9.856$ mean in terms of the original units of results? In other words, what should the results be when a few samples are tested in the optimum condition? The answer to this question is a valuable piece of information to your project team members. Let us find out if it is possible to estimate the expected result in terms of the units originally used to measure the results. The issue here is that *S/N* is known and we want to learn about the result (Y) that will make the *S/N*. The procedure is to back-transform *S/N* to find MSD, and from MSD to estimate the expected result. Since $S/N = -10 \log_{10} \text{MSD}$,

$$\text{MSD} = 10^{-(S/N)/10} \qquad \text{for known } S/N$$

$$= 10^{-(-9.856/10)} \qquad \text{for } S/N = -9.856$$

$$= 9.6738$$

Now, for QC = smaller,

$$\text{MSD} = \frac{(Y_1)^2 + (Y_2)^2 + (Y_3)^2 + \cdots + (Y_n)^2}{n}$$

With known MSD (9.6738), the equation above involves many unknowns (Y's), and thus cannot be solved for exact values of Y. An approximate solution, however, can be obtained by assuming that all results (Y) are expected to be the same as their representative value, Y_{expected}.

Using all Y's as $Y_{expected}$, the MSD equation allows calculation of an approximate value of the results expected in the units originally measured.

$$\text{MSD} = \frac{n(Y_{expected})^2}{n}$$

or

$$Y_{expected} = (\text{MSD})^{1/2} \quad (\text{approximately})$$

$$= 3.11 \quad \text{for known MSD} = 9.6738$$

You can use the *Transform* button from the optimum screen of QT4 to perform the calculation above as shown in Figure 12.7.

(c) To answer this part of the problem, you will need to perform analysis using both the standard and *S/N* analysis options of QT4 from the *Analysis* menu item. Since you already know the steps for these analyses, only the important option selection and mouse clicks are mentioned here. To compare the analyses in two methods, the *ANOVA* and *Optimum* screens obtained by both methods will be examined.

When you perform the standard analysis, your *ANOVA* screen should look like as shown in Figure 12.8. Factor *A*: spring rate is identified as the dominant factor in both analyses (compare Figure 12.8 with Figure 12.5). Standard analysis compares the influence of the other two factors (*B* and *C*) and finds their influence to be much smaller than that of the error factor and diminishes their effect (Figure 12.8). The *S/N* analysis cannot do that, as the error DOF is zero (Figure 12.5). Notice that the error DOF in standard analysis is nonzero ($8 = 12 - 3 - 1$), as it should be.

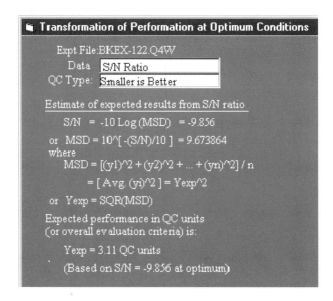

Figure 12.7 Conversion of *S/N* back to expected results.

ANOVA Table — Qualitek-4

Expt. File: BKEX-122.Q4W   Data Type: Average Value   QC Type: Smaller is Better

Col # / Factor	DOF (f)	Sum of Sqrs. (S)	Variance (V)	F - Ratio (F)	Pure Sum (S')	Percent P(%)
1 A:Spring Rate	1	6.75	6.75	3.857	5	22.471
2 B:Cam Profile	1	.75	.75	.428	0	0
3 C:Wt.of Push Rod	1	.75	.75	.428	0	0
Other/Error	8	14	1.75			77.529
Total:	11	22.25				100.00%

Figure 12.8 ANOVA with original results.

Both analyses identified the same factor levels for the optimum condition in this case (1, 1, 1). This is not always expected to be the case, particularly when a large number of factors are involved (Figures 12.9 and 12.6). The standard analysis predicts performance at the optimum condition as 3 (Figure 12.9). This number should be compared with the transformed *S/N* (Figure 12.6) expected from the *S/N* analysis, which is 3.11 (Figure 12.7). Here again, the close correlation of the expected results is coincidental and should not always be expected.

When your experimental results consist of multiple samples per trail condition, QT4 allows you options to perform analyses of the same experimental results both ways. It is recommended that you analyze your results first by *S/N* option. Then, to get a better understanding and help you in interpretation of *S/N* data, analyze the same by selecting the *standard analysis* option. Generally also, one analysis should support the other, and the conclusions drawn should be consistent. But should you encounter

Optimum Conditions and Performance — Qualitek

Expt. File: BKEX-122.Q4W   Data Type: Average Value   QC: Smaller is Better

Column # / Factor	Level Description	Level	Contribution
1 A:Spring Rate	Cur. Desi	1	-.75
2 B:Cam Profile	Type 1	1	-.25
3 C:Wt.of Push Rod	Lighter	1	-.25

Total Contribution From All Factors...	-1.25
Current Grand Average Of Performance...	4.25
Expected Result At Optimum Condition...	3

Figure 12.9 Optimum condition and performance by standard analysis.

	Factors	Level 1	Level 2	Level 3
1	A:Indexing	-30 DDEG.	0	+30DEG.
2	B:Overlap Area	-30%	0%	+30%
3	C:Spark Advance	20 DEG.	30 DEG.	40 DEG.
4	UNUSED/UPGRADED	*UNUSED*	------------	------------

Inner Array Design
Array Type: L-9
Use <ctrl> + <arrows> to move cursor.

Figure 12.10 Factor and level descriptions.

discrepancies, rely more on the *S/N* analysis. This is why QT4 expects *S/N* analysis as your default choice when there are multiple sample results in your experiment.

Example 12.3: Engine Idle Stability Study An experiment was carried out to study design settings for optimum idle performance of a newly developed engine. The factors, their levels, and the column assignment to an L-9 array for the design are shown in Figure 12.10. The results (measured in revolution per second) of three samples tested in each of the nine trial conditions and the *S/N* ratios of trial results are as shown in Figure 12.11. If the performance desired (target) is around 7.5, determine the range of performance obtainable from the optimum condition.

Solution: This experimental design and the results collected are saved in the file *BKEX-123.Q4W*. Open this file and perform analysis using the *S/N Analysis* option. When prompted, enter 7.5 as the target value for the *nominal is best quality* characteristic as shown in Figure 12.12. Proceed with analysis and go to the *Optimum* screen. Your screen showing the optimum condition $(A_1B_1C_1)$ and performance at the optimum condition (−0.586) should be that shown in Figure 12.13. You can now convert the optimum performance (−0.586), which is expressed in terms of *S/N* ratio, by click-

	Sample# 1	Sample# 2	Sample# 3	Sample# 4	Sample# 5	Sample# 6	S/N Ratio
Trial# 1	6.6	8.3	8.6				0.522
Trial# 2	11.3	12	8.6				-10.78
Trial# 3	15	11.3	8.6	S/N ratios with			-13.797
Trial# 4	4.3	7.6	7.3	Nominal/Target = 7.5			-5.353
Trial# 5	12	15	11.6				-14.929
Trial# 6	7.6	8.3	11.3				-7.016
Trial# 7	11.6	15	17.7				-17.711
Trial# 8	18.7	15.3	17.4				-19.767

Figure 12.11 Experimental results and the *S/N* ratios.

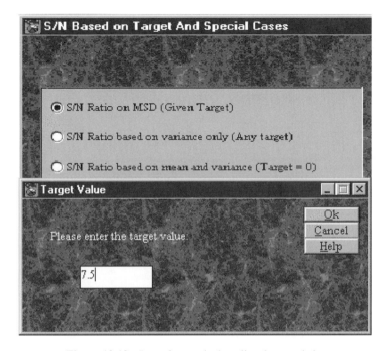

Figure 12.12   Input for nominal quality characteristic.

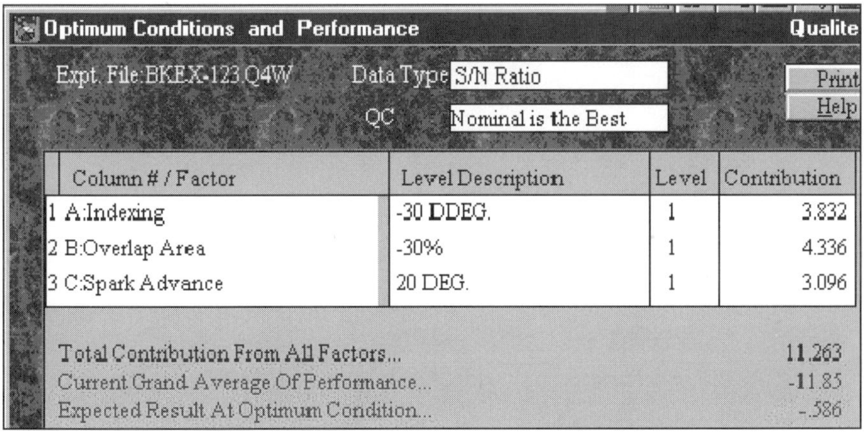

Column # / Factor	Level Description	Level	Contribution
1 A:Indexing	-30 DDEG.	1	3.832
2 B:Overlap Area	-30%	1	4.336
3 C:Spark Advance	20 DEG.	1	3.096

Total Contribution From All Factors...   11.263
Current Grand Average Of Performance...   -11.85
Expected Result At Optimum Condition...   -.586

Figure 12.13   Optimum condition and performance.

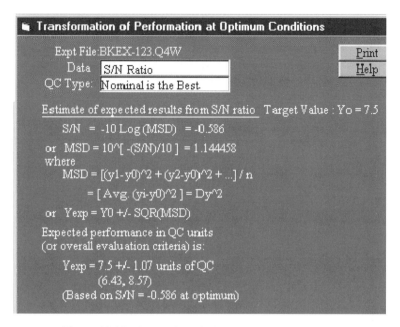

Figure 12.14   Conversion of *S/N* back to expected result.

ing on the *Transform* button. Converted results, shown in Figure 12.14, indicate that the result at the optimum condition is expected to range between 6.43 and 8.57.

Example 12.4: Blow Molding Process Optimization Study   Engineers in a plastic beverage bottle manufacturing plants identified the 6 two-level control factors and 3 two-level noise factors shown in Figures 12.15 and 12.16, respectively, for an optimization study. Among all the possible interaction pairs between the control factors, only

	Factors	Level 1	Level 2
1	A:Blow Temperatur	250 deg.F	290 deg.F
2	B:Blow Pressure	60 psi	100 psi
3	INTERACT 1 X 2	*INTER*	------------
4	C:Molder Speed	2.5 rpm	3.5 rpm
5	D:Base Wight	4.25 oz.	5.60 oz.
6	E:Preform Design	Present	Modified
7	F:Shoulder Weight	0.4 oz.	0.6 oz.

Array Type: L-8

Use <ctrl> + <arrows> to move cursor.

Figure 12.15   Control factors and their levels.

	Factors	Level 1	Level 2
	Array Type: L-4		Use <ctrl> + <arrows> to move cursor.
1	X:Machine Brand	Sidel	Sipa
2	Y:Material Source	Eastman	Nanya
3	Z:Coolant Temp.	Room Temp	Cooler

Figure 12.16 Noise factors and their levels.

that between blow temperature and blow pressure was included in the study. The plant uses several brands of machines for the same process. The raw material for the process is also supplied by different vendors and is expected to have minor difference in chemical composition. Selecting a particular brand of machine or a single source of material are not desirable choices. For these reasons, machine brand and material source are treated as noise factors along with coolant temperature.

The experiment was designed using an L-8 inner array and an L-4 outer array. It was then completed by running four samples in each of the eight trial conditions. The samples were then tested to determine the highest compressive load (measured in pounds) carried by the bottle shoulder. The sample results and the experiment configuration are shown in Figure 12.17. Perform analysis using *S/N* ratios of the results.

(a) If the interaction between blow temperature and blow pressure does exist and is found to be significant, what are the levels of the interacting factors due to interaction?

(b) What are the three most significant factors among those that have been tested?

Review Noise Factors

Review Control Factors

							Outer Array			
							3	1	2	2	1
							2	1	2	1	2
							1	1	1	2	2
	1	2	3	4	5	6	7	1	2	3	4
1	1	1	1	1	1	1	1	30	33	36	32
2	1	1	1	2	2	2	2	36	40	44	41
3	1	2	2	1	1	2	2	30	28	27	25
4	1	2	2	2	2	1	1	44	36	28	33
5	2	1	2	1	2	1	2	24	28	32	29
6	2	1	2	2	1	2	1	52	44	36	43
7	2	2	1	1	2	2	1	32	24	16	29
8	2	2	1	2	1	1	2	46	43	40	38
			Inner Array							Results	

Figure 12.17 Arrays used for the design and the experimental results.

(c) Assuming the current performance to be the same as trial 1, what gain (improvement) in *S/N* ratio (performance) is expected from the optimum condition?

(d) What performance is expected in terms of the original units (shoulder load in pounds) of the results?

Solution: This experiment has been designed and results recorded in the experiment file *BKEX-124.Q4W*. From the main screen, open this file to perform analysis. Select the *S/N analysis* option from the *Analysis* menu items and proceed with analysis by selecting the *Bigger is better* quality characteristic and be in the *Main effects* screen.

(a) From the bottom of this screen, click on the *Interactions* button and choose the *Auto* option. On selection QT4, show all 15 ($6 \times 5/2$) possible interactions in order of their severity (Figure 12.18). Notice that interaction between the factors blow temperature and blow pressure (A × B) is tenth in the list. You can double-click on the description to see the plot of interaction shown in Figure 12.19. In the interaction screen (Figure 12.19), click on the word *HERE* to see details of the combined factor effects calculated. Since the numbers are all in *S/N*, the desirable levels of the interaction factors are obtained by the highest of the four calculated values, which in this case is $X_1 Y_1 = 31.13$ (see Figure 12.19). Understand that X and Y are dummy symbols used for the plotting purposes. The factors X and Y in this case are A: blow temperature and B: blow pressure, respectively. Thus if these interaction effects were found to be significant (from ANOVA, shown next) and you decide to adjust the levels of the factors, the factor levels desirable will be A_1 and B_1. You could also find this information from the last column (1,1) in the screen listing the interactions (Figure 12.18). Recall that

Automatic Test for Presence of Interaction Qu

Number of interactions between two factors calculated = 15

#	Interacting Factor Pairs (in Order of Severity)	Columns	SI(%)	Col	Opt.
1	B:Blow Pressure x E:Preform Design	2 x 6	71.7	4	[1,2]
2	A:Blow Temperatur x D:Base Wight	1 x 5	70.35	4	[2,1]
3	A:Blow Temperatur x F:Shoulder Weight	1 x 7	53.04	6	[2,2]
4	B:Blow Pressure x F:Shoulder Weight	2 x 7	51.63	5	[1,1]
5	D:Base Wight x F:Shoulder Weight	5 x 7	48.36	2	[1,1]
6	A:Blow Temperatur x E:Preform Design	1 x 6	46.95	7	[2,1]
7	A:Blow Temperatur x C:Molder Speed	1 x 4	29.64	5	[2,2]
8	C:Molder Speed x E:Preform Design	4 x 6	28.29	2	[2,2]
9	E:Preform Design x F:Shoulder Weight	6 x 7	22.52	1	[1,2]
10	A:Blow Temperatur x B:Blow Pressure	1 x 2	15.37	3	[1,1]
11	B:Blow Pressure x D:Base Wight	2 x 5	13.95	7	[1,1]

Figure 12.18 Interactions between two factors.

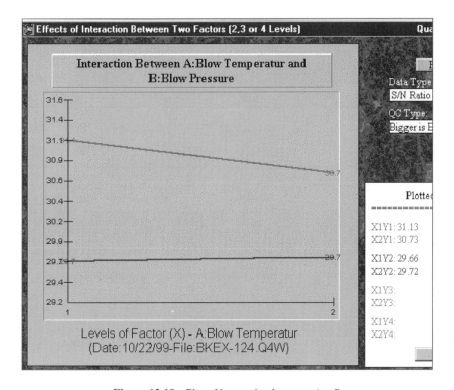

Figure 12.19 Plot of interaction between $A \times B$.

you will only change the factor levels (in optimum condition) based on the interaction information when the interaction is included in the study and subsequently found to be significant (in ANOVA).

Whether an interaction effect is significant or not is determined in ANOVA. In this example the interaction of interest, $A \times B$ (columns 1×2), is found to be insignificant and is pooled as shown in Figure 12.20. Therefore, there is no need to alter the selection of the levels of factors A and B because of the interaction between them.

(b) Also, from ANOVA (Figure 12.20), the three more significant factors, in order of their influence, are C: molder speed, D: base weight, and B: blow pressure. The other three factors, all of which happen to be insignificant and pooled, can be adjusted to levels most economical for the project.

(c) The optimum condition, and the performance at the optimum condition in terms of the S/N ratio, is shown when you proceed to the optimum screen shown in Figure 12.21. The gain refers to the increase (S/N ratio must increase for improvement) in expected performance in S/N (33.178) at the optimum condition relative to the current performance. You can find the S/N ratio (30.248) for the results in trial 1 from the result screen that follows when you proceed with analysis after selecting the quality characteristic. The improvement in performance in terms of S/N is generally expressed as

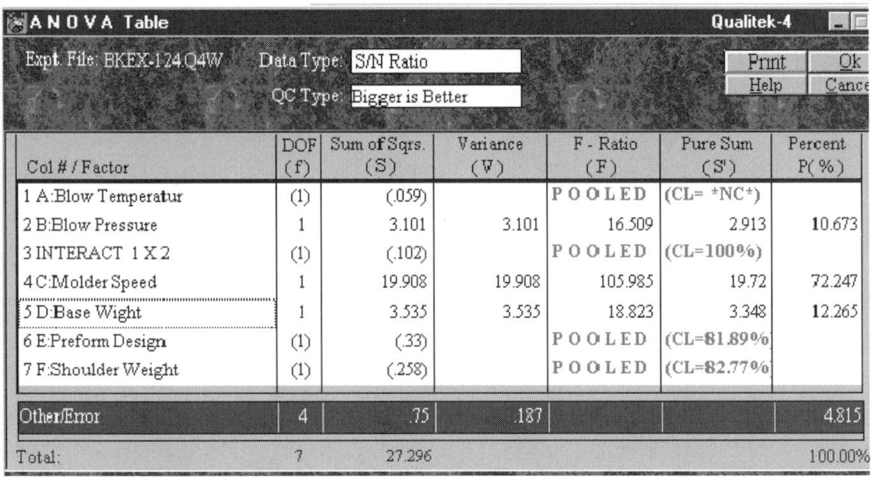

Col # / Factor	DOF (f)	Sum of Sqrs. (S)	Variance (V)	F - Ratio (F)	Pure Sum (S')	Percent P(%)
1 A:Blow Temperatur	(1)	(.059)		POOLED	(CL= *NC*)	
2 B:Blow Pressure	1	3.101	3.101	16.509	2.913	10.673
3 INTERACT 1 X 2	(1)	(.102)		POOLED	(CL=100%)	
4 C:Molder Speed	1	19.908	19.908	105.985	19.72	72.247
5 D:Base Wight	1	3.535	3.535	18.823	3.348	12.265
6 E:Preform Design	(1)	(.33)		POOLED	(CL=81.89%)	
7 F:Shoulder Weight	(1)	(.258)		POOLED	(CL=82.77%)	
Other/Error	4	.75	.187			4.815
Total:	7	27.296				100.00%

(ANOVA Table — Qualitek-4. Expt. File: BKEX-124.Q4W. Data Type: S/N Ratio. QC Type: Bigger is Better.)

Figure 12.20 ANOVA after interaction and other insignificant factors are pooled.

$$\text{gain} = 33.178 - 30.248$$

$$= 2.922$$

(d) You can also obtain the expected performance in terms of the original units of measurement by letting QT4 transform (click the *Transform* button in the *Optimum* screen) the expected *S/N* shown in Figure 12.22. From the transformed data, the expected mean performance is about 45 (45.593) pounds of shoulder loads.

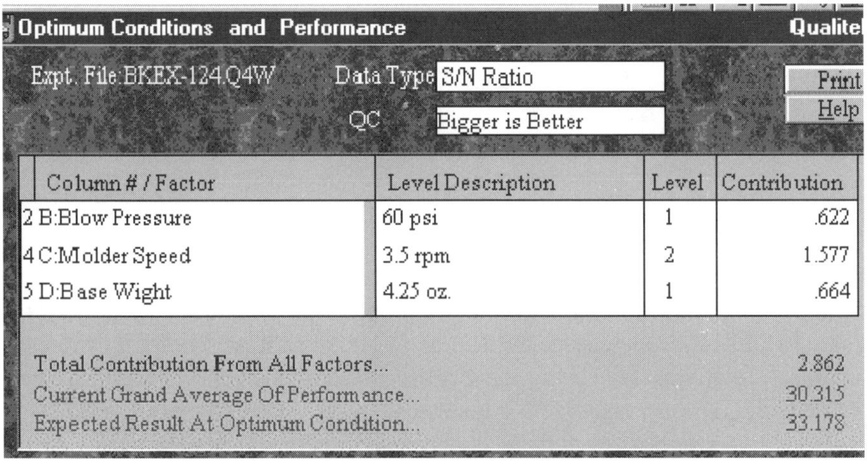

Optimum Conditions and Performance — Qualite

Expt. File: BKEX-124.Q4W Data Type: S/N Ratio QC: Bigger is Better

Column # / Factor	Level Description	Level	Contribution
2 B:Blow Pressure	60 psi	1	.622
4 C:Molder Speed	3.5 rpm	2	1.577
5 D:Base Wight	4.25 oz.	1	.664
Total Contribution From All Factors...			2.862
Current Grand Average Of Performance...			30.315
Expected Result At Optimum Condition...			33.178

Figure 12.21 Optimum condition and performance.

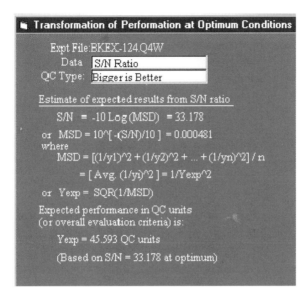

Figure 12.22 Results expected at the optimum condition.

Example 12.5: Automobile Wheel Pad Machining Process Study A supplier of automobile wheels made of aluminum ran an experiment to optimize the flatness variation in the mounting pad. The factors and interactions included in the experiment are shown in Figure 12.23. Three samples were tested in each of eight trial conditions and the results measured in terms of deviation from the ideal reference surface located at 40 units (target = 40). The sample results and the *S/N* ra-

Inner Array Design

Array Type: L-8

Use \<ctrl\> + \<arrows\> to move cursor.

	Factors	Level 1	Level 2
1	A:Chuck Speed	1000 rpm	1600 rpm
2	B:Feed Rate	0.05 in/s	0.80 in/sec
3	INTERACT AxB	*INTER*	
4	C:Rake Angle (too	60 degree	57 degrees
5	D:Tool Nose Radiu	6 mm	2 mm
6	INTERACT BxC	*INTER*	------------
7	E:# of Finish Pas	One Pass	Three passes

Figure 12.23 Factor and level descriptions.

tios for the trial conditions are shown in Figure 12.24. Perform sample calculations to describe the key analysis steps.

Solution: The experiment design and the results for this example are saved in the file *BKEX-125.Q4W*. You can review and compare the sample calculations performed below with that displayed by QT4. The *S/N* ratio for the results of trial condition 1 is calculated using the target value, $Y_o = 40$, (see Figure 12.24, row 1 for the results and *S/N* ratio):

$$\text{MSD} = \frac{(y_1 - y_o)^2 + (y_2 - y_o)^2 + \cdots}{n}$$

$$= \frac{(38 - 40)^2 + (42 - 40)^2 + (46 - 40)^2}{3}$$

$$= 14.667$$

$$S/N = -10 \log_{10} \text{MSD} = -10 \log_{10} 14.667 = -11.67$$

Similarly, the *S/N* ratios for all other trial conditions are calculated. The *S/N* ratios calculated are shown in the far-right column of Figure 12.24. Using the column of *S/N* ratios (just like the average of trial results), the factor average effects are calculated. Sample calculations are

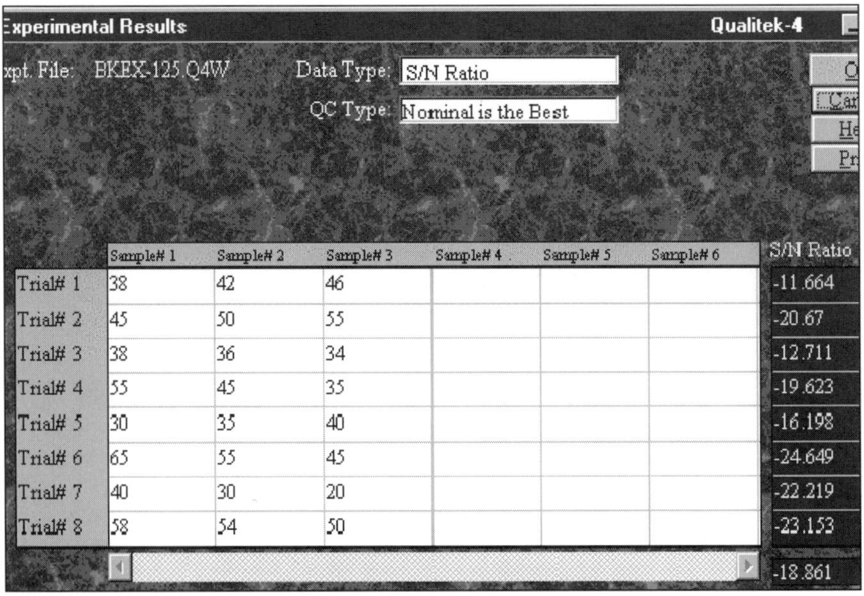

	Sample# 1	Sample# 2	Sample# 3	Sample# 4	Sample# 5	Sample# 6	S/N Ratio
Trial# 1	38	42	46				-11.664
Trial# 2	45	50	55				-20.67
Trial# 3	38	36	34				-12.711
Trial# 4	55	45	35				-19.623
Trial# 5	30	35	40				-16.198
Trial# 6	65	55	45				-24.649
Trial# 7	40	30	20				-22.219
Trial# 8	58	54	50				-23.153
							-18.861

Experimental Results — Qualitek-4
Expt. File: BKEX-125.Q4W   Data Type: S/N Ratio   QC Type: Nominal is the Best

Figure 12.24   Results and the *S/N* ratios.

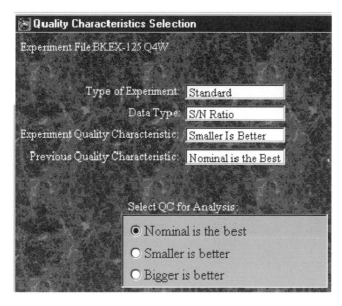

Figure 12.25 Quality characteristic selection option.

$$\overline{A}_1 = \frac{-11.67 + (-20.67) + (-12.72) + (-19.63)}{4} = -16.17$$

$$\overline{A}_2 = \frac{-16.21 + (-24.65) + (-22.22) + (-23.16)}{4} = -21.55 \quad \text{etc.}$$

To see how QT4 performs the S/N analysis with the *Nominal* quality characteristic and what the average effects are, follow these steps:

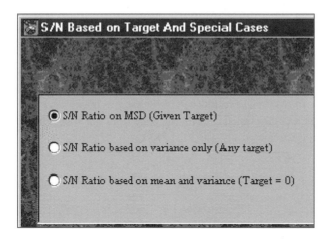

Figure 12.26 Type of nominal S/N ratio selection option.

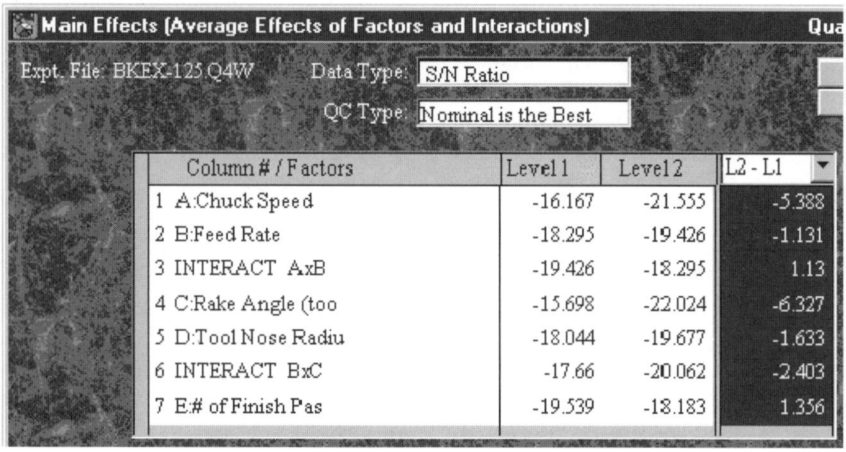

Main Effects (Average Effects of Factors and Interactions)			Qua

Expt. File: BKEX-125.Q4W Data Type: S/N Ratio
QC Type: Nominal is the Best

Column # / Factors	Level 1	Level 2	L2 - L1 ▼
1 A:Chuck Speed	-16.167	-21.555	-5.388
2 B:Feed Rate	-18.295	-19.426	-1.131
3 INTERACT AxB	-19.426	-18.295	1.13
4 C:Rake Angle (too	-15.698	-22.024	-6.327
5 D:Tool Nose Radiu	-18.044	-19.677	-1.633
6 INTERACT BxC	-17.66	-20.062	-2.403
7 E:# of Finish Pas	-19.539	-18.183	1.356

Figure 12.27   Average effects of factors.

1. Run QT4 and select the *BKEX-125.Q4W* file from the main screen.
2. Select the *S/N analysis* option from the *Analysis* menu items.
3. Check the box for *Nominal is best* and click *OK* (Figure 12.25).
4. Check the *S/N ratio on MSD (given target)* and click *OK* (Figure 12.26).
5. Enter the target value for analysis in the input field (Figure 12.27).

As you proceed with analysis, you will find the average factor effects displayed as in Figure 12.27.

CALCULATION OF ANOVA TERMS

The formulas used to calculate ANOVA terms are the same as used in standard analysis. The only difference is that all numbers are in *S/N* ratios (can be positive or negative). Also, for purposes of calculating degrees of freedom (DOFs), the number of *S/N* ratios of the original sample results are considered as the number of results. Therefore, in this analysis, the number of results will be the same as the number of trial conditions regardless of the samples tested in each trial condition. The total number of DOFs for the experiment is calculated as

$$f_T = 8 - 1 = 7$$

(Each *S/N* ratio is counted as 1 DOF, regardless of the number of repetitions). The correction factor (C.F.) is

$$C.F. = \frac{T^2}{n}$$

$$= \frac{[(-11.67) + (-20.67) + (-12.72) + \cdots + (-23.16)]^2}{8}$$

$$= 2847.4831$$

where T is calculated by summing the S/N column, n is the total number of S/N values, and f is the number of degrees of freedom (DOFs). The *sums of squares* are (for known factor A)

$$S_T = [(-11.67)^2 + (-20.67)^2 + (-12.72)^2 + (-19.63)^2 + (-16.21)^2 + (-24.65)^2 + (-22.22)^2 + (-23.16)^2] - C.F.$$

$$= 163.77$$

$$S_A = \frac{A_1^2}{n_{A1}} + \frac{A_2^2}{n_{A2}} - C.F. \text{ (general formula)}$$

$$= 1046 + 1859.33 - 2847.4831$$

$$= 58.063$$

The sums of squares for other terms are calculated similarly.

The variance, F-ratios, pure sums of squares, and the percent influences for factor A are calculated as shown. When you pool the insignificant factors, QT4 displays the original and revised ANOVA results (Figures 12.28 and 12.29). The variance for factor A is

$$V_A = \frac{S_A}{f_A}$$

$$= \frac{58.063}{1}$$

$$= 58.063$$

The error variance

$$V_e = 0 \qquad \text{(since the error DOF is zero)}$$

The pure sum of squares

$$S_A' = S_A$$

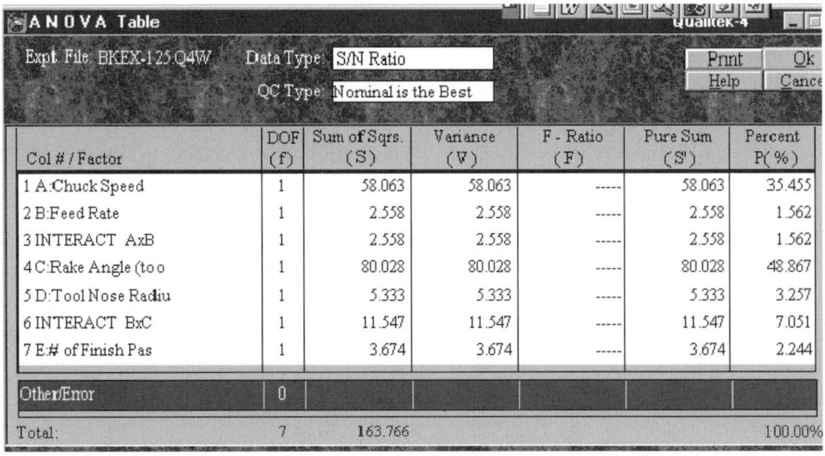

Col # / Factor	DOF (f)	Sum of Sqrs. (S)	Variance (V)	F - Ratio (F)	Pure Sum (S')	Percent P(%)
1 A:Chuck Speed	1	58.063	58.063	-----	58.063	35.455
2 B:Feed Rate	1	2.558	2.558	-----	2.558	1.562
3 INTERACT AxB	1	2.558	2.558	-----	2.558	1.562
4 C:Rake Angle (too	1	80.028	80.028	-----	80.028	48.867
5 D:Tool Nose Radiu	1	5.333	5.333	-----	5.333	3.257
6 INTERACT BxC	1	11.547	11.547	-----	11.547	7.051
7 E:# of Finish Pas	1	3.674	3.674	-----	3.674	2.244
Other/Error	0					
Total:	7	163.766				100.00%

Figure 12.28 ANOVA of S/N results.

The percent influence of factor A,

$$P_A = \frac{100 \times S'_A}{S_T}$$

$$= \frac{100 \times 58.063}{163.77}$$

$$= 35.45\%$$

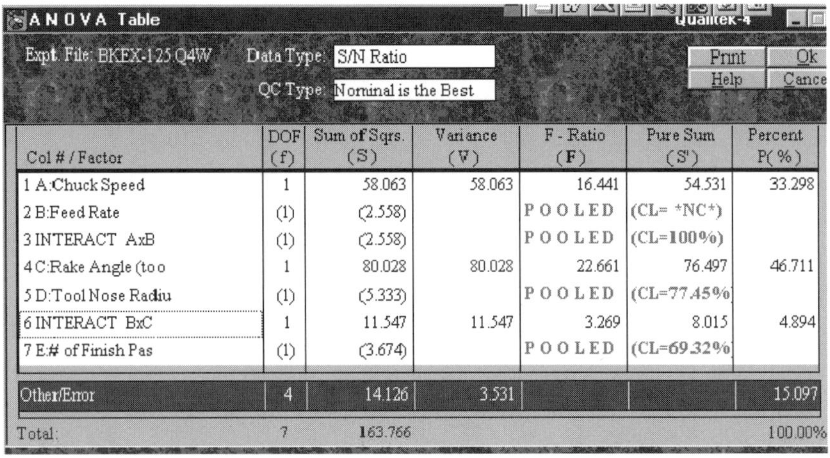

Col # / Factor	DOF (f)	Sum of Sqrs. (S)	Variance (V)	F - Ratio (F)	Pure Sum (S')	Percent P(%)
1 A:Chuck Speed	1	58.063	58.063	16.441	54.531	33.298
2 B:Feed Rate	(1)	(2.558)		POOLED	(CL= *NC*)	
3 INTERACT AxB	(1)	(2.558)		POOLED	(CL=100%)	
4 C:Rake Angle (too	1	80.028	80.028	22.661	76.497	46.711
5 D:Tool Nose Radiu	(1)	(5.333)		POOLED	(CL=77.45%	
6 INTERACT BxC	1	11.547	11.547	3.269	8.015	4.894
7 E:# of Finish Pas	(1)	(3.674)		POOLED	(CL=69.32%	
Other/Error	4	14.126	3.531			15.097
Total:	7	163.766				100.00%

Figure 12.29 Pooled ANOVA.

The same formula as used earlier for the optimum condition also applies for the S/N ratio. The S/N ratio expected at the optimum condition is calculated by using the average effects of significant factors (A and C) and interaction ($B \times C$; see Figure 12.28) shown in Figure 12.30.

$$Y_{opt} \text{ (in } S/N) = -18.861 + (-16.167 + 18.861) + (-15.698 + 18.861) + (-17.66 + 18.861)$$

$$= -18.861 + 2.694 + 3.162 + 1.201$$

$$= -18.861 + 7.056$$

$$= -11.804$$

In S/N analysis, the optimum performance (-11.804) is always expressed in terms of the S/N ratio of the performance expected tested at the optimum condition. To understand this number you should consider expressing this $S/N = -11.804$ in an equivalent number in the original units of measure of results (microinches). Since S/N is related to MSD, and MSD is expressed in terms of results (Y), it is possible to calculate the expected result (Y) from a known S/N ratio. This calculation is done by QT4 with a click on the *Transform* button in the *Optimum* screen shown in Figure 12.31. The steps in the transformation are repeated here for clarity.

From a known value of optimum performance (-11.804), MSD is calculated as

$$MSD = 10^{-[(S/N)/10]}$$

$$= 10^{1.1804}$$

$$= 15.14956$$

Optimum Conditions and Performance				Qualitek
Expt. File: BKEX-125.Q4W	Data Type: S/N Ratio			Print
	QC: Nominal is the Best			Help
Column # / Factor	Level Description	Level	Contribution	
1 A:Chuck Speed	1000 rpm	1	2.694	
4 C:Rake Angle (too	60 degree	1	3.162	
6 INTERACT BxC	*INTER*	1	1.201	
Total Contribution From All Factors...			7.056	
Current Grand Average Of Performance...			-18.861	
Expected Result At Optimum Condition...			-11.804	

Figure 12.30 Optimum condition and performance.

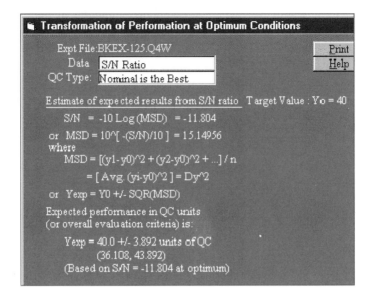

Figure 12.31   Transformation of *S/N* into original units of measured results.

For QC = N,

$$MSD = \frac{(Y_1 - Y_0)^2 + (Y_2 - Y_0)^2 + (Y_3 - Y_0)^2 + \cdots}{n}$$

or

$$15.14956 = \frac{(Y_1 - Y_0)^2 + (Y_2 - Y_0)^2 + (Y_3 - Y_0)^2 + \cdots}{n}$$

The expression above represents a single equation with n unknown expected performances (Y). To make an estimate of performance, the assumption that all performances are equal to a single expected performance ($Y_{expected}$). This allows us to make a logical estimate:

$$15.14956 = \frac{(Y_{expected} - Y_0)^2}{n}$$

or

$$Y_{expected} = Y_0 \pm (15.14956)^{1/2}$$

$$= 40 \pm 3.892$$

which expresses the performance expected within a boundary around the target value.

ALTERNATIVE FORM OF "NOMINAL IS BEST" *S/N* RATIO

When it comes to *nominal is best* quality characteristics, several expressions for *S/N* ratios are used. For those experimenters who are interested in adjusting the factors based selectively on influence on the mean and variation (around the mean), they may prefer the following formulation. Using the notation

$$T = \text{sum of all data} = Y_1 + Y_2 + Y_3 + \cdots + Y_n$$

$$Y_a = \text{mean (average) of results} = \frac{T}{n}$$

$$S_m = \text{sum of squares due to mean} = \frac{T^2}{n}$$

$$\sigma_{n-1}^2 = \text{mean square (variance, } V_e) =$$
$$\frac{(Y_1 - Y_a)^2 + (Y_2 - Y_a)^2 + (Y_3 - Y_a)^2 + \cdots + (Y_n - Y_a)^2}{n - 1}$$

$$S/N = \text{signal-to-noise ratio} = 10 \log_{10} \frac{S_m - V_e}{nV_e}$$

The preceding expression can be simplified as

$$S/N = 10 \log_{10} \frac{S_m - V_e}{nV_e}$$

$$= 10 \log_{10} \frac{T^2/n - \sigma_{n-1}^2}{n\sigma_{n-1}^2}$$

$$\approx 10 \log_{10} \frac{T^2/n}{n\sigma_{n-1}^2}$$

or

$$S/N \approx 10 \log_{10} \left(\frac{Y_a}{\sigma_{n-1}} \right)^2$$

This definition expresses *S/N* as a ratio between response due to signal factors (Y_a) and that due to noise factors (σ_{n-1}^2). The multiplier 10 and the transformation of the ratio into a logarithmic scale are done for convenience, as before. Should you prefer to use

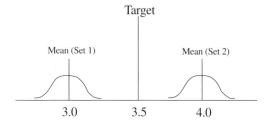

Figure 12.32 Quality characteristic: nominal is best; target = 3.5.

this definition of *S/N* for the *nominal is best* quality characteristic, be aware that there are some special situations when *S/N* alone may not help you identify the desirable condition. Compare and determine better among the following two sets of data (Figure 12.32):

  Data set 1: 2, 3, and 4
  Data set 2: 3, 4, and 5

Based on the target value of 3.5, both sets of data are equally desirable or undesirable. In such cases, you would expect the yardstick of measurement, like the *S/N*, should facilitate differentiating the two performances. As you can see from the calculations shown in Table 12.2, the two definitions of *S/N* may lead you to conclude differently.

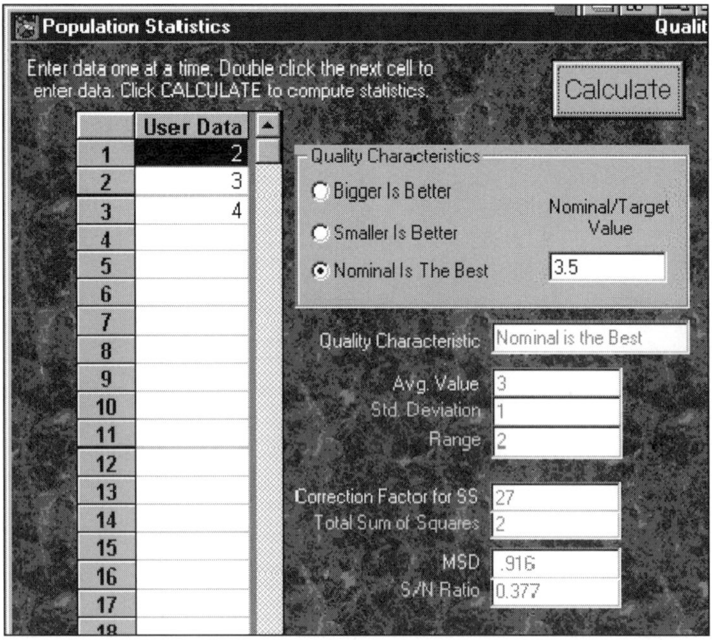

Figure 12.33 MSD and *S/N* for data set 1.

TABLE 12.2

Data Set	Mean, Y_a	Std. Dev., σ_{n-1}	$[Y_a/\sigma_{n-1}]^2$	MSD, $Y_a = 3.5$	Alternative Expression of S/N, $S/N \simeq 10\log_{10}(Y_a/\sigma_{n-1})^2$	S/N Based on MSD, $S/N \simeq -10\log_{10}\text{MSD}$
Set 1: 2, 3, and 4 (Figure 12.32)	3.0	1.0	9.0	0.91666	$(10 \times \text{Log}9 =)$ 9.54	0.3779
Set 2: 3, 4, and 5 (Figure 12.33)	4.0	1.0	16.0	0.91666	$(10 \times \text{Log}16 =)$ 12.04 (larger than set 1)	0.3779 (same as set 1)

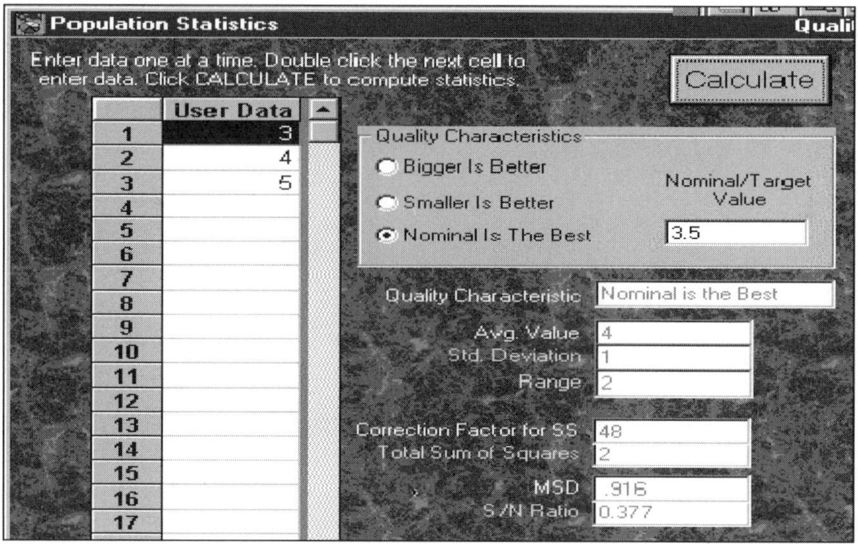

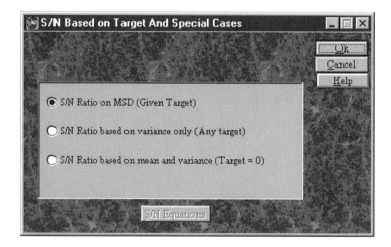

Figure 12.34 MSD and *S/N* for data set 2.

To calculate data set statistics like those shown in Figure 12.33 and 12.34, select the *Population statistics* option from the *Loss/savings* menu item from the main screen of QT4. You can use this expression of the *S/N* ratio to analyze your experiments by selecting the appropriate option. When selecting the quality characteristic (in *S/N* analysis) in QT4, check the *Nominal is best* option and then

Figure 12.35 Quality characteristic selection option.

check the last option as shown in Figure 12.35. Notice that when you check this option, you will not be asked to input the target value, as the expression for S/N in this case does make use of it.

SUMMARY

In this step you have learned how to compare population behavior using the mean-squared deviation (MSD) of the results and its logarithmic transformation (S/N). You understand that for convenience of linearity and to accommodate wide-ranging data, a log transformation of MSD (called the S/N) is recommended for analysis of the experimental results. You have also learned how the MSD is calculated for various quality characteristics and how analysis using S/N ratios differs from standard practice.

REVIEW QUESTIONS

Q. When should you perform analysis using S/N ratios?

A. You should analyze results using S/N ratios whenever you test more than one sample in each trial condition (two or more columns of results).

Q. If there are multiple columns of results, can I still perform standard analysis?

A. Yes.

Q. If analyses are performed by both methods, should I get the same factor level for the optimum condition?

A. Not necessarily. Factor influences in ANOVA will be different and the trend of factor influences (main effects) may also be different.

Q. When the conclusions about factors are different from the two analyses, which should I trust?

A. Rely more on S/N analysis. A good practice is to perform standard analysis first and identify factors with a major influence on the mean. Then perform S/N analysis, identify factors that are significant, and adjust their levels. Now find the factors that have no or the least influence on S/N but a greater effect on the mean. Adjust their levels based on standard analysis.

Q. The three basic steps in analysis are (1) average column effects, (2) ANOVA, and (3) optimum performance calculations. Which steps must you follow to determine optimum condition?

A. Steps 1 and 3.

Q. How should you confirm the results when you analyze it using S/N ratios?

A. Run multiple samples at the optimum condition as usual. Calculate S/N for the results of the confirmation tests and compare with the S/N estimated at the optimum condition.

Q. Should the S/N value expected be a negative number?

A. Not always. Whether *S/N* is a negative or positive number depends on whether MSD is a number greater or smaller than 1. The magnitude of MSD, however, depends on the quality characteristic.

EXERCISES

12.1 Design experiments to suit the following experimental study objectives. Determine the appropriate inner and outer arrays.

    (a) 4 two-level factors and 3 two-level noise factors

    (b) 5 two-level factors and 3 three-level noise factors

    (c) 10 two-level factors and 5 two-level noise factors

12.2 Select the answers that most closely match yours in the following situations.

    (a) Why do we need to consider running multiple samples for each trial condition? **(1)** To obtain better representative performance, **(2)** to reduce trials, or **(3)** to reduce experimental error.

    (b) While repeating experiments, what was the objective or purpose that Taguchi wanted to satisfy? **(1)** To study more factors, **(2)** to learn about noise factors, or **(3)** to design robustness.

12.3 Answer the following questions as they relate to projects in which you are involved.

    (a) What are noise factors? Give an example of the noise factors related to a project with which you are familiar. What does robust design mean as applied to your project?

    (b) Can averages be used to compare two data sets? Discuss the limitations.

    (c) Why is MSD preferred as a better representative of a data set over the standard analysis method?

    (d) What are the advantages of transforming MSD to an *S/N* ratio?

    (e) If set *A* has MSD = 5 and set *B* has MSD = 6, would the *S/N* for set *B* be higher than that for set *A*?

12.4 An L-16 array was used to design an experiment to study 15 two-level factors. Determine the degree of freedom of an error factor when:

    (a) Each trial condition is tested once.

    (b) Each trial condition is repeated three times and standard analysis is carried out.

    (c) Each trial condition is repeated five times and the *S/N* ratio of the result is used for analysis.

12.5 What does *zero error term* ($f_e = 0$, $S_e = 0$) mean? Select all appropriate answers.

    (a) It indicates a poorly run experiment.

    (b) It represents a very well run experiment.

    (c) *Zero error term* does not mean that there is no experimental error. It simply means that information concerning the error sum of squares cannot be determined specifically.

12.6 Compare the *S/N* ratios of the following two sets of data and determine which set is more desirable. Assume that the target/nominal value = 12.

    Set 1: 11, 9, 12, 10, and 9

    Set 2: 10, 12, 8, 14, and 16

12.7 In an experiment, the estimate of performance at optimum conditions is expressed as *S/N* = 22.5. If the quality characteristic is *Bigger is better*, determine the expected performance in terms of the measured value.

12.8 Determine the performance at the optimum condition in terms of the *S/N* ratio of the results when the quality characteristic is *Smaller is better* (Table 12.3).

    Trial 1:

$$\text{MSD} = \frac{4^2 + 6^2 + 5^2 + 7^2}{4} = 31.5 \qquad S/N = -10 \log \text{MSD} = -10 \log 31.5 = -14.98$$

    Trial 2:

$$\text{MSD} = \frac{7^2 + 8^2 + 6^2 + 7^2}{4} = \qquad S/N = -16.94$$

    Trial 3:

$$\text{MSD} = \frac{5^2 + 6^2 + 4^2 + 3^2}{4} = \qquad S/N = -13.32$$

    Trial 4:

$$\text{MSD} = \frac{3^2 + 2^2 + 4^2 + 2^2}{4} = \qquad S/N = -9.16$$

TABLE 12.3 Array for Exercise 12.8

Trial	A	B	C	Results				S/N
1	1	1	1	4	6	5	7	−14.98
2	1	2	2	7	8	6	7	−16.94
3	2	1	2	5	6	4	3	−13.32
4	2	2	1	3	2	4	2	−9.16

Computing the grand average,

$$\bar{T} = \frac{-14.98 - 16.94 - 13.32 - 9.16}{4} = -13.60$$

$$\bar{A}_1 = \frac{-14.98 - 16.94}{2} = -15.96 \qquad \bar{B}_1 = \frac{-14.98 - 13.32}{2} = -14.15$$

$$\bar{A}_2 = \frac{-13.32 - 9.16}{2} = -11.24 \qquad \bar{B}_2 = \frac{-16.94 - 9.16}{2} = -13.05$$

$$\bar{C}_1 = \frac{-14.98 - 9.16}{2} = -12.07$$

$$\bar{C}_2 = \frac{-16.94 - 13.32}{2} =$$

Optimum condition: $A_2 B_2 C_1$

$$Y_{opt} = -13.06 + (-11.24 + 13.60) + (-13.05 + 13.6) + (-12.07 + 13.6)$$

$$= -13.06 + 2.36 + 0.55 + 1.53$$

$$=$$

EXERCISE ANSWERS

Use QT4 capability to solve exercises when appropriate.

12.1 (a) L-8 and L-4; (b) L-8 and L-9; (c) L-12 and L-8.

12.2 (a) Answer (1) is correct; (b) Answer (3) is correct.

12.3 Answers will vary.

12.4 (a) 0; (b) 32; (c) 0.

12.5 Answer (c) is correct.

12.6 Set 1.

12.7 13.33 (MSD = 0.005623).

12.8 Optimum performance = −9.166. Refer to file *BKEX-129.Q4W*.

Results Comprising Multiple Criteria of Evaluations

What You Will Learn in This Step

Although analyzing results to satisfy one objective (criterion) at a time is a common practice, most DOE applications involve more than one objective. Generally, different objectives are also evaluated by different criteria of evaluations, each of which has different units of measurement and different relative weighting. Combining the different evaluation criteria into a single quantity desirable for determining the optimum factor combination that best satisfies all the objectives requires devising a special scheme. This step will allow you to see how the overall evaluation criterion (OEC) is calculated and how its quality characteristic is assigned for several application examples.

Thought for the Day

Give a man a fish and you feed him for a day. Teach a man to fish and you feed him for a lifetime.

—Chinese proverb

By now you have learned many different ways of designing experiments and analyzing the results of such experiments. So far, you only know how to analyze results based on evaluation of only one objective at a time. This way, if there were multiple objectives, you could analyze results from each criterion evaluation (result) separately. But a better way to analyze results of multiple criteria of evaluations will be first to combine the different evaluations into a single index, which incorporates the relative weights of all criteria and their individual quality characteristics. The concept of combining different criteria into overall evaluation criteria (OECs) has been described in Step 2. In this step you learn how to create the OECs of evaluations for analysis purposes through a few application examples. Before proceeding with this step, you will find it helpful to review Step 2. For the sake of completeness, the examples used in this step will contain detailed descriptions of the experimental design and analysis involved. Since methods for accomplishing design and analysis using QT4 have already been covered in earlier steps, the emphasis in this step is focused on creating OECs from evaluations of experimental samples.

OVERALL EVALUATION CRITERION

Whenever there is more than one performance objective, formulation of the OEC is recommended. The method of OEC formulation and computation is given in Step 2. Once the OEC for a sample is formulated, it becomes the result for the sample. When OECs are calculated for all samples, these become the results and subject of analysis of the experimental results. For experiments with multiple objectives, it is a good practice first to perform separate analysis for each criterion of evaluation (objective). Then use the OEC from all evaluations of samples to perform an additional analysis. In the event that all individual analysis identifies the same levels of the factors as the desirable ones, OEC analysis becomes redundant.

Recommended Analysis Strategy for Multiple Objectives

1. Analyze results by taking evaluations of a single objective one at a time.
2. Compare optimum factor levels from each analysis. If all the analyses identify the same factor level, no further analysis is needed. If analyses were not able to identify the factor levels, try to determine them intuitively such that they produce overall better performance for all objectives. As a last resort, perform OEC analysis.
3. To perform OEC analysis, first combine different sample evaluations into a single index. Then use the sample OEC and analyze results based on the quality characteristic of the OEC (*bigger* or *smaller*, depending on the formulation).

Example 13.1: Study of a Magnet Subassembly for Adhesion and Placement
Production quality assurance personnel of an audio speaker manufacturer conducted an experiment to determine the best process parameters for the magnet assembly process. For superior performance, earlier tests have indicated that the assembly needs to satisfy three objectives: (1) high impact strength, (2) minimum squeeze-out adhesive, and (3) proper positioning of the magnet. Based on the factors that influence these objectives (Figure 13.1), an experiment designed using an L-8 array was carried out. The description, relative weight, and quality characteristics of the criteria of evaluation of the three objectives are shown in Table 13.1. Three samples were tested in each of the eight trial conditions. Evaluations of each sample under the three criteria of evaluation are shown in Table 13.2. Analyze results and determine the level of factor D: cure temperature that is desirable for all three objectives.

Solution: The results of this experiment have been analyzed in four different ways. In the first three analyses, evaluations (Table 13.2) for each criterion (impact strength, placement, etc.) are used as the result. In the fourth analysis, the evaluations are first combined into an OEC, then analyzed.

(a) *Analysis with results of impact strength* (C1). The experiment designs and results appropriate for impact strength (C1) have been saved as experiment file *BKEX131A.Q4W*. You may re-create the design and enter the results or review this file. After you are done designing the experiment and entering results, your result

	Factors	Level 1	Level 2	Level 3	Level 4
1	D:Cure Temp.	100 deg.F	125 deg.F	150 deg.F	175 deg.F
2	UPGRADED	*UNUSED*	--------------	--------------	--------------
3	UPGRADED	*UNUSED*	--------------	--------------	--------------
4	A:Cure Time	Current	Longer	------------	------------
5	B:Glue Viscosity	40000 cps	50000cps	------------	------------
6	C:Tooling Pressur	2 kgf	4.2 kgf	------------	------------
7	COLUMN UNUSED	*UNUSED*	--------------	--------------	--------------

Figure 13.1 Factor and level descriptions.

screen should look as shown in Figure 13.2. The results shown are for three samples tested in each trial condition. The evaluations for impact strength, which make up the results for analysis, are the same under column C1 in Table 13.2. You will see this screen (Figure 13.2) when you proceed with the *Standard analysis* option in QT4. As higher impact strength is desirable, you will need to select the *Bigger is better* quality characteristic to perform this analysis. At the end of the analysis, you will find that the desirable level for cure temperature based on impact strength is level 1 (100°F), as shown in Figure 13.3.

(**b**) *Analysis with results of placement accuracy.* The experiment design and results appropriate for placement accuracy (C2) have been saved as experiment file *BKEX131B.Q4W.* You can create this file simply by replacing the results of the previous experiment file with that for the placement accuracy, which are listed under columns C2 in Table 13.2. Update your own file or work with the *BKEX131B.Q4W* experiment file. When you are finished entering the evaluations for placement accuracy, your experimental results should look like Figure 13.4. Proceed with analysis by selecting the *Standard analysis* option from the *Analysis* menu item and the quality characteristic as *Nominal is best.* Set the target value as 90 and proceed with the analysis. For analysis purposes in the case of *Nominal is best,* the results are converted to deviation prior to analysis as shown in Figure 13.5. At the conclusion of analysis, you

TABLE 13.1   Evaluation Criteria for Example 13.1[a]

	Criterion	Worst Reading	Best Reading	QC	Relative Weight (%)
C1:	impact strength (grams)	20	40	Bigger	60
C2:	accuracy of placement	85–96	90	Nominal/smaller deviation)	25
C3:	squeeze-out glue	10	0	Smaller	15

[a]Impact strength: QC = bigger with observed range between 20 and 40 grams of force; accuracy of placement: QC = nominal with target value at 90 (the worst deviant value is 96); squeeze-out glue: QC = smaller with expected range between 0 and 10; relative weight: subjective emphasis is determined by the group consensus during the planning meeting.

TABLE 13.2 Sample Evaluations for Example 13.1[a]

	Sample								
	1			2			3		
Trial	C1	C2	C3	C1	C2	C3	C1	C2	C3
1	27	88	6	33	88	5	29	84	8
2	34	86	4	28	86	7	35	83	9
3	22	96	8	25	96	5	21	95	4
4	30	92	1	24	92	2	31	93	0
5	28	89	3	25	89	1	34	94	4
6	32	92	3	29	92	3	32	95	5
7	25	90	5	27	90	3	26	92	5
8	20	89	3	23	89	2	24	91	4

[a]C1, C2, and C3 are the three criteria of evaluation.

will find that this analysis also shows that level 1 (100°F) of cure temperature is the desirable condition (Figure 13.6; same as with impact strength).

(c) *Analysis with results of squeeze-out glue.* The experiment design and results appropriate for squeeze-out glue (C3) have been saved as experiment file *BKEX131C.Q4W*. You can create this file simply by replacing (select *Edit/results*) the results of the previous experiment file with that for the squeeze-out glue, which are

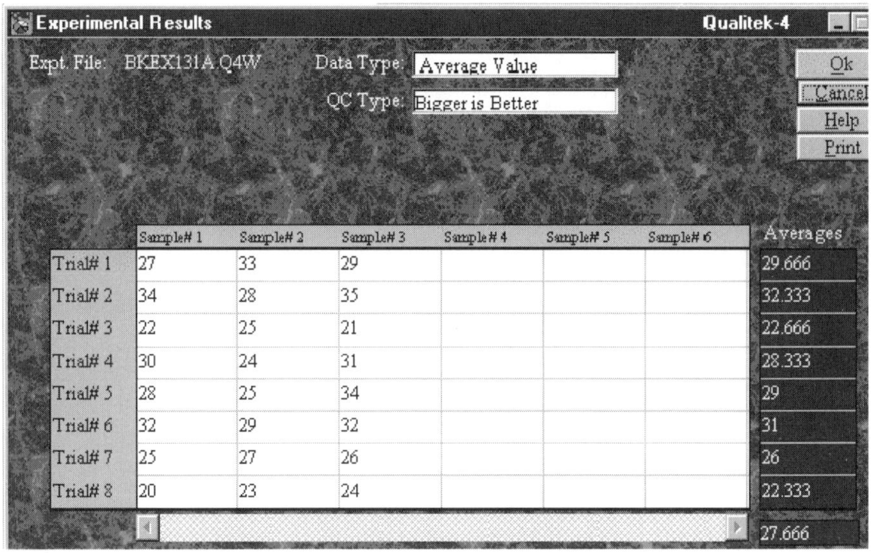

Figure 13.2 Results based on impact strength.

Optimum Conditions and Performance			Qualitek-

Expt. File:BKEX131A.Q4W   Data Type Average Value   Print

QC   Bigger is Better   Help

Column # / Factor	Level Description	Level	Contribution
1 D:Cure Temp.	100 deg.F	1	3.333
4 A:Cure Time	Longer	2	.833
5 B:Glue Viscosity	50000cps	2	1.249
6 C:Tooling Pressur	4.2 kgf	2	.333

Total Contribution From All Factors...   5.748
Current Grand Average Of Performance...   27.666
Expected Result At Optimum Condition...   33.414

Figure 13.3   Optimum level for cure temperature.

listed under columns C3 in Table 13.2. Update your own file or work with the *BKEX131C.Q4W* experiment file. When you are finished entering the evaluations for squeeze-out glue, and you proceed with standard analysis by selecting the *Smaller is better* quality characteristic, your experimental results should look like Figure 13.7. Notice that the optimum condition (Figure 13.8) in this case shows a different level of cure temperature (level 3, 150°F) than the previous two cases. So what value of cure temperature should you specify? To resolve such conflicts with one or more factors, you can rely on analysis using the OEC of the original evaluations.

(d) *Analysis with OEC of original evaluations.* The experimental design and OEC of the original evaluation as results have been saved as experiment file *BKEX131D.Q4W*. Review this file or create your own experiment file following these steps.

	1	2	3	4	5	6	7		1	2	3
1	1	0	0	1	1	1	0		86	66	84
2	1	0	0	2	2	2	0		82	86	83
3	2	0	0	1	1	2	0		94	96	95
4	2	0	0	2	2	1	0		90	92	93
5	3	0	0	1	2	1	0		88	89	94
6	3	0	0	2	1	2	0		94	92	95
7	4	0	0	1	2	2	0		89	90	92
8	4	0	0	2	1	1	0		93	89	91
			Inner Array								Results

Figure 13.4   Results of placement accuracy.

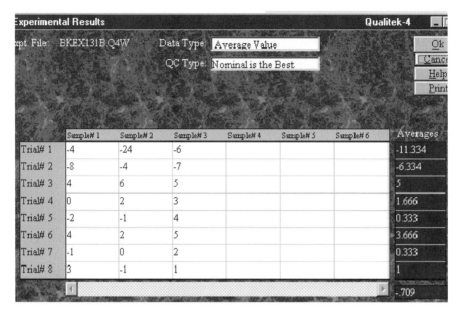

Figure 13.5 Placement results reduced to deviation from the target.

1. Run QT4 and open one of the last three files (say, *BKEX131C.Q4W*).

2. From the main screen, select the *Results* option from the *Edit* menu item. Place the cursor on top of each column and click the *Reset column* button. This way, you remove all existing results. Click *OK* and return to the main screen by naming the file *BKEX131D.Q4W* (or *BKEX131Y.Q4W*) or any other of your

Optimum Conditions and Performance				Qualitek
Expt. File: BKEX131B.Q4W	Data Type Average Value			Print
	QC Nominal is the Best			Help
Column # / Factor	Level Description	Level	Contribution	
1 D:Cure Temp.	100 deg.F	1	-8.125	
4 A:Cure Time	Current	1	-.709	
5 B:Glue Viscosity	50000cps	2	-.292	
6 C:Tooling Pressur	2 kgf	1	-1.375	
Total Contribution From All Factors...			-10.501	
Current Grand Average Of Performance...			-.709	
Expected Result At Optimum Condition...			-11.21	

Figure 13.6 Desirable cure temperature based on placement accuracy.

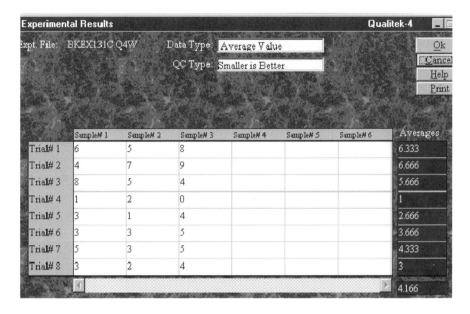

Figure 13.7   Results of squeeze-out glue.

liking. You are now ready to combine all evaluations into OECs and use that as the results for analysis.

3. Select *Multiple criteria (OEC)* from the *Results* menu item. First work with the lower left box (Figure 13.9) to define the criteria. Describe one row for each criterion of evaluation. Worst and best values refer to the sample readings under this criterion. Click on the *QC* box and set it until the correct description is

Optimum Conditions and Performance				Qualitek
Expt. File:BKEX131C.Q4W	Data Type Average Value		Print	
	QC Smaller is Better		Help	

Column # / Factor	Level Description	Level	Contribution
1 D:Cure Temp.	150 deg.F	3	-1
4 A:Cure Time	Longer	2	-.584
5 B:Glue Viscosity	50000cps	2	-.501
6 C:Tooling Pressur	2 kgf	1	-.917

Total Contribution From All Factors...	-3.002
Current Grand Average Of Performance...	4.166
Expected Result At Optimum Condition...	1.164

Figure 13.8   Desirable cure temperature based on squeeze-out glue.

Used	Criteria Description	Worst	Best Value	QC	Rel Wt
X	C1 Impact Strength	20	40	B»	40
X	C2 Placement	82	90	«N»	35
X	C3 Squeeze Out	10	0	«S	25

QC of the overall evaluation criterion | Bigger is better | 100

Figure 13.9 Criteria descriptions for OEC.

displayed. Enter the relative weight as percent number (40 for 40%) and see that 'X' appears on the left of the description, indicating that this criterion is counted. Complete the description of the other two criteria in the same manner. Make sure that the relative weights add up to 100 as shown at the bottom of Figure 13.9. Make a note of the QC description at the bottom of the screen. It represents the QC of the overall evaluation criteria. You will need this for analysis using OEC, which you will perform after you are finished entering all the evaluations.

4. After you are done describing the criteria of evaluation, you are ready to enter sample evaluations for all criteria (Table 13.2). In this experiment, there are eight trial conditions with three samples in each condition, and there are three numbers (evaluations) for each sample. The challenge for you is to enter the evaluations in the right input field in QT4. You may find that a few simple guidelines make it easier to place them.

   a. Select the trial number by clicking the scroll bar on top left box (Figure 13.10). This box displays only the calculated values of OECs.

   b. Select the sample number by clicking on the scroll bar of the bottom right box (Figure 13.10).

   c. Enter evaluations by double-clicking on the appropriate space in the bottom right box (Figure 13.10).

   d. Start entering evaluations from the last sample (sample 3) of the first trial of your experiment.

5. After you are done describing the criteria of evaluation, you are ready to enter sample evaluations. Set up your input condition to enter evaluations (29, 84, and 8) for sample 3 of trial 1 as shown in Figure 13.11. Double-click on the space for evaluation of impact strength (29). Enter 29 in this space. Use the arrow key (or double-click) to move the cursor down and enter the other two evaluations (84 and 8). You are now done entering evaluations for sample 3 of trial 1.

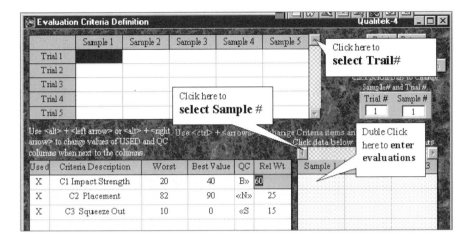

Figure 13.10   Description of criteria and input of evaluations.

6. Now click on the scroll bar to change to trial 2 (stay at sample 3). Enter evaluations (35, 83, and 9) for sample 3 of trial 2 as shown in Figure 13.12. Enter evaluations for all other trial conditions in the same manner. After all evaluations for sample 3 are finished, repeat the process for samples 2 and 1 for all trial conditions.

7. After you have entered all evaluations, click on the *Return* button in the OEC screen (Figure 13.12). All OECs calculated by QT4 are now shown as the results shown in Figure 13.13. You are now ready to perform analysis using the OECs

Evaluation Criteria Definition   **Qualitek-4**

	Sample 1	Sample 2	Sample 3	Sample 4	Sample 5
Trial 1	41.5	64.75	31.75		
Trial 2	43	41	36.87		
Trial 3	26.5	31.25	36.87		
Trial 4	77.5	54.25	68.87		
Trial 5	59.75	63.12	60.5		

Print   Return
Tips   Cancel
Default Data   Reset Data
Click Scroll Bars to Change Sample# and Trial #

Trial #   Sample #
1   3

Use <alt> + <left arrow> or <alt> + <right arrow> to change values of USED and QC columns when next to the columns. Use <ctrl> + <arrows> to change Criteria items and Sample #. Click data below to activate Scroll Bars

Used	Criteria Description	Worst	Best Value	QC	Rel Wt	Sample 3	Sample 4	Sample 5
X	C1 Impact Strength	20	40	B»	40	29		
X	C2 Placement	82	90	«N»	35	84		
X	C3 Squeeze Out	10	0	«S	25	8		

Figure 13.11.   Evaluations input.

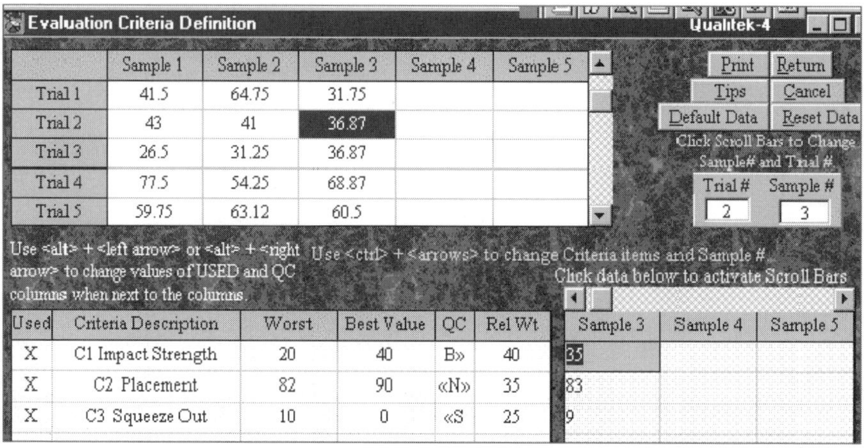

Figure 13.12 Evaluation for sample 3 of trial 2.

as results. Save the experiment file by selecting the *Save* option from the *File* menu item.

8. Analyzing results using the OEC is the same as any other form of results. You have the option to perform analysis using the *Standard analysis* or *S/N* option. Since all other analysis in this example is done using the *Standard analysis*

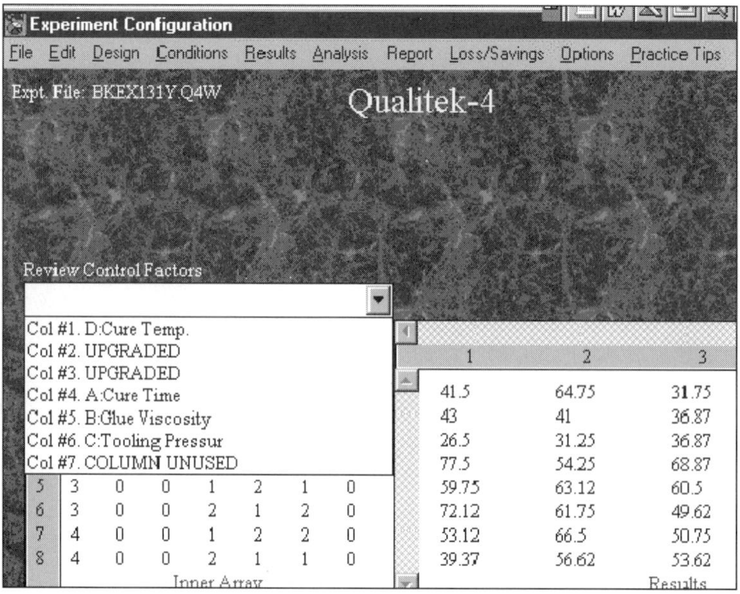

Figure 13.13 OEC as experimental results.

option; you should analyze OEC results by selecting the *Standard analysis* option from the *Analysis* menu items. Recall that the quality characteristic (QC) for OEC is *Bigger is better.* So you should check *Bigger is better* as the QC to proceed with analysis. Upon completion of analysis, you will obtain the optimum condition, indicating the desirable level for cure temperature as the third level (150°F) shown in Figure 13.14.

So as you can see, the dilemma about the level of cure temperature is now resolved. Based on the three criteria of evaluation [i.e., analyses (a), (b), and (c) discussed above], different levels of cure temperature were identified as the desirable selections. An analysis using OEC in such situations can include an objective determination of the optimum levels. In summary, when different criteria indicate different factor levels, perform OEC analysis and let it determine the factor level. You can extend the same approach to any or all other factors included in the experiment.

Example 13.2: Refrigeration Fluid Carrying Hybrid System Design Study
Guha Hydraulics is a major supplier of automotive and refrigeration fluid-carrying tubes. Its product line is primarily small-diameter metal- and nylon-based (hybrid) tubing used to transfer fuel, brake, or refrigeration fluids. Recent introduction of government regulations dictate a higher standard of performance and durability life (10 years) for the product. To meet and exceed the performance requirements, the responsible design group within the company launched a study to optimize the design. A large portion of Guha's fuel-carrying systems includes joints made up by inserting steel tubes into the inner diameter of nylon tubes. For lasting performance, the tensile strength of the joints and hydrocarbon permeation were among the major concerns. The objectives of the experiment were to optimize the design based on these two criteria described in Table 13.3. The factors and their levels (6 two-level factors and an

Optimum Conditions and Performance			Qualitel
Expt. File:BKEX131Y.Q4W DataType Average Value			Print
QC Bigger is Better			Help

Column # / Factor	Level Description	Level	Contribution
1 D:Cure Temp.	150 deg.F	3	9.437
4 A:Cure Time	Longer	2	2.842
5 B:Glue Viscosity	50000cps	2	4.562
6 C:Tooling Pressur	2 kgf	1	4.26

Total Contribution From All Factors...	21.101
Current Grand Average Of Performance...	51.706
Expected Result At Optimum Condition...	72.807

Figure 13.14 Optimum condition with OEC.

TABLE 13.3 Description of Evaluation Criteria[a]

	Criterion	Worst Reading	Best Reading	QC	Relative Weight (%)
C1:	permeation rate (grams/day)	5	0	Smaller	60
C2:	tensile strength (newtons)	25	85	Bigger	40

[a]Permeation rate of hydrocarbon: QC = smaller, with observed range between 0 and 5 grams/day; tensile strength of joint: QC = bigger, with observed range between 25 and 85; relative weight: a subjective emphasis is determined by the group consensus during the planning meeting.

interaction) that were tested as part of the experiments are shown in Figure 13.15. One sample was tested in each of the eight trial conditions. The evaluations for each sample under the two criteria of evaluations are shown in Table 13.4. This experimental study had two specific objectives: (1) to determine three factors that have most influence on the variability of the two criteria of evaluations, and (2) to determine the desirable levels of the six factors studied for overall best performance.

Solution: The experiment design for this example and the results has been saved as file *BKEX132C.Q4W*. As the objectives include finding factor influence and levels based on both criteria of evaluation, analysis using the OEC of the original evaluations will be preferred. You may either use this file containing the original evaluations for analysis or create your own experiment file for the purpose. (Files *BKEX132A.Q4W* and *BKEX132B.Q4W* contain individual criteria evaluations and may be used for analysis specifically for the criteria.)

	Factors	Level 1	Level 2
1	A:3rd Member ID	Low Dia.	High Dia.
2	B:Multi-layer OD	Lower Lim	Higher Limit
3	INTER COLS 1 x 2	*INTER*	------------
4	C:Insertion Speed	Current	10% Higher
5	D:Tube End OD	Present D	Expanded
6	E:Tube End Angle	Narrow	At Spec.
7	F:Clamp Loading	Old Sprin	New Spring

Figure 13.15 Factor and level descriptions.

TABLE 13.4   Test Sample Evaluations[a]

Trial	Sample 1	
	C1 (0–5)	C2 (25–85)
1	4	79
2	3	80
3	3	49
4	5	23
5	4	36
6	1	27
7	3	53
8	2	13

[a]C1 and C2 are two criteria of evaluation.

1. Run QT4 and open file *BKEX132C.Q4W* (or the file you created to design this experiment). Review the factor and level description (*Edit/Factor & levels*) to make sure that the description looks like that in Figure 13.15.

2. Review the description of criteria of evaluations of test samples (*Edit/Evaluation criteria*) and see that they are as shown in Figure 13.16.

3. Examine the OEC values used as results when you return to the main screen as shown in Figure 13.17. (When entering evaluation criteria for the first time, be sure to update/save your experiment file by selecting the *Save* option in the *File* menu.)

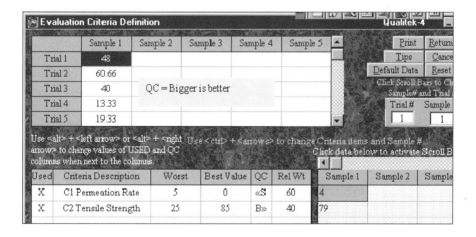

Figure 13.16   Criteria descriptions and evaluations.

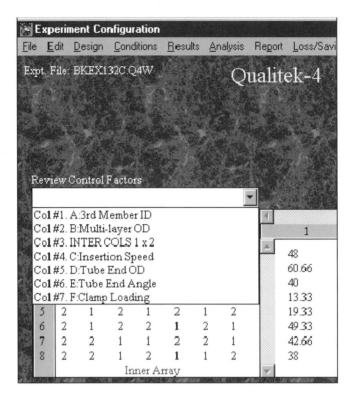

Figure 13.17   OEC results for analysis.

4. Perform analysis (*Standard analysis/Analysis* menu) by selecting the *Bigger is better* quality characteristic, which is the QC for the OECs in this case. Review the ANOVA output (Figure 13.18) and identify factors in order of their percent influence to the variation of results (OEC).

5. Review the pie diagram of factor influences (Figure 13.19) and identify the ranks of the factors in terms of their influence. Note that factor E: tube end angle has the most influence and factor F: clamp loading has the least influence. This information is recorded (ranks) in the first row (OEC) in Table 13.5 under the corresponding factors. From the pie diagram, the first three most influential factors are E, B, and D. This answers part (a) of the question.

6. Proceed with analysis and obtain the optimum condition shown in Figure 13.20. Note down the levels of the factors that are considered better for both criteria of evaluation. The desirable level for the factors are A_1, B_1, C_2, D_1, E_2, and F_2 (answer to the second part/objective). The optimum factor levels are also recorded in Table 13.5 (first row for OEC) as a numeric number under the factor influence ranking.

Expt. File: BKEX132C.Q4W   Data Type: Average Value   Print   Qk
                            QC Type: Bigger is Better   Help   Cance

Col # / Factor	DOF (f)	Sum of Sqrs. (S)	Variance (V)	F - Ratio (F)	Pure Sum (S')	Percent P(%)
1 A:3rd Member ID	1	20.066	20.066	-----	20.066	1.168
2 B:Multi-layer OD	1	234.687	234.687	-----	234.687	13.659
3 INTER COLS 1 x 2	1	566.667	566.667	-----	566.667	32.982
4 C:Insertion Speed	1	16.047	16.047	-----	16.047	.934
5 D:Tube End OD	1	193.552	193.552	-----	193.552	11.265
6 E:Tube End Angle	1	684.316	684.316	-----	684.316	39.83
7 F:Clamp Loading	1	2.727	2.727	-----	2.727	.158
Other/Error	0					
Total:	7	1718.063				100.00%

Figure 13.18   ANOVA showing factor influences.

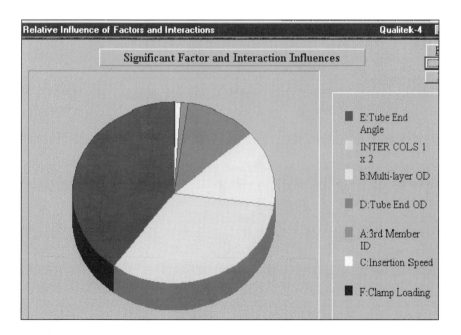

Figure 13.19   Pie diagram of factor influences.

TABLE 13.5 Comparison of Individual Criteria Analysis with OEC Analysis[a]

| | Rank ( ) of Factor Influence and Optimum Levels | | | | | | |
Type of Results	A	B	A × B	C	D	E	F
OEC	(5th)	(3rd)	(2nd)	(6th)	(4th)	(1st)	(7th)
(BKEX-132C)	1	1	N/A	2	1	2	2
C1:	(1st)	(5th)	(6th)	(4th)	(2nd)	(3rd)	(5th)
permeation	2	1	N/A	2	1	2	2
(BKEX-132A)							
C2:	(1st)	(3rd)	(2nd)	(4th)	(5th)	(6th)	(7th)
tensile strength	1	1	N/A	1	2	2	1
(BKEX-132B)							

[a]Effect of interaction not considered for selection of optimum levels.

7. You can pursue similar analysis using only a single criterion of evaluation at a time. Observations from such analysis are recorded in the second and third rows of Table 13.5. For better understanding of the process, you may compare the optimum level and relative influence of the factors with the corresponding value obtained by OEC analysis. Here are a few key observations:

   a. Optimum levels for a few factors are different for the two criteria analyses (second and third row data: factor levels 2 1 2 1 2 2 and 1 1 1 2 2 1). Here OEC analysis presents compromised levels (1 1 2 1 2 2), which coincidentally, is the same as that for criteria C1 except the level for factor A.

Optimum Conditions and Performance			Qualitek
Expt. File:BKEX132C.Q4W	Data Type Average Value		Print
	QC Bigger is Better		Help

Column # / Factor	Level Description	Level	Contribution
1 A:3rd Member ID	Low Dia.	1	1.583
2 B:Multi-layer OD	Lower Lim	1	5.416
3 INTER COLS 1 x 2	*INTER*	1	8.416
4 C:Insertion Speed	10% Higher	2	1.416
5 D:Tube End OD	Present D	1	4.918
6 E:Tube End Angle	At Spec.	2	9.248
7 F:Clamp Loading	New Spring	2	.583

Total Contribution From All Factors...	31.579
Current Grand Average Of Performance...	38.913
Expected Result At Optimum Condition...	70.493

Figure 13.20 Optimum condition and performance.

b. In terms of the influence of the factor, analyses for both criteria showed factor A (first rank) to be most influential. But when analyzed using the OEC, factor E was identified as the most influential variable. This is an example where common intuition will fail to identify the factor that is most influential (6 two-level factors and an interaction).

Example 13.3: Gear Distortion Study The task force in a division of a local automotive manufacturer undertook a study to investigate the parameters involved in distortion of a transmission gear. The plan for a formal design of experiment was initiated in a meeting of the project team members. In this meeting, several factors with potential influence on the distortion and objective characteristics (evaluation criteria) indicative of the distortion were determined. Based on the information from the planning session, an experiment that prescribed 16 separate sample preparation schemes were laid out and the data collection procedure established.

Following the specification prescribed by the designed experiment, two gears for each of the 16 separate conditions were fabricated and heat-treated. Upon completion of the heat treatment, the desired characteristics were measured at several participating facilities. The results were then reduced to a format suitable for analysis. The experiment investigated 11 factors (see Figure 13.21) that are considered to influence the

Inner Array Design

Array Type: L-16   Print   Help

Use \<ctrl\> + \<arrows\> to move cursor.

	Factors	Level 1	Level 2	Level 3	Lev
1	J:Type of quench	Fast Oil	Slow Oil	Marquench	------------
2	UNUSED/UPGRADED	*UNUSED*	------------	------------	------------
3	UNUSED/UPGRADED	*UNUSED*	------------	------------	------------
4	K:Material(alloy)	8620-L	8620-H	5130	9310
5	A:Initial Conditi	Unreleave	Relieved		------------
6	B:Quench Temp.	Low	High	------------	------------
7	C:Case Depth	Low	High	------------	------------
8	UNUSED/UPGRADED	*UNUSED*	------------	------------	------------
9	D:Grain Flow(tex)	Bar Stock	Forging	------------	------------
10	F:Surface Carbon	Low	High	------------	------------
11	G:Carburizing Mtd	Direct	Re-Heat	------------	------------
12	UNUSED/UPGRADED	*UNUSED*	------------	------------	------------
13	I:Type of Cycle	Boost	Straight	------------	------------
14	E:Part Location	Center	Edge	------------	------------
15	H:Part Orientatio	Vertical	Horizonta	------------	------------

Figure 13.21 Factor and level descriptions.

characteristics, which are used as measures of distortion. All but two factors were two-level: 1 four- and 1 three-level factors were also included in the study.

The experiment was designed using a Taguchi L-16 orthogonal array, which has 15 two-level columns. Columns 1, 2, and 3 were combined, upgraded, and then downgraded to a three-level column for the three-level factor. Columns 4, 8, and 12 were used to create a four-level column, which accommodated the four-level factor. The remaining 9 two-level factors were then assigned to the remaining nine columns.

As indicators of distortion, test samples were evaluated under nine separate criteria of evaluation (Table 13.6). These characteristics were all evaluated in terms of linear distance (measured in microinches) and each carried a different relative weight as determined by consensus of the group involved in the experiment planning. Two samples were tested in each of the 16 trial conditions. The samples were then evaluated nine different ways, one for each of the nine criteria of evaluations. The readings for the samples are shown in Table 13.7. Based on the sample evaluations, determine the combination of the process parameters that produce the best overall improvement, and identify five factors with most significant influence on all the criteria.

Solution: This example experiment and its results (OEC from all criteria evaluations) are saved in experiment file *BKEX133A.Q4W*. You should review this file to make sure that the experiment design (shown in Figure 13.21) is correct and that the evaluations (shown in Table 13.7) are represented properly. Because you already know how to design experiments (from earlier examples in this and other steps), and also how to enter multiple evaluation criteria (for OEC) information, major focus in this example is on the key observations.

The experiment design was accomplished by modifying an L-16 array. To accommodate 1 three-level factor and another four-level factor, columns 1 (using columns 1, 2, and 3) and 4 (using columns 4, 8, and 12) were modified as shown in Figure 13.22. The results were analyzed using *S/N* ratios, as they consisted of two samples (two columns) for each trial condition. The results which formed the *S/N* were created

TABLE 13.6 Description of Evaluation Criteria[a]

Criterion	Worst Reading	Best Reading	QC	Relative Weight (%)
1: Change in runout	62	0	Smaller	20
2: Lead	25	0	Smaller	20
3: Involute	65	18	Nominal	15
4: Flatness	125	0	Smaller	10
5: Size of tooth	24	16	Nominal	10
6: Bore shape (ovality)	60	0	Smaller	10
7: Bore size in plane 1	104	35	Nominal	5
8: Bore center shift in plane 1	270	0	Smaller	5
9: Bore center shift in plane 2	25	0	Smaller	5

[a]Worst and best values are determined by considering the readings from all test samples.

TABLE 13.7 Evaluation of Performance for Two Samples in Each of 16 Trial Conditions[a]

	Sample 1									Sample 2								
Trial	C1	C2	C3	C4	C5	C6	C7	C8	C9	C1	C2	C3	C4	C5	C6	C7	C8	C9
1	9	12	13	48	15	16	14	0	12	32	11	16	33	15	15	18	18	13
2	24	5	0	110	17	3	20	17	25	62	8	5	125	17	10	25	14	18
3	29	3	2	56	20	50	20	60	16	30	8	5	72	20	28	12	74	8
4	10	10	47	72	17	14	98	58	17	37	18	50	116	17	8	99	31	15
5	45	18	44	19	22	15	52	75	15	14	19	38	27	22	20	70	80	10
6	23	10	12	31	19	13	28	19	11	11	9	12	26	19	10	35	24	17
7	5	0	0	38	14	6	12	62	16	4	0	0	36	14	54	4	55	16
8	15	5	11	97	10	9	80	97	15	28	3	28	53	10	10	90	91	14
9	3	18	17	41	19	11	46	25	22	37	17	16	55	19	3	46	20	22
10	0	14	39	39	23	35	48	56	14	8	14	33	70	9	33	44	112	19
11	15	0	2	22	19	20	10	0	20	32	2	0	42	20	8	18	4	16
12	19	3	28	47	9	23	47	168	9	34	4	25	18	13	10	40	157	2
13	52	18	9	123	20	5	40	27	23	0	13	8	35	20	17	22	3	15
14	10	1	13	78	13	9	10	48	10	26	2	2	91	18	12	12	74	17
15	6	17	10	105	20	3	19	96	24	25	0	3	118	23	24	75	82	21
16	26	4	62	47	18	20	82	256	17	20	9	55	114	19	30	46	265	19

[a]C1, C2, C3, . . . are the criteria of evaluation. All evaluations are measurements in millionths of an inch.

Edit Inner Array

Array Type: L-16

	1	2	3	4	5	6	7	8	9	10	11	12	13	14	15
1	1	0	0	1	1	1	1	0	1	1	1	0	1	1	1
2	1	0	0	2	1	1	1	0	2	2	2	0	2	2	2
3	1	0	0	3	2	2	2	0	1	1	1	0	2	2	2
4	1	0	0	4	2	2	2	0	2	2	2	0	1	1	1
5	2	0	0	1	1	2	2	0	1	2	2	0	1	2	2
6	2	0	0	2	1	2	2	0	2	1	1	0	2	1	1
7	2	0	0	3	2	1	1	0	1	2	2	0	2	1	1
8	2	0	0	4	2	1	1	0	2	1	1	0	1	2	2
9	3	0	0	1	2	1	2	0	2	1	2	0	2	1	2
10	3	0	0	2	2	1	2	0	1	2	1	0	1	2	1
11	3	0	0	3	1	2	1	0	2	1	2	0	1	2	1
12	3	0	0	4	1	2	1	0	1	2	1	0	2	1	2
13	1	0	0	1	2	2	1	0	2	2	1	0	2	2	1
14	1	0	0	2	2	2	1	0	1	1	2	0	1	1	2
15	1	0	0	3	1	1	2	0	2	2	1	0	1	1	2
16	1	0	0	4	1	1	2	0	1	1	2	0	2	2	1
Total	28	0	0	40	24	24	24	0	24	24	24	0	24	24	24

Figure 13.22   Modified orthogonal array.

by combining evaluations from all nine criteria. When combining different criteria of evaluation, their relative weights (see Table 13.6), as determined by the group consensus, were maintained.

While performing analysis using *S/N* ratios, larger values are always preferred regardless of the quality characteristics of the OEC or the individual criteria. Therefore, in determining the desirable levels of the factor, greater (algebraic) values of the factor averages were chosen.

1. Review (select *OEC* from the *Edit* menu) the description of evaluation criteria and evaluations. Check to see that all evaluations are entered correctly and that the corresponding OECs are calculated. Take note of the quality characteristic of OEC as shown at the bottom of the OEC screen (Figure 13.23).

2. Proceed with analysis by selecting the *S/N analysis* option from the *Analysis* menu. Check *Smaller is better*, which is appropriate for the OEC for the evaluations. When at the *Results* screen, scroll the results to verify that all trial conditions have the same OEC values as those shown on the *OEC* screen (Figure 13.22). The results of all trial conditions are shown (pasted together) in Figure 13.24. Review the factor influences from average effects of factors at their

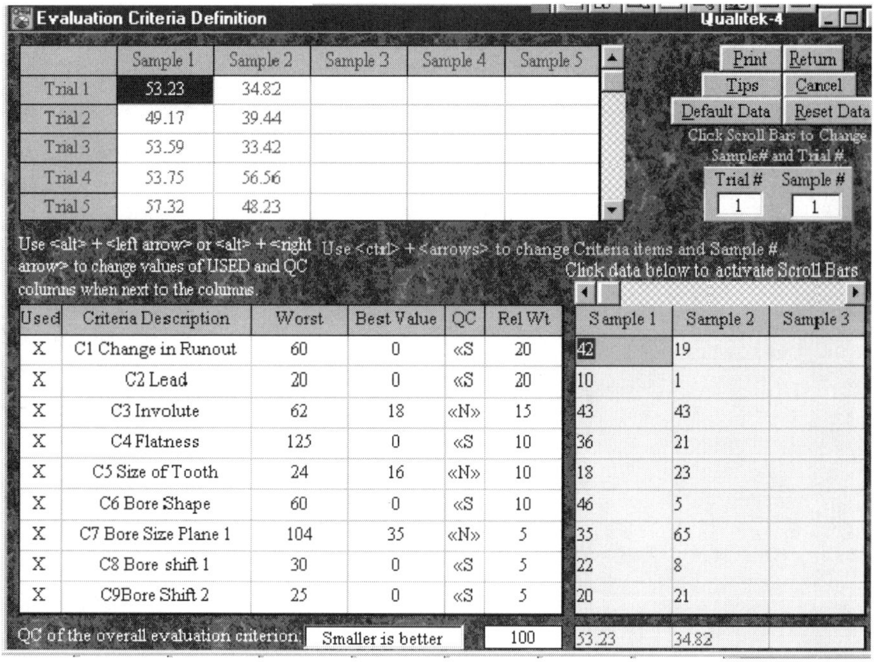

Figure 13.23 Criteria descriptions and the OEC.

respective levels in the *Main effects* screen as shown in Figure 13.25. Display and review plots if necessary and identify the desirable factor levels based on higher values (for *S/N* analysis).

3. While in the *ANOVA* screen, pool factors that are found to have a confidence level below 95%. After pooling, your *ANOVA* screen should look as shown in Figure 13.26. Make note of the relative influences of the six significant factors.

4. From the *Optimum* screen (Figure 13.27), note the levels of the factors desirable for improved performance. Notice that, as before, only the significant factors are used to calculate the performance expected at the optimum condition, as it is expected to produce a more conservative estimate.

5. As always, you will need to transform the performance expected at the optimum condition in *S/N* to a number in the original units of the results. Since the results in this case are OECs, the transformed value of *S/N* shown in Figure 13.28 is also in OEC (has no units). Based on the performance in *S/N* of −30.158, the OEC expected from samples tested at the optimum condition will be around 32.203.

By comparing ANOVA and optimum level values in the OEC analysis done above, you can put together a summary of factor influences and their levels, shown in Table 13.8. Such a list of factor influences can serve as a quick reference of influences of the

	Sample# 1	Sample# 2	Sample# 3	Sample# 4	Sample# 5	Sample# 6	S/N Ratio
Trial# 1	53.23	34.82					-33.06
Trial# 2	49.17	39.44					-32.982
Trial# 3	53.59	33.42					-32.999
Trial# 4	53.75	56.56					-34.835
Trial# 5	57.32	48.23					-34.481
Trial# 6	29.54	58.81					-33.356
Trial# 7	48.03	53.59					-34.132
Trial# 8	36.85	47.7					-32.593
Trial# 9	53.52	33.75					-33.015
Trial# 10	65.93	52.24					-35.488
Trial# 11	37.3	23.8					-29.908
Trial# 12	35.97	55.35					-33.383
Trial# 13	59.55	44.24					-34.396
Trial# 14	26.46	55.78					-32.801
Trial# 15	48.5	35.88					-32.601
Trial# 16	57.8	48.16					-34.519
							-33.409

Experimental Results — Qualitek-4
Expt. File: BKEX133A.Q4W   Data Type: S/N Ratio   QC Type: Smaller is Better

Figure 13.24 *S/N* ratios made of OEC results.

Main Effects (Average Effects of Factors and Interactions) — Qualitek-4
Expt. File: BKEX133A.Q4W   Data Type: S/N Ratio   QC Type: Smaller is Better

Column # / Factors	Level 1	Level 2	Level 3	Level 4	L2 - L1
1 J:Type of quench	-33.524	-33.641	-32.948		-.117
4 K:Material(alloy)	-33.738	-33.657	-32.41	-33.832	.081
5 A:Initial Conditi	-33.036	-33.782			-.746
6 B:Quench Temp.	-33.549	-33.27			.278
7 C:Case Depth	-32.907	-33.912			-1.006
9 D:Grain Flow(tex)	-33.858	-32.961			.896
10 F:Surface Carbon	-32.781	-34.037			-1.257
11 G:Carburizing Mtd	-33.485	-33.334			.15
13 I:Type of Cycle	-33.221	-33.598			-.378
14 E:Part Location	-33.398	-33.421			-.023
15 H:Part Orientatio	-33.712	-33.107			.605

Figure 13.25 Average effects of factors.

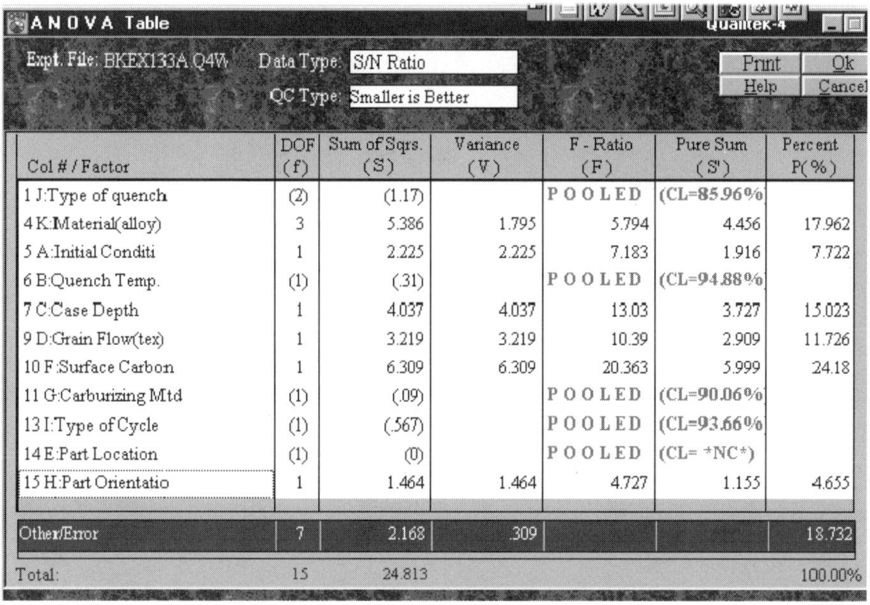

ANOVA Table — Qualitek-4

Expt. File: BKEX133A.Q4W Data Type: S/N Ratio QC Type: Smaller is Better

Print Ok Help Cancel

Col # / Factor	DOF (f)	Sum of Sqrs. (S)	Variance (V)	F - Ratio (F)	Pure Sum (S')	Percent P(%)
1 J:Type of quench	(2)	(1.17)		POOLED	(CL=85.96%)	
4 K:Material(alloy)	3	5.386	1.795	5.794	4.456	17.962
5 A:Initial Conditi	1	2.225	2.225	7.183	1.916	7.722
6 B:Quench Temp.	(1)	(.31)		POOLED	(CL=94.88%)	
7 C:Case Depth	1	4.037	4.037	13.03	3.727	15.023
9 D:Grain Flow(tex)	1	3.219	3.219	10.39	2.909	11.726
10 F:Surface Carbon	1	6.309	6.309	20.363	5.999	24.18
11 G:Carburizing Mtd	(1)	(.09)		POOLED	(CL=90.06%)	
13 I:Type of Cycle	(1)	(.567)		POOLED	(CL=93.66%)	
14 E:Part Location	(1)	(0)		POOLED	(CL= *NC*)	
15 H:Part Orientatio	1	1.464	1.464	4.727	1.155	4.655
Other/Error	7	2.168	.309			18.732
Total:	15	24.813				100.00%

Figure 13.26 ANOVA statistics from *S/N* analysis.

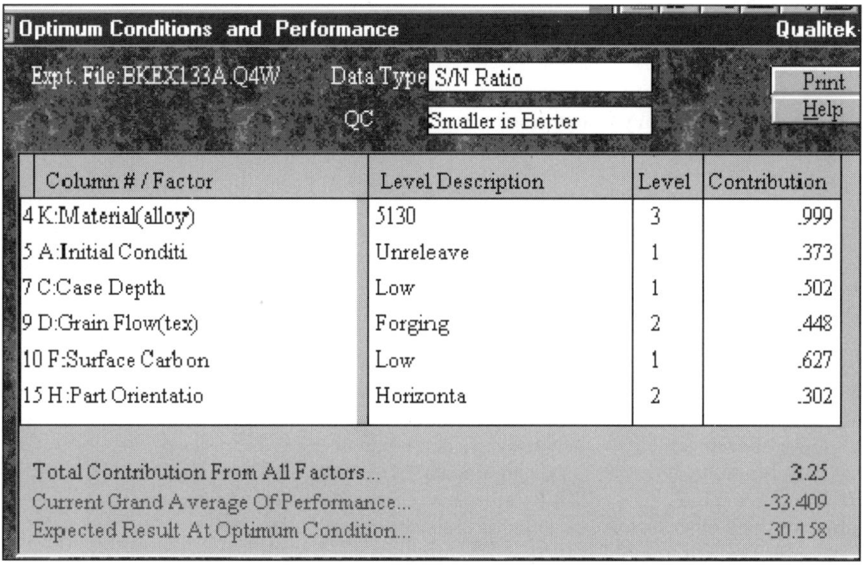

Optimum Conditions and Performance — Qualitek

Expt. File: BKEX133A.Q4W Data Type S/N Ratio QC Smaller is Better

Print Help

Column # / Factor	Level Description	Level	Contribution
4 K:Material(alloy)	5130	3	.999
5 A:Initial Conditi	Unreleave	1	.373
7 C:Case Depth	Low	1	.502
9 D:Grain Flow(tex)	Forging	2	.448
10 F:Surface Carbon	Low	1	.627
15 H:Part Orientatio	Horizonta	2	.302

Total Contribution From All Factors...	3.25
Current Grand Average Of Performance...	-33.409
Expected Result At Optimum Condition...	-30.158

Figure 13.27 Optimum condition and performance.

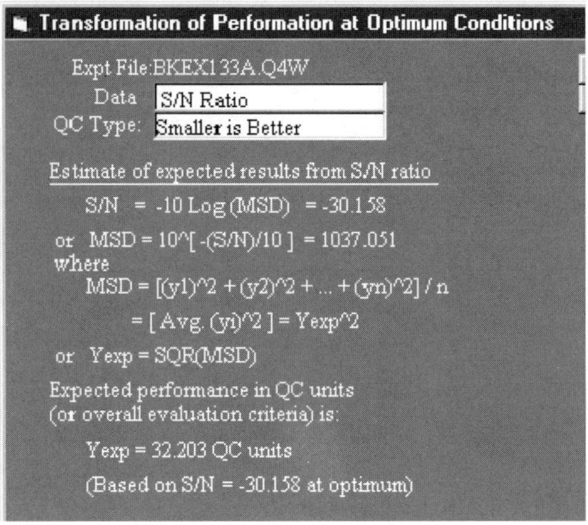

Figure 13.28 Expected OEC obtained by transforming *S/N*.

factors studied. For detailed analysis, you could also perform nine additional such analyses by taking the results of each criterion. Should you decide to do that, you could compile nine additional tables, such as the one for OEC shown in Table 13.8. Your ability to decide which factor has what influence on which criteria now becomes a difficult task. But since no additional experiments are necessary, it is a good idea to perform the individual criteria analyses (not done here) that help to understand the process better.

TABLE 13.8 Summary Conclusions

Factor	Overall Influence of Factor and Desirable Level[a]
J: type of quench	Not significant; use least costly factor level
K: material (alloy)	17.96; rank = 2; desirable level = 3 (5130)
A: initial condition	7.72%; rank = 5; desirable level = 1
B: quench temperature	Not significant; use least costly factor level
C: case depth	15.02%; rank = 3; desirable level = 1 (low)
D: grain flow (texture)	11.72%; rank = 4; desirable level = 2 (forging)
F: surface carbon	24.18%; rank = 1; desirable level = 1 (low)
G: carburizing MTD	Not significant; use least costly factor level
I: type of cycle	Not significant; use least costly factor level
E: part location	Not significant; use least costly factor level
H: part orientation	4.65%; rank = 6; desirable level = 2 (horizontal)

Influence of all other factors not included in the experiments: 18.73%
Total influence = 100% (always)

[a]Compare the *ANOVA* and *Optimum* screens.

SUMMARY

In this step you have learned how to handle a situation where results need to satisfy more than one objective. Although you were familiar with multiple evaluation criteria from Step 2, you had the opportunity to see how the concept is utilized to evaluate the results of designed experiments.

REVIEW QUESTIONS

Q. How is the quality characteristic (QC) of OEC determined?

A. The quality characteristic (QC) of an OEC is determined (automatically by QT4) based on the presence of the *bigger is better* QC for the constituent criteria. QT4 prefers to set it as *Bigger is better* as long as there is a criterion in this category with large relative weight (over 30%). You should pay special attention to what is set as the quality characteristic and be sure to use it while performing the analysis.

Q. What is the unit of OEC?

A. OEC is a dimensionless (ratio) measurement. It has no units.

Q. Should I always combine results into OEC whenever I have multiple criteria of evaluations?

A. Not necessarily. You should first analyze each criterion of evaluation (taking it as results) separately. You should go for OEC only when there are conflicts in the desirable factor levels.

Q. What does the estimated performance at the optimum condition mean when OEC is used for analysis?

A. The value predicted is an estimate of OEC expected when several samples are tested at the optimum condition.

Q. Is there a way to estimate the performance under individual criterion from the OEC predicted?

A. To determine performance at optimum condition under different criteria of evaluation, test samples at the optimum condition and compute the average performance under the criteria of interest.

Q. How is the relative weight of an individual criterion determined?

A. Relative weights of the criteria play an important role in analyses and formulation of OEC results. Generally, the relative weights are subjective and determined by consensus of the project team.

Q. How are worst and best performances determined?

A. Criteria (table) can be described completely only after the experiments are done. The best and worst readings are obtained by examining the evaluations of all samples tested under the planned experiments. Note that the best and worst are always relative to the quality characteristic of the criteria under consideration, not the highest or lowest magnitudes necessarily. These are the limits within which all evaluations in the criterion must lie.

Q. Is the OEC applicable when the experiment has an outer array?

A. As long as you have multiple objectives that are evaluated by multiple criteria of evaluations, you may consider analyzing results by combining evaluations into OEC. It does not matter whether there are multiple samples in each trial condition or noise factors are present in the design.

EXERCISES

13.1 A team of manufacturing engineers and quality professionals of an automotive supplier company performed an experiment to optimize an interior trim panel molding process that is suspected to be influenced by four control factors (Table 13.9) and three noise factors (Table 13.10). The experiment used an L-8 inner array and an L-4 outer array to assign the factors in the manner indicated in the tables. Upon completion of the test, the test samples were evaluated using the three evaluation criteria shown in Table 13.11. The quality characteristic (QC) for the OEC becomes *Bigger is better.* The evaluations under each criterion and for each of the four samples tested in all eight trial conditions were recorded as shown in Table 13.12. Run the QT4 program and open the experiment file *BKEX134A.Q4W*. Review:

- Experiment design
- Noise factors/outer array
- Trial and noise conditions
- Results created from evaluations

and carry out the analysis.

(a) Determine the main effect of the factor *cure time*.

(b) Determine the optimum condition.

TABLE 13.9 Control Factors and Levels for Exercise 13.1

Column	Factor	Level			
		1	2	3	4
1	A: Fixture type (D)	Fixture 1	Fixture 2	Fixture 3	Fixture 4
2, 3	Used for upgrading column 1 to four levels				
4	B: heat (A)	Low heat	High heat		
5	C: pressure (B)	Standard	Stepped up		
6	D: cure time (C)	Current	Increased		

TABLE 13.10   Noise Factors (L-4 Outer Array) for Exercise 13.1

	Factor	
Column Factor	1	2
1 Room temperature	Cold	Warm
2 Humidity	Low	High
3 Maintenance	Frequent	Random

(c) Determine the performance at optimum condition due to all factors having influence over 1%.

(d) Determine the range of performance expressed in terms of *S/N* and OEC at the 80% confidence interval.

(e) Determine the average effect of the noise factor *room temperature*.

(f) If trial condition 1 represents the current status, calculate the savings expected by adjusting the process to the optimum condition. Assume that the loss associated with current production of 15,000 parts per month is $2.50 per part.

(g) If you were to control the specification of two factors only, which factors will you control?

(h) If you were to relax tolerance of one of the factors studied, which will it be?

(i) If for some reason you are forced to hold *Fixture* at level 3 (i.e., go for fixture 3), what will be performance in *S/N* and the OEC?

TABLE 13.11   Evaluation Criteria Description for Exercise 13.1

Criterion	Worst Value	Best Value	QC	Relative Weighting (%)
1: Shape factor	2	9	Bigger	40
2: Surface texture	5	1	Smaller	35
3: Dimensional conformance	0.05	0.35	Nominal	25

TABLE 13.12 Evaluation of Samples for Exercise 13.1

	Run			
Trial	1	2	3	4
1	8, 4, 0.15	9, 5, 0.05	8, 4, 0.10	9, 5, 0.07
2	8, 3, 0.24	7, 4, 0.15	8, 3, 0.12	7, 4, 0.22
3	6, 4, 0.25	5, 5, 0.21	6, 5, 0.20	5, 4, 0.12
4	9, 2, 0.05	8, 1, 0.10	9, 2, 0.07	7, 2, 0.15
5	7, 3, 0.26	6, 4, 0.26	5, 3, 0.17	6, 5, 0.16
6	2, 2, 0.18	3, 1, 0.28	3, 1, 0.25	3, 1, 0.20
7	4, 3, 0.12	4, 4, 0.05	6, 3, 0.03	4, 5, 0.28
8	6, 4, 0.25	2, 2, 0.18	2, 2, 0.15	4, 2, 0.25

EXERCISE ANSWERS

13.1 Before proceeding with analysis, review experiment file *BKEX134A.Q4W* and see that its configuration is the same as shown in Figure 13.29. Check that first-row OECs are 51.36, 40, 47.2, and 41.66 with overall QC = bigger.
(**a**) 32.099 – 33.645 *S/N*; (**b**) $A_1B_2C_2D_1$; (**c**) *S/N* = 36.415; (**d**) *S/N*: 35.609 – 37.221; (**e**) 46.605 – 48.31 average value, OEC; (**f**) *S/N* from 32.947 to 36.415, \$20,624 per month; (**g**) Fixture and heat; (**h**) Pressure; (**i**) *S/N* = 35.891, OEC = 62.309.

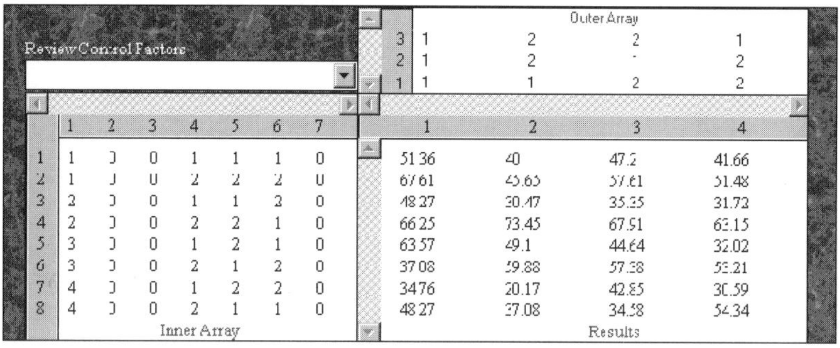

Figure 13.29 Experiment configuration.

Quantification of Variation Reduction and Performance Improvement

What You Will Learn in This Step

Most of your DOE applications allow you to determine optimum design that is expected to produce an overall better performance. The improvement in performance means that either average or variations have improved, or both have done so. When the new design is put into place, it is expected to reduce the volume of scraps and warranty. The reductions in the number of scrapped parts, amount of reworks, and incidents of warranty returns in turn off set the cost of the new design. The savings expected from the improved design in terms of dollars can be calculated by using the loss function proposed by Taguchi. In this step you will learn how to estimate the expected savings from the improvements predicted by the experimental results. Furthermore, you will also learn how the expected improvement in performance from the new design is expressed in terms of popular capability indices such as C_p and C_{pk}.

Thought for the Day

Consider the postage stamp, my son. It secures success through its ability to stick to one thing till it gets there.

—Josh Billings

One of the main reasons for completing a designed experiment is always to expect performance improvement from the new (optimum) design. Fortunately, you do not need to make any special preparation to achieve this objective. As long as you perform the experiments as planned and analyze the results correctly, you may expect an improvement in performance most of the times (over 95%). But of what kind of improvements are we speaking?

Performance improvement will generally be seen in two areas. First the average of the results will move toward the desirable value. This means that the average will be closer to the target when there is a target (QC = nominal). If, on the other hand, smaller (QC = smaller) or bigger (QC = bigger) results are desirable, the average performance will be closer to zero or a higher value, respectively. The second element of improvement is expected in terms of variation reduction of the distribution of results around the average. The variation of the expected results is commonly measured in terms of

433

its standard deviation (range and variance are also used). Improvement in performance may therefore mean improvement in either the average or standard deviation or both.

There would always be some cost (time and material) involved in carrying out planned experiments. The cost involved will depend on the scope of the study and the cost and availability of the hardware. Although cost of materials is minimum when running experiments using analytical simulations, the labor invested may not be so. Naturally, your management or your employer has the right to ask you about the benefits derived from such studies. Unfortunately, while you and your project team members may be satisfied learning about the performance improvements (average and standard deviation), they may expect you to estimate numbers in more common, and industrial, standard terms. A common presentation of performance improvement is done by drawing a distribution (assume normal) with respect to the target value and the upper and lower specification limits (USL and LSL). Such graphical presentation can also allow calculation of capability statistics such as C_p and C_{pk}, which are commonly used in the industry to compare the performance of two sources of the same product. Another interpretation of improvement, which most of your company's financial management would appreciate, is to express it in terms of dollar savings that the new design could potentially produce. How does a better design produce savings? Naturally, improved design means that more parts are made within the specification limits, and fewer parts are made outside the limits. This means reduced rejects and warranty callbacks, which reduces costs.

So how do you tell the whole world what improvement you expect from the new design? People who would need to be convinced of the expected improvement are those who will decide to incorporate the changes necessary to implement the new design. Such people would probably like to know the potential benefits of the improvement expected in terms of dollar savings. They may also appreciate the improvement when it is expressed in terms of commonly used process capability indices such as C_{pk} or C_p. Expressing savings in terms of these quantities is simple when you understand the relationships among the various distribution statistics (average, standard deviation).

HOW TO MEASURE AND EXPRESS IMPROVEMENT

Improvement expected from the new design is established by comparing it with a reference status. Commonly, the current condition is used as a basis for comparing improvement. In the simplest case, an improvement can be measured by comparing (taking the difference of) how close the average performances between the expected value and the current performance are to the desirable (depending on the QC) value. A better indication of improvement is a measure of how much the variability (standard deviation) around the average (or mean) is reduced. A third indicator, which combines both the effects of average and standard deviation, is the S/N ratio of the performance. It combines the effects of both the average and standard deviation. Here is how much benefit each of these basic measures of improvement offers:

- *Average.* Improvement is indicated by closeness to the desirable goal. Use of the average alone, however, cannot help us calculate any of the common capability indices and cost.

- *Standard deviation.* Reduction of standard deviation represents improvement. This and the average combined help us calculate the cost and capability indices.

- *S/N ratio.* This ratio combines the effects of the average and standard deviation. Since the *S/N* ratio is a preferred method of analyzing results of experiments with multiple samples tested under each trial condition, it is a natural way to measure performance improvement (larger *S/N*). Improvement measured in terms of *S/N* is necessary for calculation of savings from the improved design using the loss function proposed by Taguchi.

Interrelationships among Common Distribution Statistics

The relationships among different distribution statistics (average, standard deviation, *S/N*, loss, etc.) are slightly different for individual quality characteristics. To minimize the confusion, all calculations in this step will be done assuming that the *nominal is best* quality characteristic applies. In a simplistic way, all quality characteristics can be viewed as a form of *nominal* case where the averages are at different locations depending on the QC. The average may be viewed as being at zero when QC is *smaller is better* and a larger value when the QC is *bigger is better*.

Example 14.1: Evaluation of a Population Performance Determine the common distribution statistics for the following readings from a lot of 9-volt batteries: 9.5, 8.3, 8.6, 9.2, 9.6, 10, 8.8, and 9.3. Show the distribution of data expected from the population of batteries.

Solution: As part of the common distribution statistics, you will need to calculate average (y_a), standard deviation (σ), mean-squared deviation (MSD), and *S/N* ratio (*S/N*). Other performance measures mentioned earlier would all be calculated using these numbers. You can calculate all such calculations using special calculation options in QT4. Be aware that QT4 does not save calculations of this nature in any experiment file. To perform calculations for this example:

1. Run QT4 and go to the main screen (you need not work with any particular experiment file).
2. Select the *Population statistics* option from the *Loss/savings* menu items.
3. While at the *Population* screen (Figure 14.1), enter the observations (9.5, 8.3, etc.) under the *User data* box, one number in each row. When you have entered all data, check the *Nominal is best* quality characteristic option and enter the target value as 9.0 in the input box (Figure 14.1). You need not *Order* (press this button) data before calculation.
4. Click the *Calculate* button and obtain the statistics desired (Figure 14.1).

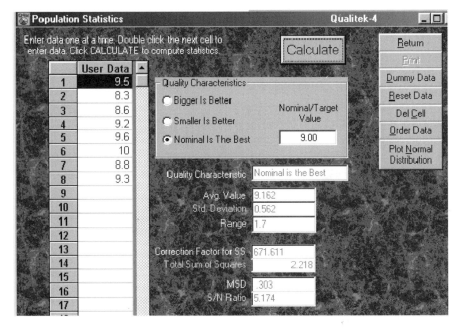

Figure 14.1 Calculation of distribution statistics.

Average: $y_a = 9.162$

Standard deviation: $\sigma = 0.562$

Mean-squared deviation: $MSD = 0.303$

S/N ratio: $S/N = 5.174$

5. Click the *Plot normal distribution* button to see the plot of the population distribution shown in Figure 14.2. (Normal distribution is defined mathematically using only the average and the standard deviation.)

Example 14.2: Comparison of Population Performances Heat-treated samples from two furnaces showed the following Rockwell C hardness readings. If the desired hardness is 60, compare the performances of the two furnaces and identify the better of the two.

Furnace 1: 55, 54, 56, 57, 54.5, 55.3, 57, and 56.5
Furnace 2: 64.5, 59, 58.4, 66, 63, 65, 56, 64, and 69

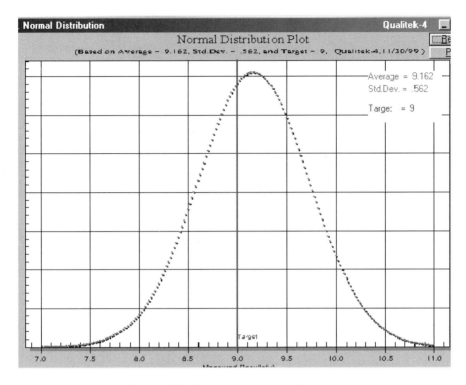

Figure 14.2   Plot of population distribution.

Solution: Follow the steps outlined in Example 14.1 to calculate the population sta-tistics for furnaces 1 and 2 as shown in Figures 14.3 and 14.4. List the distribution pa-rameters for the two furnaces with hardness values shown in Table 14.1. This example will show you how the *S/N* ratio can work as a better distribution statistic for compari-son purposes. If you look at the average performance, furnace 1 (average 55.662) ap-pears worse than furnace 2 (62.766), as it is farther away from the target (60). If, on the other hand, you compare the variability indicated by the standard deviations (1.138 and 4.152), furnace 1 looks better than furnace 2. In situations like this, *S/N* can help you make your final decision. It combines the effect of both mean and variability with equal emphasis. Comparing *S/N* ratios, you will find that furnace 1 is slightly better, as −13 is greater than −13.614 (larger *S/N* is better). You can compare the plots of the two distributions in Figure 14.5.

Expressing Improvements in Terms of Dollars

Taguchi has proposed a mathematical formulation, called the *loss function* to quantify the effect of poor quality in terms of monetary units. In his formulation, lack of qual-ity, which is measured by variation around the target value, is held directly responsible for additional cost (*loss*) of production per unit of product. Since the *S/N* ratio repre-

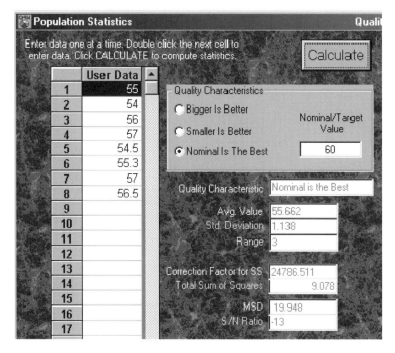

Figure 14.3 Population statistics for samples from furnace 1.

sents the status of performance with respect to the variation around the target, an improvement in S/N can be translated in reduction of loss in dollar amounts. The difference between the loss before and after the experiments can be shown to produce savings.

The loss function and its history of development have been well described in most texts on the subject. For theory and extended background on the subject, you should review the texts listed in the References in the Appendix. Here we simply concern ourselves with the method of manipulating the relationship to express dollar saving from design improvements. *Loss* is expressed as an additional cost of production (overhead) incurred and expressed in terms of dollars per unit of product (Figure 14.6). Two expressions are commonly applicable:

$$L = \begin{cases} K(y - y_0)^2 & \text{for a single product performing at } y \\ K(\text{MSD}) & \text{for multiple products with the sample performance of a given MSD} \end{cases}$$

where L is the loss/unit product in dollars, K a constant, y_0 the target value, and y the performance measure.

Before you attempt to calculate loss, it is important to understand clearly how it is calculated. For a sample calculation, assume the following production data:

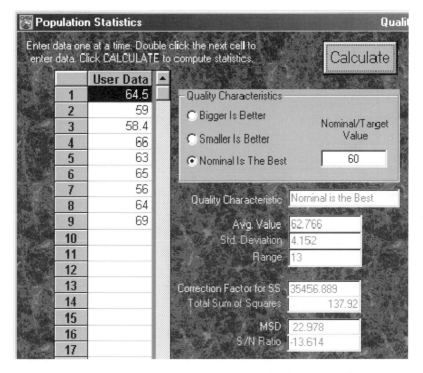

Figure 14.4   Population statistics for samples from furnace 2.

monthly production = 2000 units

monthly rejection or rework = 50 units per month

cost of rejection or rework= $8 per unit (same as cost of production)

From the data above,

TABLE 14.1   Furnace Distribution Statistics[a]

Distribution Statistic	Furnace	
	1	2
Average y_a	55.662	62.766
Standard deviation, σ	1.138	4.152
S/N ratio	−13.0	−13.614

[a]Desired performance target is 60.

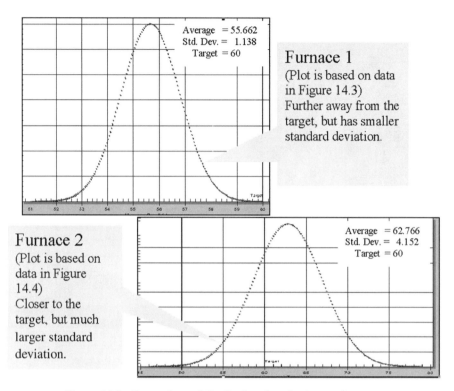

Figure 14.5   Comparison of distribution plots for the two furnaces.

total additional cost of production = number of reject × cost of rejects

$$= 50 \text{ units} \times \$8/\text{unit}$$

$$= \$400$$

Thus the loss is

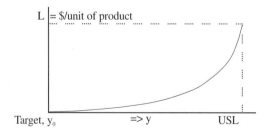

Figure 14.6   Loss function (one-sided graph).

$$L = \frac{\$400}{2000}$$

= $0.20 per unit of part

Following the same logic, you can see that $L = 0$ when the number of rejects is zero, and $L = \$8$ (cost of rejection) when the number of rejects is 2000 (all parts are made out of specification in that month). Losses at all other points between these two boundaries ($0 and 8) are assumed to follow a parabolic curve.

Example 14.3:   Expected Savings from Improved Designs Process development engineers of an automotive battery supplier completed an optimization study, which identified an improved design. The expected performance from the improved design (optimum performance) is expected to be $S/N = 18.3863$. Based on an evaluation of samples from current production, the current status is found to be $S/N = 13.233$. The target voltage of the battery is 12 volts with an acceptable tolerance of ±0.35 volt. The plant produces 15,000 batteries a month at a cost of $20 per unit. Determine the saving expected by implementing the new design (ignore the cost of implementation and training).

Solution: Saving is computed by subtracting loss at the new design from the current condition. But to calculate loss using the loss function, you will need to know a minimum amount of production data. To calculate savings using QT4, run the program and be in the main screen. Select the *Loss–nominal case* from the *Loss/savings* menu item. Enter the input data, starting with the target value, one at a time (Figure 14.7). Type your input over the default value shown. Here are explanations on the input required.

- *Target value.* This is the nominal value of performance. If the nominal or target value is unimportant, enter the average of performance. This can be the expected value or the observed current performance.

- *Tolerance of quality characteristic.* This is the limit of performance that a customer would tolerate. In many instances, this will be the specification limit (±tolerance value).

- *Cost of rejection.* This is the cost incurred to replace the item rejected. In most cases it is the cost to rework a part and is assumed to be the same as the cost of production per part.

- *Units of production.* This represents the production volume in the number of parts produced in a month.

- *S/N of Current design.* This is a number representing the status of current design in terms of S/N ratio. You will need this to compare improvement and express it in terms of dollars. If your application deals with a new product and you do not have a current design to compare with, consider the average S/N of all trial conditions as the reference value for an estimation of improvement. In case the first trial condition (all factors at level 1) of your experiment is set such

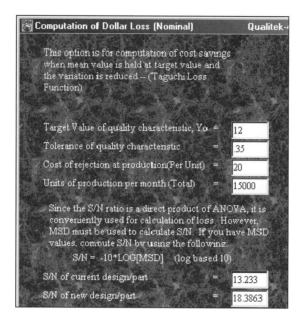

Figure 14.7   Input for savings calculation.

that it matches closely the current condition, take its *S/N* (assuming that you carried out multiple sample tests) as the current status.

- *S/N of new design.*   This is the performance expected from the new design. You will get this number in one of two ways. If you completed your experiments and performed confirmation tests, calculate the *S/N* from samples tested at the optimum condition as part of the confirmation test. If you have not yet conducted confirmation tests, use the calculated value of the expected performance (at the optimum condition, *S/N* analysis) to estimate the improvement.

Once you are finished inputting all the data, click *OK* to have QT4 perform calculations and display savings as shown in Figure 14.8. QT4 follows these steps for savings calculation. From expression of the loss function:

$$L = K(y - y_0)^2 \qquad \text{for a single unit}$$

When $y = y_0 \pm$ Tol., *loss L* = \$20 (all parts are rejected). Thus $20 = K(y_0 \pm \text{Tol.} - y_0)^2$ or $20 = K(\text{Tol.})^2$ or

$$K = \frac{20}{(0.35)^2}$$

$$= 163.265$$

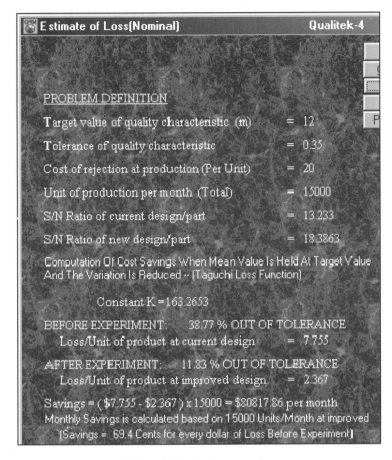

Figure 14.8   Savings expected.

With a known value of the production dependent constant K, the loss for a population of parts can be calculated using the equation

$$L = K(\text{MSD}) \qquad \text{for the population of parts}$$

The loss in the current design is

$$L = K(\text{MSD}) = 163.265 \times 0.0475 = 7.755$$

since MSD = 0.0475 (calculated from known S/N = 13.23307). The loss after experiment in the new design is

$$L = K(\text{MSD}) = 163.265 \times 0.0145 = 2.367$$

since MSD = 0.0145 (calculated from known S/N = 18.38631). The savings per month is

$$7.755 - 2.367 = \$5.388 \text{ per unit of product}$$

and the total savings per month is

$$15{,}000 \times 5.388 = \$80{,}820 \text{ (close to the QT4 value of } 80{,}817.86)$$

The saving calculated by QT4 is shown in Figure 14.8.

ESTIMATION OF VARIATION FROM KNOWN S/N RATIOS

The average and standard deviation are the two basic distribution statistics. Other distribution statistics, such as MSD, S/N, C_p, C_{pk}, and so on, are all dependent on these two statistics. But there are times when the improvement is estimated only in terms of S/N, such as the estimate of performance at optimum from S/N analysis. In such cases it is necessary to transform S/N into the standard deviation and/or mean, so that the distribution can be plotted and capability indices (C_p and C_{pk}) can be calculated.

In a typical DOE with multiple samples tested in each trial condition, you will analyze the results using S/N ratios and your estimated performance at the optimum condition will be expressed in terms of S/N. You will then need to transform the S/N value into the standard deviation and average of performance expected. Because, mathematically, it is not possible to determine two unknowns from one equation (one known value), you may need to make use of other experimentally determined data or make some assumptions. Once the average and standard deviation are found, the normal distribution can be plotted (Gaussian equation) and capability indices can be computed.

When S/N is given, MSD can be calculated as

$$\text{MSD} = 10^{-[(S/N)/10]}$$

Also, since

$$\text{MSD} = \sigma^2 + (y_a - y_0)^2$$

where y_a is the sample average and y_0 is the target value, standard deviation σ can also be calculated when y_a and y_0 are known or assumed. Note that σ used in the expression of MSD is for population, which is calculated by dividing deviation squares by N and not $N - 1$, as done for sample standard deviation. Thus, when the number of samples is known, σ must be modified using the relation

$$\sigma_{msd} = \sigma_{conventional} \left(\frac{N - 1}{N} \right)^{0.5}$$

Example 14.4: Finding Variation from Known *S/N* Value A study involving designed experiments predicted $S/N = 4.92$ as the optimum performance. If the performance at optimum condition is expected to be at the target ($y_a = y_0$), calculate the standard deviation of the performance and plot distribution with target value as 9.

Solution: QT4 has a calculator option that allows you to compute this and many such items. Once you know how to manipulate the relationships among various distribution parameters, you will be able to compute it yourself or rely on the software. Here is how you can use your scientific calculator to convert *S/N* into standard deviation:

$$MSD = 10^{-[(S/N)/10]}$$

$$= 10^{-[(4.92)/10]}$$

$$= 0.322$$

Also, since $MSD = \sigma^2 + (y_a - y_0)^2$,

$$MSD = \sigma^2 + (y_0 - y_0)^2$$

since $y_a = y_0$. Therefore, $\sigma^2 = MSD$ or

$$\sigma = (0.322)^{1/2}$$

$$= 0.567$$

You can use QT4 to perform these calculations and based on the average and standard deviation calculated, plot the expected population distribution.

1. Run QT4 and go to the main screen.
2. Select *Loss and capability statistics* from the *Loss/savings* menu item.
3. Use the middle portion of the screen to enter *S/N* on the left, and click *Calculate* to see the MSD value on the right as shown in Figure 14.9. You could also perform the reverse operation by entering MSD to find the equivalent *S/N*.
4. Now, to calculate standard deviation, enter the input parameters in the top part of the screen (Figure 14.9). Enter *S/N* under *S/N ratio* field. Enter a number, say 9, for average and target (equal value). Set the quality characteristic as *Nominal is best*, and assign two numbers as the LSL and USL. You are not interested in these limits, but since QT4 computes a few other quantities in this option, it requires these limits (any arbitrary value will not affect answers for standard deviation). Now click *Calculate* and find the standard deviation as 0.567 (Figure 14.9).

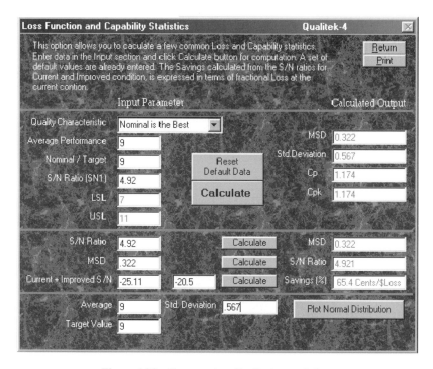

Figure 14.9 Computation distribution statistics.

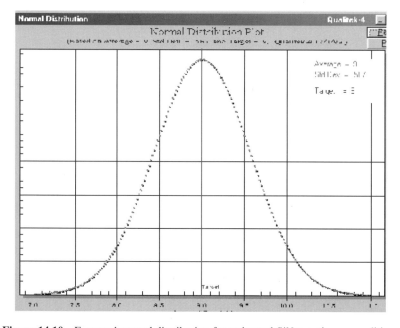

Figure 14.10 Expected normal distribution for estimated *S/N* at optimum condition.

5. Use the bottom part of the screen to plot the expected distribution. Enter the standard deviation calculated above and the average and target (see the bottom of Figure 14.9). Click *Plot normal distribution* to obtain the plot shown in Figure 14.10. (The number of samples for the distribution is assumed to be large such that the standard deviation is assumed equal for sample and population.)

RELATIONSHIP BETWEEN GAIN IN *S/N* AND STANDARD DEVIATION

Assuming that the performance is on target (QC = N), a simple relationship between standard deviation and *S/N* ratio can be established. For arbitrary σ and y_a,

$$MSD = \sigma^2 + (y_a - y_0)^2$$

when the average performance is unchanged and is at the target ($y_a = y_o$), suppose that the standard deviation at the current condition is $\sigma = \sigma_{current}$ and that at the improved design condition it is $\sigma = \sigma_{improved}$. Then from the expression for MSD above,

$$\sigma_{improved} = \sigma_{current} \left(\frac{MSD_{improved}}{MSD_{current}} \right)^{0.5}$$

$$= \sigma_{current} \; 10^{-[(S/N)improved - (S/N)current]/20}$$

or

$$\sigma_{improved} = \sigma_{current} \; 10^{-(gain/20)}$$

or

$$\frac{\sigma_{improved}}{\sigma_{current}} = 10^{-(gain/20)}$$

TABLE 14.2 Relationship between Gain and Reduction in Standard Deviation

Gain $\Rightarrow$ (*S/N*)	1	2	3	4	5	6	7	8
Reduction of σ (%)	11	21	29	37	44	50	55	60
$\dfrac{\sigma_{improved}}{\sigma_{current}}$	89	79	71	63	56	50	45	40

Sample calculation: let gain = 6. Then $\sigma_{improved}/\sigma_{current} = 10^{-gain/20} = 0.50$; reduction = $100 - 50\%$

where gain = $(S/N)_{\text{improved}} - (S/N)_{\text{current}}$ is the difference between the S/N at the improved condition and that at the current condition. This relationship can be used to establish a set of numbers that indicates percent reduction of standard deviation for different levels of gains, as shown in Table 14.2. The numbers in the third row indicate the percentage by which the standard deviation at the current condition must be multiplied to obtain the same for the improved design. The second row in the table shows that standard deviation can be reduced by half for a gain of 6 (S/N).

RELATIONSHIP BETWEEN GAIN IN S/N AND LOSS IN DOLLARS

Using the notation (subscripts) 1 for current condition and 2 for improved condition, the loss relationships can be written [L_1, $(MSD)_1$, $(S/N)_1$, L_2, $(MSD)_2$, and $(S/N)_2$, are distribution statistics]; using the general expression of loss for population of parts,

$$L = K(\text{MSD})$$

Substituting $L = L_1$ and MSD = MSD_1, the constant K can be found as

$$K = \frac{L_1}{(\text{MSD})_1}$$

Therefore,

$$L = \frac{L_1(\text{MSD})}{(\text{MSD})_1}$$

Substituting $L = L_2$ and MSD = MSD_2 in the equation above,

$$L_2 = \frac{L_1[(\text{MSD})_2]}{(\text{MSD})_1}$$

$$= \frac{L_1\left[10^{-(S/N)_2/10}\right]}{10^{-(S/N)_1/10}}$$

$$= L_1 \times 10^{[-[(S/N)_2 - (S/N)_1]/10}$$

or $L_2/L_1 = 10^{-[\text{gain}/10]}$, where gain = $(S/N)_2 - (S/N)_1$. The savings that resulted from the gain in S/N can now be expressed conveniently as a percentage of the current *loss* L_1, as

TABLE 14.3 Relationship between Gain and Savings in Terms of Percentage of Current Loss

Gain $\Rightarrow$ (S/N)	1	2	3	4	5	6	7	8
Saving (% of L_1)	21	37	50	60	68	75	80	84
L_2/L_1 (%)	79	63	50	40	32	25	20	16

Sample calculation: Let gain = 4. Then $L_2/L_1 = 10^{-4/10} = 0.40$; savings = $1 - 0.40$ or 60%

$$\text{savings} = 100 \times \frac{L_1 - L_2}{L_1}$$

$$= 100 \times \frac{L_1 - L_1 \times 10^{-[\text{gain}/10]}}{L_1}$$

$$= [1 - 10^{-(\text{gain}/10)}] \times 100\% \text{ of } L_1$$

The relationships above can be used to establish and calculate the percent savings from the improved design in terms of the loss in the current condition. By expressing the savings in terms of the loss at the current condition, the need for calculation of the constant (K) in the loss function is avoided. The numbers in the second row of Table 14.3 indicate the percent savings for gain in S/N values from the improved design. The numbers in the third row indicates percent reduction of loss in the improved condition. The numbers in the fourth column verifies that loss can be reduced to half its original value by making a 3-point gain in the S/N ratio, which is expected to result in savings of 50 cents for every dollar spent (loss, L_1) in the current condition.

RELATIONSHIP BETWEEN CAPABILITY INDICES AND STANDARD DEVIATION

The common two capability indices, C_p and C_{pk}, are easily calculated from known values of average and standard deviation. You will also need the specification limits (LSL and USL) for both indices:

$$C_p = \frac{USL - LSL}{6\sigma_{n-1}}$$

$$C_{pk} = \frac{D_L}{3\sigma_{n-1}}$$

where D_L is the smaller of $y_{\text{avg}} - LSL$ and $USL - y_{\text{avg}}$.

The capability index, C_p, depends strictly on the magnitude (inversely proportional) standard deviation for a fixed set of specification limits. C_{pk}, on the other hand, depends both on the location of the average and the magnitude of standard deviation. However, for distribution, which is centered on the target, both indices are inversely proportional to the standard deviation. Both C_p and C_{pk} will double when the standard deviation is reduced to half. In the next examples, performance improvement is calculated in terms of savings and increase in capability indices. You can use the relationships above to study and verify the results displayed by QT4.

Example 14.5: Calculation of Capability Indices In an effort to evaluate the production quality of a friction-lining material, the hardness of the material was tested and the following BHN readings were recorded: 35, 36, 31, 38, 33, 32, 34, and 36. The material has a target value of 35 BHN with LSL and USL as 28 and 41, respectively. Determine the capability indices (C_p and C_{pk}).

Solution: From the original observations you will first need to calculate the standard deviation to find the capability indices.

1. Run QT4 and go to the main screen.
2. Select *Population statistics* from the *Loss/savings* menu items.
3. Enter the hardness readings, target value (35), and click *Calculate* to obtain distribution statistics (includes *S/N* and standard deviation) as shown in Figure 14.11.

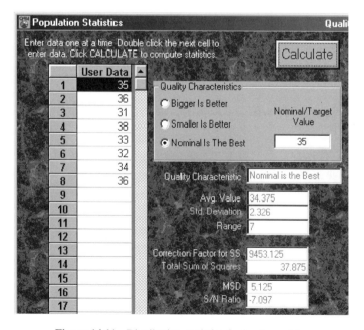

Figure 14.11 Distribution statistics for hardness data.

Using the *S/N* and average calculated, you can now proceed to calculate the capability indices.

4. Go to the main screen.

5. Select the *Loss and capability statistics* option from the *Loss/savings* menu item.

6. In the top portion of the screen, enter the average, target, *S/N*, LSL, and USL data as shown in Figure 14.12. Make sure that you have checked the *Nominal is best* quality characteristic. Now click *Calculate* to obtain the capability statistics on the right.

$$C_p = 0.995 \qquad C_{pk} = 0.976 \quad \text{(Figure 14.12)}$$

Note: Capability calculations require standard deviation only. But since QT4 is set up to work with *S/N*, standard deviation was calculated from *S/N* using an arbitrary value of number of samples ($N = 8$). As a result, the standard deviation calculated (2.175) differs slightly from the value (2.326) calculated from the original observation.

Example 14.6: Measurement of Improvement and Variation Reduction In an experiment with different industrial grinding wheel materials, surface roughness (rev/min, microinches) was used as the criterion of evaluation. The process parameter that produced a roughness value of 15 (target) was considered most desirable. The roughness values below 10 (LSL) and above 18 (USL) are considered unacceptable. Upon analysis of experimental results, performance at the optimum condition (improved design) was estimated to produce $S/N = -2.5$. A check of current production

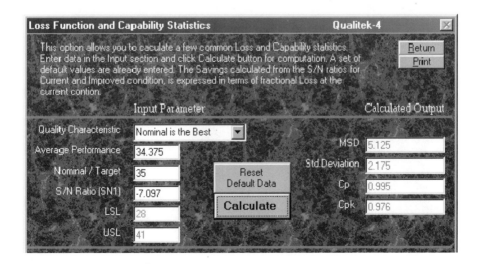

Figure 14.12   Capability statistics for hardness data.

samples showed the average performance to be 13.5 with an of $S/N = -5$. If the performance from the proposed design (optimum condition) is assumed to be on target:

(a) Compare the performance distributions.

(b) Determine the expected increase in C_{pk}.

(c) Determine the expected savings as a percentage of the loss incurred in the current design.

Solution: You can solve this problem by using the variation reduction option of QT4.

1. Run QT4 and go to the main screen.

2. Select the *Variation reduction* option from the *Loss/savings* menu item. Click *OK* at the reminder prompt (ignore this message).

3. Enter the values of the average, target, *S/N*, and so on, in the appropriate input box (Figure 14.13). Notice that there is no input field for average performance in the improved condition, as it is always assumed to be on the target.

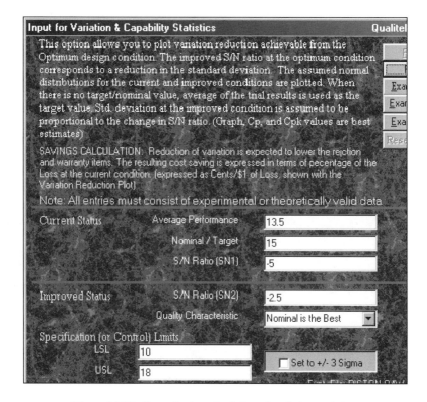

Figure 14.13 Input for the calculation of variation statistics.

4. Click *Plot* when finished entering data. Review plots of the distributions at the two conditions and find answers to the questions.

   (a) The current performance is not on target and has greater variability, as indicated by the standard deviation (0.955). The distribution in the improved condition is on target and is narrower.

   (b) The C_{pk} is increased from 1.221 at the current condition to 1.396 in the improved condition.

   (c) If the new design were implemented, 43.6% of the current loss could be eliminated. This means that the new design can potentially save 43.6 cents of every dollar spent, due to the poor quality of the current design.

Example 14.7: Quantifying Improvement in Terms of Dollar Savings, Nominal QC In the engine idle stability study of Example 12.3, determine the savings expected from the optimum (improved status) condition determined based on a nominal (target) result of 12. Assume that the current status is represented by the average of all trial results and that the LSL and USL are 7 and 20, respectively.

   Solution: While carrying out an analysis of results using QT4, you might have noticed a special function button in the *Optimum* screen labeled *Variation*. You will now find how to use this button to display the variation plot and determine savings from the improved design condition. But before using that option, you will need some information about the current status of performance (average). Proceed as follows.

1. Run QT4 and go to the main screen.
2. Open the experiment file *BKEX-123.Q4W*.
3. Select *Standard analysis* from the *Analysis* menu item. Click *Yes* at prompt and check the *smaller* or *bigger* quality characteristic. Note the value of the grand average (11.688). You will need this input for the variation plot. Click *Cancel* and return to the main screen.
4. Select the *S/N analysis* option from the *Analysis* menu item. Check the *Nominal* QC and enter 12 as the target/nominal value.
5. Proceed with the analysis and go to the optimum screen. Note that the current grand average of performance is −10.583 (*S/N*) and the expected result at the optimum condition is −4.58 (*S/N*).
6. Click on the *Variation* button. In the variation input screen (Figure 14.14), enter the data (average = 11.688, current *S/N* = −10.583, etc.) in the appropriate field. You will notice that QT4 automatically sets some of the input data. Type over them if they are not correct. Be sure to uncheck *Set to ± 3 sigma* box before entering the LSL (7) and USL (20) values.
7. Click on the *Plot* button when done inputting in all fields. The variation plot comprising distributions in the current and improved conditions is shown in Figure 14.15. Confirm that the savings indicated is 74.9 cents for every dollar spent correcting quality problems in the current condition.

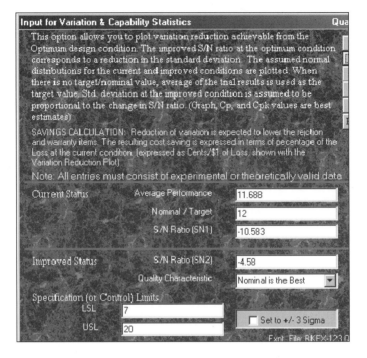

Figure 14.14 Input for variation reduction plot.

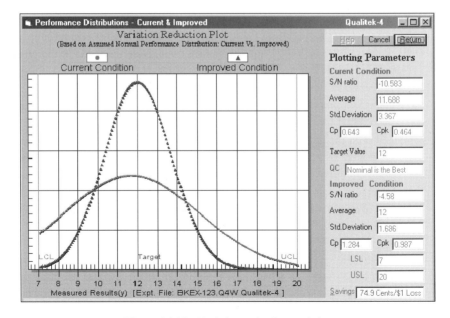

Figure 14.15 Variation reduction statistics.

Example 14.8: Quantifying Improvement in Terms of Dollar Savings, Smaller QC In the cam design study of Example 12.2, analysis (QC = S) of experimental results shows slight improvement (−12.647 to −9.856) in *S/N*. Assuming that the current average performance is 4 and the limits of results are within 2 and 8 (LSL and USL), determine the saving expected from the optimum (improved condition) design.

Solution: The variation plot option you used earlier can also be used for *smaller* and *bigger* QC's. In these cases, however, savings and other distribution statistics are only rough estimates, as they are calculated with the assumption that the average is on target even though no target is present. To determine savings, follow these calculations. This assumption of target location is required for two reasons. First, it makes calculation of standard deviation from *S/N* simpler. Second, it facilitates calculation of capability indices available as QT4 options.

1. Run QT4 and go to the main screen.
2. Open the experiment file *BKEX-122.Q4W*.
3. Select the *S/N analysis* from the *Analysis* menu item. Check the *smaller* quality characteristic.
4. Proceed with the analysis and go to the *Optimum* screen. Note that the current grand average of performance is −12.647 (*S/N*) and the expected result at the optimum condition is −9.856 (*S/N*).
5. Click on the *Variation* button. In the variation input screen shown in Figure 14.16, enter the data (average = 4.0, current *S/N* = −12.647, etc.) in the appropriate field. Uncheck *Set to* ± 3 *sigma* box before entering the LSL (2) and USL (8) values.
6. Click on the *Plot* button when done inputting in all fields. The variation plot which displays distributions in the current and improved conditions is shown in Figure 14.17. Confirm that the savings is 47.4 cents per $1 loss, which indicates that 47.4% of the amount of current loss dollars can be saved with the new design.

Example 14.9: Estimating Future Loss in Terms of Current Loss Production engineering personnel of a pump manufacturer ran an experiment to study the influence of design variables and to optimize the design. Based on the analysis of results (pump output), the proposed design was estimated to produce *S/N* = 18.5. The current status of the production pump is 15.4 and the loss is $7.50 per production item. If the monthly production is 50,000 units, determine the savings expected from the new design.

Solution: The situation presented in this example can be true for any of the three quality characteristics. To estimate savings, all you need is the gain in *S/N* and the loss at current conditions. The production volume is needed only to express the monthly savings. To solve problems of this type, you will use the QT4 option as indicated here.

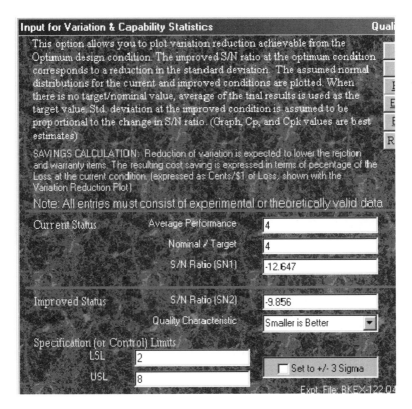

Figure 14.16 Input for variation plot.

1. Run QT4 and go to the main screen. You do not need to work with any particular experiment file.

2. Select the *Loss–general case* option from the *Loss/savings* menu item.

3. Enter the given input (loss, monthly production, *S/N*, etc.) in the appropriate boxes as shown in Figure 14.18. Click *OK* when done. Based on the given values of *S/N* and loss at the current condition, the constant for the *loss* function, *L* = *K*(MSD), is first calculated. With the known constant, the loss at the proposed (improved) design condition is calculated. The savings obtainable is shown by QT4 in the screen shown in Figure 14.19.

4. This screen (Figure 14.19) allows you to plot the loss function applicable for the production process. But for plotting purposes you need to know the target value, cost of rejection, and tolerance of the quality characteristic. For a theoretical scenario, QT4 can plot the loss diagram by assuming some arbitrary values of these parameters (try the loss diagram you should use *Loss–nominal* option under the *Loss/savings* menu item). You can get such a representation of the *loss* diagram when you click the *Plot loss* button in this screen (Figure 14.19) and see that the plot is as shown in Figure 14.20.

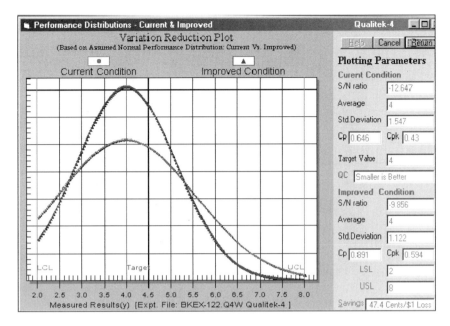

Figure 14.17 Variation plot for smaller QC.

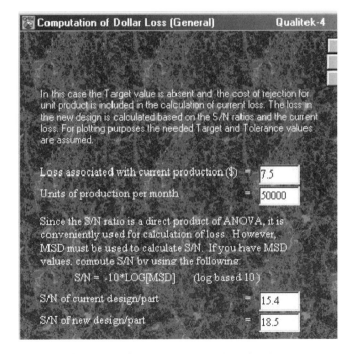

Figure 14.18 Input for calculation of saving.

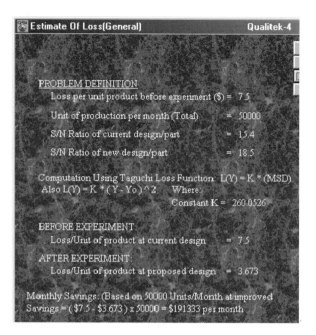

Figure 14.19   Savings from improved design.

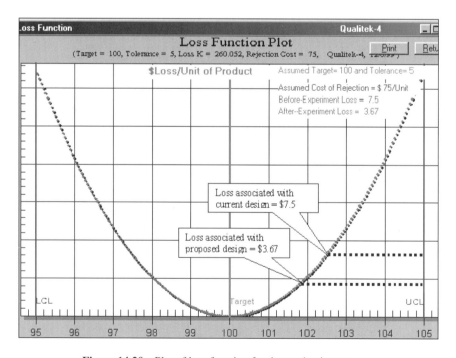

Figure 14.20   Plot of loss function for the production process.

SUMMARY

In this step you have learned how to estimate the savings expected from an improvement predicted by the experimental results. You have also learned how the expected improvement in performance from the new design is expressed in terms of popular capability indices such as C_p and C_{pk}.

REVIEW QUESTIONS

Q. What does *status* mean?

A. *Status* refers to the condition of the distribution of results (performance). A distribution is defined completely by two statistical characteristics, average and standard deviation. A single quantity such as MSD or *S/N* is made up of both average and standard deviation. When designed experiments are completed, you should use a *S/N* ratio to represent status.

Q. What is meant by *improved status*?

A. It refers to the performance at the optimum condition expressed in terms of either *S/N* or average and standard deviation.

Q. How do you know the improved status?

A. The direct way to determine status of an improved condition is to collect sample performance by running tests at that condition. When you complete a designed experiment and analyze its results, this would mean that you run the confirmation tests and express the results in terms of the *S/N* ratio. If you have not had a chance to run the confirmation tests, the performance expected at the optimum condition, in terms of *S/N*, will represent the status in the improved designed.

Q. How is the status at the current design determined?

A. You can determine the current status of your design several ways. A direct way is to collect a number of production samples, measure their performances, and calculate *S/N*. Of course, you can do that only when production samples reflecting the current design are available. If it is a new design, your project will fall under the heading of research and development. Here you will have to make an effort to define a reference for comparison purposes. The reference or current status is needed only to compare a new design and express the improvement. If your project involves a designed experiment, you can check to see if trial 1, for which all factors have a level 1, represents the current status of your design. If not, a common practice is to assume the average of all trial results (average of all trial *S/N* values) as the reference point from which all improvements can be estimated.

Q. Does an increase in *S/N* always indicate improvement?

A. Yes. Regardless of the magnitudes, if the *S/N* value of the proposed (optimum condition) design is greater than that at the current condition, you are guaranteed to reduce variation and save costs.

Q. Is there a rule of thumb to relate savings to gain in S/N?

A. You can reduce loss by half when you make a 3-point gain in S/N.

EXERCISES

14.1 Match the symbols of this formula to their proper definitions:
$L(y) = K(y - y_0)^2$
  (a) Measured value of quality characteristic.
  (b) Target value of quality characteristic.
  (c) Constant which depends on the cost structure of the manufacturing operation.

14.2 Consider a typical project in your own area and see if you can apply Taguchi's loss function to quantify the savings that may result from an improved design. Discuss with your group how you would determine the following items.
  (a) The current status of design (performance) in terms of the S/N ratio
  (b) The target value for *nominal is best* performance
  (c) The customer tolerance
  (d) The cost of rejection per unit
  (e) The rate of production per month
  (f) The expected performance (S/N ratio) of the design proposed

14.3 Taguchi's loss equation helps you calculate the dollar loss associated with a production process. Discuss how can you use it to determine the savings expected.

14.4 For the given manufacturing conditions, calculate the dollar savings per month.

Target = 5.00, customer tolerance = 0.05, cost of rejection = $21.00

Production = 18,000/month (rolled-steel machine part)

S/N before = 32.00 S/N after = 39.00

14.5 (a) For the given production parameters and Figure 14.21, determine the savings from the improved design condition.

Target/nominal value = 9 volts

Tolerance = 0.25 volt

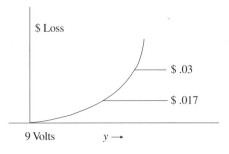

Figure 14.21   Graph for Exercise 14.5.

Cost of rejection at production = $0.15

Production per month = 450,000 units

S/N before the experiment = 19.0

S/N in the improved design = 21.25

Partial solution:

$$L = K(y - Y_0)^2$$

when $y = y_0 \pm$ tol. Then

$$K = \frac{\$0.15}{(0.25)^2} \text{ or } K = 2.40$$

Thus, $L = 2.40$(MSD). Before the experiment,

$$L = 2.40 \times \quad = \$0.03 \text{ per unit; after the experiment,}$$

$$L = 2.40 \times \quad = \$0.017 \text{ per unit}$$

Therefore

savings = $(0.03 - 0.017) \times 450,000 = \$5850$ per month

(b)   Calculate the savings resulting from the new process design, which is expected to produce $S/N = -12.60$. The current performance is at 20,000 units per month at a cost of $65 per unit. The reject rate is 400 pieces per month, and the current S/N is estimated to be at -15.0. Compute the saving expected from the new design.

Partial solution:

$$\text{Current loss} = \frac{400 \times \$65}{20{,}000} = \$1.30$$

$$L = K(\text{MSD}) \qquad \text{MSD} = 10^{-(S/N)/10} \text{ or } K = \frac{1.30}{31.62} = 0.0411$$

Therefore

$$\text{new loss } L = 0.0411(\text{new MSD}) = \$0.75 \text{ per unit}$$

Thus

$$\text{Savings} = (1.30 - 0.75) \times 20{,}000 = \$11{,}000 \text{ per month}$$

14.6   An experiment used an L-9 to study 3 three-level factors (*A*, *B*, and *C* in columns 1, 2, and 3) and 1 two-level factor (*D* in column 4, with level 3 changed to 1). Each of the nine trials was repeated three times and the results were recorded as shown in Table 14.4. The target value of the performance is 50 (*nominal is best*). The current production and cost data are as follows:

Performance: 52, 54, 45, 43, 57, 51, 53, 44, 34, 39, and 62
(measure data from 11 samples)

(Based on inspection of current production parts, calculate *S/N* for the current status.)

target = 50   tolerance = 15 (UCL = 65 and LCL = 35)

monthly production = 20,000 units

cost of repair/rejection by manufacturer = $6.5 per unit

TABLE 14.4   Results for Exercise 14.6 Trials

Trial	Results		
1	59	58	60
2	48	49	52
3	45	44	41
4	78	71	65
5	67	69	65
6	72	67	69
7	45	54	59
8	55	56	57
9	65	66	68

Analyze results using the *S/N* ratio and determine:

(a)   The *S/N* of trial 1

(b)   The effect of factor *D* (good or bad)

(c)   The optimum condition

(d)   The optimum performance in terms of *S/N*

(e)   The status (*S/N*) of the current design

(f)   The savings expected from the new design

(g)   The difference in C_{pk} (variation reduction)

EXERCISE ANSWERS

Use QT4 capability to solve Exercises 2.1 to 2.3: *Population statistics* option of *Loss/savings*.

14.1   (a) y; (b) y_0; (c) *K*.

14.2   Answers will vary.

14.3   Answers will vary.

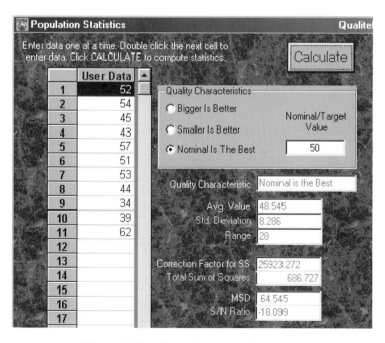

Figure 14.22   Status of current performance.

14.4 K (constant) = 8400
Before experiment: loss = $5.3 per unit
After experiment: loss = $1.05 per unit
Savings = $76,374

14.5 **(a)** $5850; **(b)** $11,000.

14.6 Run QT4 and analyze the results of experiment file *BKEX147E.Q4W* to obtain answers to questions (a) to (d). Use the *Population statistics* option in the *Loss/savings* menu item to determine the current status in *S/N* as shown in Figure 14.22. For the calculation of savings, use the *Loss–nominal* case option from the *Loss/savings* menu items.
(a) −19.121; **(b)** good; **(c)** $A_1B_2C_2D_2$; **(d)** *S/N* = −4.774 (50 ± 1.73); **(e)** −18.099; **(f)** $24,617; **(g)** will depend on the current *S/N*, UCL, LCL, etc.

Effective Experiment Preparation and Planning

What You Will Learn in This Step

Planning is necessarily the first and most important step in application of the DOE/Taguchi technique. It requires structured brainstorming with project team members. The nature of discussions in the planning session is likely to vary from project to project and is best facilitated by one who is expert in the technique. Because good planning draws on an in-depth knowledge of the subject, it is introduced as one of the last steps in this book. A recommended agenda for a typical project application planning session is discussed in this step.

Thought for the Day

The man who does not read good books has no advantage over the man who can't read them.

—Mark Twain

PREPARATION FOR EXPERIMENTATION

Two activities must precede actual conduct of the experiment: selection of the project and planning for the experimental study. After the project is selected, the information necessary to lay out the plan and conduct the experiment is found by conducting a planning session. Now that you know something about the technique, it is appropriate that we talk about elements that should be included in these preparatory stages.

As mentioned in earlier steps, your experiment will have a better chance of success when you conduct it as a project team, particularly when you work in a production environment for a large manufacturer. The only exception will be when you apply the technique to the research project and you alone have control over the destiny of the project. As a professional employee of an organization, you belong to one of these two groups. You either have project (part or process design) responsibility, or you belong to one or more groups that have responsibilities for many projects. Being acquainted with the experimental design technique and its benefit, you can now be proactive and initiate the project to optimize the designs. If you own the project, you need to explain and convince all the group members. If you are not the project owner, you can be a good team member and convince the project owner to undertake such optimization studies.

No matter who initiates the experimental study, careful planning often controls the outcome. In a broader sense, planning would include project selection and everything that must be known before the experiments can be run. As the project is launched, experiment planning must be the first and most important phase in the application process.

Project Selection

Selecting the right project for your first few experimental studies is crucial. The correct project may assure you a success or guarantee an opportunity to try it again the second time. Often, you will end up using the technique for problem-solving exercises, in which case the problem is thrust upon you. You will not have the option to decide whether to apply the technique for design optimization or problem-solving projects. But for the vast majority of cases, you will be proactive and would want to select the project that can return the most. In those situations, you may keep some of these considerations in mind.

- *Customer satisfaction.* Select the project that has the highest potential to satisfy customers (internal and external). To find such a project, ask your customers where they want improvement. Review some known problem areas. Look into where the warranty dollars are going.

- *Management support.* Ask your management about the areas they feel need improvement. Asking them is a great way to involve them in the project right from the start. After all, you would need their support, if not in carrying out the experiments, when you want to implement the design improvement determined from the project.

- *Return on investment.* When you have a number of projects for immediate application, consider the return on investment. Savings due to reduced warranty returns and increased customer satisfaction may produce a higher return in the long term. Go for a project that will give you the "biggest bang for the buck."

- *Scope of project.* Analyze the target system. Break down your product/process into a system and subsystems. Generally, the smaller the system, the better defined it is and the better the outcome will be. Working with each subsystem means that there will be too many projects. So you must make a practicable compromise. Working toward improving automobile performance, don't go for optimizing the entire vehicle, go after optimizing the suspension or braking system. If you can afford to, break the braking system down into many subsystems, and seek to optimize the surface finish of the rotor disk instead of the entire system at a time.

- *Completion possibility.* Undertake projects that have a good potential for completion. Too many projects with all good intentions remain unfinished. Anticipate all potential roadblocks. Prefer the one that has the fewest obstacles to completion. An unfinished DOE (all planned experiments need to be completed) is wasteful and detrimental to future initiatives. Think it through as much as you can, and secure funding and approval for all phases of the study.

- *Implementation of the design.* You may have found the best design or may have discovered a great new way of improving the present design, but it will be of no value if no one is asking for it. Who benefits from the new design? What do they want? If your project delivers what they want, chances are they would incorporate the improvement you would determine from the project. It would definitely pay to select a project that the project beneficiaries wanted in the first place.

- *Value to company.* You should select a project that adds long-term value to your company's future worth (stockholders' perception of the company). This would be an area that you need to explore when your products are market leaders and your company is currently enjoying higher customer satisfaction. Undertake projects to achieve levels of product/process performance that your customers are not asking for now but may ask for soon. Define your objectives such that your products perform with a wider gap from the competition.

Team Selection

Two or more members involved in a study make a project team. Six to 12 people make a good team. When the number of members in a team is larger (over 12), you have to be extra careful, for it is likely to become a burden on the facilitator. With too many people in a team, "people problems" can overwhelm the main business of a meeting. But who should form the team? Who make good team members?

The project owner should form the team. In selecting team members, the following criteria should be considered:

- *Direct involvement.* Invite everyone who has firsthand knowledge of the project. Select people who can provide input in the planning session based on experience. When planning a study of a plastic molding process, don't forget the technician who operates the machine. For a typical product, personnel involved in design, test, validation, quality, sales, and similar categories are likely candidates.

- *Your customer.* Your internal (and external) customers make good team members. What they want does truly count. This will be a great way to involve them in the project and develop a better understanding of your performance target.

- *Customer support personnel.* The customer relations and sales personnel in your organization generally have a better understanding of what end-users want. They are able to offer a perspective that engineering personnel may overlook.

- *Project beneficiaries.* Consider the beneficiaries of your project. These are people who must accept your recommendation for improvement. If you cannot include them as participants, try to bring them in the loop somehow.

- *Implementation considerations.* Think about people who would be involved in the implementation of changes that your study may recommend. Having one or more such people in your project team may save a lot of time in the end.

- *Parts/materials supplier.* Conducting experiments definitely will require time and materials, which someone will have to approve. If some of these people have a direct interest and involvement in the subject product/process, it will be good to have them in the team.

- *Involvement in planning.* Although all the qualifications noted above make one a potential team member, what is immensely important is that the person is able to offer at least a day (generally, an 8-hour day) dedicated to planning the experiment. You will not need all team members to be involved in conducting the experiment or analyzing results, nor will they all be interested in these activities. But your project will depend on the participation and input from all during the experiment planning session (brainstorming). It is also important that you secure commitment from all team members in participation in the planning session.

PLANNING AND EXECUTION OF THE EXPERIMENT

Understanding of the experimental design technique and analysis of results is necessary for successful applications. But what yields the most benefit is the way the experiment planning process is carried out. More often than not, the effort put into the planning session determines the benefits obtained. Experiment planning is one of five phases of application. You are already familiar with the other five phases of application, introduced in Step 1 and shown again in Figure 15.1. These phases are described briefly again here for your benefit.

- *Phase I: Planning.* Everything regarding the experiment's purpose and scope is determined here. A typical project should require a full 8-hour day for a planning session. Preferably, planning an experiment should be carried out in one long session rather than many short ones. The content and nature of the discussion in a planning meeting is discussed in detail in the remainder of this step.

- *Phase II: Designing.* In Steps 1 through 14 you have learned the technique of experimental design. Specifically, you perform the following tasks in this phase:
  - Design experiments and describe trial conditions.
  - Determine the order of running the experiment.

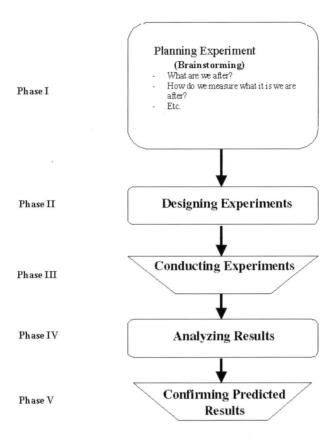

Figure 15.1   Five phases in application of DOE.

- Describe noise conditions for testing samples if the design includes an outer array.
- *Phase III: Conducting (Doing).*   You are now ready to carry out the planned experiment and evaluate the results. Specifically, you must perform the following tasks in this phase.
  - Carry out experiments in random trial order (when possible).
  - Record readings. Calculate and record averages if multiple readings of the same criteria are taken.
  - Calculate OEC (result) using the formula defined in the planning session when multiple objectives are involved.
- *Phase IV: Analyzing.*   You already know how the results of the experiments are analyzed. The specific tasks performed as part of the analyses are as follows:
  - Analyze the results.
  - Determine factor influence (main effects).
  - Identify significant factors (ANOVA).

- Determine optimum conditions and estimate performance.
- Calculate the confidence interval for optimum performance.
- Adjust design tolerances based on ANOVA.
- *Phase V: Confirming.*   In this concluding phase, you should do the following:
  - Run confirmation tests with multiple samples at the optimum condition.
  - Compare the average performance with the confidence interval determined from DOE.

By now you are familiar with the details of all the phases above except experiment planning. Following are the topics of discussion in the planning meeting with the project team.

1. Project objectives and methods of evaluation
2. Control factors and their levels
3. Interaction studies (if any)
4. Noise factors (if any)
5. Number of samples to be tested in each trial condition
6. Experiment logistics (parts/material and machines)

A formally convened planning session should be called by the project leader (project owner) and preferably, facilitated by someone else. Ideally, a facilitator must be one who has a working knowledge of the technique and is skilled in working with people. Knowledge of the product/process is not necessary. In reality, you should let someone with proven background to do the job. If you are the project owner and you end up facilitating the planning session, take extra measures to demonstrate that things are decided by consensus. To decide things under individual topics of discussion, a facilitator may consider asking questions that provoke response and participation by team members. A number of typical questions that should be asked are listed under each topic.

1. Project Objectives and Methods of Evaluation (2–4 hours). This should
be the first item of discussion. Here, as a group, you will agree on what the objectives are and their order of priority. It is quite common to identify more than one objective that members are after. It is important to assign relative weight to the objectives and establish ways that each will be evaluated. You should not be surprised if discussions on this topic take 2 to 4 hours.

- What is the quality characteristic?
- How do we evaluate the objective?
- How do we measure the quality characteristic?
- What are the units of measurement?

- What are the criteria (attributes) of evaluation for the quality characteristic?
- When there is more than one criterion of evaluation involved, do we need to combine them into a single numerical index such as an overall evaluation criterion (OEC)?
- How are the various quality criteria weighted?
- What is the sense of the quality characteristic of the OEC?

2. Control Factors and Their Levels (1–2 hours). Efficient application logistics will dictate that you complete discussions of objectives before you get to this topic. Experience shows that when objectives are discussed and agreed upon, the tasks of identifying factors and selecting a few among them for the study moves quickly. The approach here should be to list all factors identified by everyone, then reduce the list scrutinizing and qualifying each factor. Finally, select the number of factors for the study based on the size of the orthogonal array that the time and budget would allow. The factors levels are discussed and determined only after the factors to be included in the study are selected.

- What are all the possible factors?
- Do all potential factors suggested qualify as factors (must be capable of being input and be controllable)?
- Which factors are more important than others (obtained by creating a parrot diagram)?
- How many factors should be included in the study?
- How to select levels for the factors?
- How many levels of the factor should be studied?
- What is the trade-off between levels and factors?

3. Interaction Studies (1/2–1 hour). Thoughts of interaction and selection of one or more likely pairs of interaction will be burdensome to most of your project team members. You should initiate discussion on the topic and allow opportunities for the team to offer input. As you are aware, for the number (N) of factors included in the study, the number of possible interactions (among two factors) are of square order (N^2). Unless someone offers past experience and insists on inclusion of certain interactions, it will be extremely difficult to select interactions for study arbitrarily and be correct. A recommended strategy is to accommodate all factors first, then if columns are available, select a few interactions for the study.

- Which two factors are most likely to interact?
- How many pairs of 2 two-level factor interactions are possible?
- How many interactions can be included?
- Should we include an interaction or an additional factor?

- Do we need to study the interaction(s) at all?

4. Noise Factors (1/2–1 hour). As you are familiar with by now, noise factors play an important role in your experiment, particularly when the objective includes reducing variability. You will need them if you plan on running experiments that will include an outer array. You would also need to know the noise factors when you want to test multiple samples under random noise conditions. Some of the noise factors may already have been listed during the discussion of factors. This is the time to discuss their true nature and determine if they could be included in the experiment. Bear in mind that incorporating the effects of noise factors would invariably increase the time to carry out the experiment, which could make people reluctant to go for robust design strategy (with outer array). Whether or not you go for robust design, a discussion and identification of noise factor will be helpful in deciphering the results of the experiments.

- What factors are likely to influence the objective function but cannot be controlled in real life?
- How can the product under study be made insensitive to the noise factors?
- How are these factors included in the study?

5. Number of Samples to Be Tested in Each Trial Condition. By the time you are at this point of discussion, you (and the facilitator) should have a clear mental picture of what the experiment will look like. You should be able to describe the scope of the experiment to all project team members and make an effort to establish the number of samples that you want to test under each trial condition. Your discussion should reveal all matters that deals with the time and materials required for completion of the project. Questions related to test samples and a few other items that should be discussed are of the types shown here.

- How many samples can we afford to test in each trial and/or cell?
- If an outer array is utilized, should we be interested in learning about the effects of noise factors?
- What should we do with the levels of the factors not included in the study?
- How should we simulate the experiments to represent the customer/field applications?
- In what order should we run the experiments?

6. Experiment Logistics (Parts/Material and Machines). Your planning meeting should be such that you will be able to design and carry out the experiments the next day. Therefore, it will not be complete unless you decided who will perform the remaining phases (phases II to V) of application. Usually, the facilitator is the person

Project Title _____   Date & Location _____

Objectives: 1._____   2._____

            3._____   4._____

Criteria Description	Worst Value	Best Value	QC	Rel. Weighting
1.				
2.				
3.				
.				
.				

OEC/RESULT = (   ) x + (   ) x + (   ) x
               + (   ) x

Example:

FACTORS	Level 1	Level 2	Level 3	Level 4
1.				
2.				
3.				
4.				
etc.				

Interactions selected for study:

Noise factors and Outer array::

Number of samples to be tested:

Tasks and responsibilities:

Figure 15.2   Experiment planning summary.

who designs the experiment. If he or she is not the one, designate another person to do the same and ask others to identify where the necessary support will come from.

- Who will supply the parts for the test samples?
- Who will run the experiments?
- Who will collect data?
- Who will analyze the data?

You may consider keeping records of items decided in the planning session in a form similar to the one shown in Figure 15.2. Such records can serve as summary information you will need when you begin to design your experiments and lay out the trail conditions.

APPLICATION AND ANALYSIS CHECKLIST

How do you feel about tackling one of your projects? Are you ready for applications? Could you facilitate the planning session? If you read all steps and review example experiments, you should feel ready to help others to apply the technique. To test how prepared you are, here is a short checklist to confirm your skills. You should feel ready if you:

1. Know how to *plan experiments* and conduct a brainstorming session. Understand basic principles of a consensus decision-making process while working as a team. Plan your discussions to:
   a. Select criteria and prepare an evaluation table.
   b. Prepare data collection tables.
   c. Select factors for experiments.
   d. Decide on the number and values of the levels.
   e. Compromise on the number of factors and interactions.
   f. Allocate the work assignments for the project (tasks).
2. Know how to *design experiments* when (see Table 15.1 for common design solutions):
   a. All factors are at the same level (two, three, or four levels).
   b. You wish to study interactions between 2 two-level factors.
   c. Factors are at mixed levels (two, three, or four levels).
3. Can *compute* factors' *average effects* and determine:
   a. The main effects.
   b. The optimum condition.
   c. Performance at the optimum condition.
4. Understand the philosophy behind *robust design* and know how to include *noise factors* in the outer array.
5. Be familiar with the *transformation of results to MSD and S/N*, and know how to interpret the conclusions from the analysis.
6. Understand the purpose and mechanics of *ANOVA calculations* and know:
   a. How to interpret the error term.
   b. How to determine the implications of experimental results on your future efforts.
   c. Why to pool factors and when to *pool* them.
   d. How to test for significance and pool factors.
   e. How to calculate the confidence interval (C.I.) and what it means.

TABLE 15.1 Experiment Design Tips[a]

| Design # | Experiment Design Requirements | | | | Design Practices |
	2-level Factors	3-level Factors	4-level Factors	Interactions	Orthogonal Array and Column Assignments[b]
1	2–3	—	—		L-4; factors assigned to columns arbitrarily.
2	2	—	—	1: $A \times B$	L-4; factors A in col. 1, B in col. 2, and interaction $A \times B$ in col. 3.
3	4–7	—	—	—	L-8; factors cols. 1, 2, 4, and 6; remaining columns are left empty.
4	4	—	1	—	L-8; 4-level factor in col. 1, 2-level factors in cols. 4, 5, 6, and 7.
5	1–4	1	—	—	L-8; 3-level factor in col. 1, 2-level factors in cols. 4, 5, 6, and 7 as appropriate.
6	1–3	—	1	—	L-8; 3-level factor in col. 1, 2-level factors in cols. 4, 5, 6, and 7, as appropriate.
7	3–6	—	—	1: $A \times B$	L-8; factor A in col. 1, B in col. 2, and interaction of $A \times B$ in col. 3; other 2-level factors in the remaining columns.
8	3–5	—	—	2: $A \times B, B \times C$	L-8; factors A in col. 1, B in col. 2, and C in col. 4; interactions $A \times B$ in col. 3 and $B \times C$ in col. 6.
9	3–4	—	—	3: $A \times B, B \times C$, and $C \times A$	L-8; factors A in col. 1, B in col. 2, and C in col. 4; interactions $A \times B$ in col. 3, $B \times C$ in col. 6, and $C \times A$ in col. 5.
10	4	—	—	3: $A \times B, A \times C$, and $A \times D$	L-8; factors A in col. 1, B in col. 2, C in col. 4, and D in col. 7; interactions $A \times B$ in col. 3, $A \times C$ in col. 5, and $A \times D$ in col. 6.
11	4–13	—	—	2: $A \times B, C \times D$	L-16; factor A in col. 1, B in col. 2, and int. $A \times B$ in col. 3; factors C in col. 4, D in col. 8, and int. $C \times D$ in col. 12.
12	8–11	—	—	Present but ignored	L-12; assign factors to columns arbitrarily.
13	12–15	—	—	—	L-16; assign factors to columns arbitrarily.
14	—	2–4	—	—	L-9; factors are assigned arbitrarily.
15	1–2	2	—	—	L-9; dummytreat columns for the 2-level factors.

[a]Hundreds of such common experiment designs can be conceived and proposed for everyday use by experim enters. A large set of such designs is available at the Web site: http://www.rkroy.com/wp-tip.html
[b]Other possible solutions exist.

475

7. Know how to apply the *principles of brainstorming and DOE* in problem-solving projects.

SUMMARY

In this step you have learned the considerations for project and team selections. You also know how to conduct an experiment planning session with project team members. Below are a set of guidelines on experiment planning, reporting, and presenting results for quick reference.

Project Application Guidelines

1. Select a project, describe the objective, and:
   a. Establish what you are after (better looks, better performance, improved quality, etc.).
   b. Decide how you will evaluate what you are after and how many evaluation criteria you will have.
   c. When you have more than one criterion for evaluations, determine the relative weightings for each and how you will combine them into a single quantifiable number.
2. Determine factors, factor levels, interactions, noise factors, number of samples, and so on.
3. Design experiments and describe trial and noise conditions.
4. Carry out experiments and collect results.
5. Analyze results and present findings to team members.
6. Run confirmation tests.

Content of Study Report

- Project title, date, and name of participants
- Brief description of the project and objectives (about 100 words)
- Evaluation criteria table containing criteria description, QC, weighting, and so on (if more than one criterion)
- Factors, levels, and column assignments
- Results, optimum, percent influences, C.I.
- Dollar savings and other optional items, and variation reduction diagram
- Outline of actions based on findings

Presentation Guidelines

Distribute report and present findings that explains:

- How each factor influences the results

- Which factors are significant and which are not
- The optimum condition and how much improvement is expected from the optimum design
- The potential dollar savings from the new design
- How much improvement there was in terms of variation reduction (C_{pk}, C_p, etc.).

REVIEW QUESTIONS

Q. Is experiment planning or brainstorming necessary?

A. Yes. Except when you alone are applying the technique to your own project.

Q. What is the purpose of a brainstorming (planning) session?

A. The main purpose is to decide on project objectives and determine factors and levels included in the study.

Q. Who should conduct the planning session?

A. The session should be facilitated by a person who has a good working knowledge of Taguchi methodologies. Engineers or statisticians dedicated to helping others apply this tool will make better facilitators. A facilitator need not have knowledge of the subject product/process or participate in decision making.

Q. Who should host the session?

A. The team/project leader should host the brainstorming session.

Q. Who should attend?

A. All in the project team should attend the entire session. In the absence of a project team, you should invite all who have direct knowledge and interest in the project.

Q. What is the agenda for the planning session?

A. The topics of discussion are as discussed in this step.

Q. Should the host (project leader) send out the meeting agenda with an invitation for the planning session?

A. It should not be sent out with the invitation. Instead, the invitation should include a rationale for the project and establish a need for team members to participate in the entire meeting (1/2–1 day).

Q. Where should you hold the planning meeting?

A. Hold it in a place where you can expect uninterrupted attention from team members. An off-site location has been shown to work most favorably.

EXERCISES

15.1 What are the major steps in application of the Taguchi method?

15.2 If you were to arrange a brainstorming session for an important project:

    (a) Who would you ask to facilitate the session?

    (b) Who would you invite?

    (c) Where would you hold such a session?

    (d) Would you want to distribute the discussion items beforehand?

15.3 To carry out an effective brainstorming session, the following conditions are favorable:

- The project leader is open to input from all participants and is willing to accept a consensus decision.
- All participants agree to the principle of "one person, one vote" and are willing to exchange ideas.
- In all items of group decisions, the suggestions are first accepted without criticism, then scrutinized for qualification.

Which, if any, is most difficult to obtain in your work environment? Discuss.

EXERCISE ANSWERS

15.1 Planning, designing, conducting, analyzing, and confirming.

15.2 **(a)** A person from outside if available; **(b)** all project team members; **(c)** preferably off-site; **(d)** no.

15.3 Answers will vary.

Case Studies

What You Will Learn in This Step

The knowledge you gathered so far can be a lifelong asset if you put it to practice. This step includes a wide variety of examples to cover all types of experiments. You will have a chance to practice designing them and review the experiment files already created. Use them as your guide to put into practice what you already know.

Thought for the Day

I feel that the greatest reward for doing is the opportunity to do more.

—Jonas Salk

In this step we present some case studies for your reference. After you have completed steps 1 through 15 you will be able to review these examples and identify the experiments that closely match the design that can satisfy your project application needs. Also, if you didn't have the time to review the earlier steps but are familiar with the Taguchi experimental design technique, you may jump into this step and find an experiment that may be similar to the one your project needs.

From the design point of view, experiments you conduct may be different at different times because of size and complexity. Depending on your interest in information obtainable from the experimental results and the type of analysis you wish to perform, you can add another level of complexity in your experiment. Generally, however, size and complexity in the design of an experiment determine the level of difficulty in setting up and carrying out the experiments. Once an experiment is carried out, no matter how simple or complex it is, the results could be analyzed and reanalyzed to satisfy certain information demands without running additional experiments.

Practicing engineers, quality professionals, client companies, and students of the author provided all the experimental studies included in this step. To conceal the proprietary information contained in the studies, in most cases the factors and level descriptions have been altered and the results scaled up or down. Complete descriptions including analyses (experiment files included when needed) are presented for some case studies. Others contain brief descriptions of the project. All case studies presented in this step include most of the following general contents.

Case Study Content

- Project description and participants
- Objectives and evaluation criteria
- Factors and levels
- Experiment design
- Collection of data and preparation of results
- Analysis of results (main effects, ANOVA, and optimum condition)
- Conclusions and recommendations
- Variation reduction (reduction of standard deviation) and savings (relative to trial 1 performance)
- Confirmation results (if performed)

CASE STUDY 16.1: BODY PANEL THICKNESS VARIATION REDUCTION STUDY[*]

For one of the popular passenger cars, the left-hand quarter-panel thickness variation was found to be causing interference between the seal molding and the deck lid. It was also determined to be the cause for a gap and flushness variation in the interface between the deck lid and the rear quarter panel (see Figure 16.1). The deck lid is made of a fiberglass material made by a process known as resin transfer molding. Resin transfer molding (RTM) is a low-pressure molding process in which dry fiber reinforcement, or fiber preform, is packed into a mold cavity, which has the shape of the part desired. The mold is then closed and a mixed resin and catalyst are injected under pressure into the mold, where it impregnates the preform (see Figure 16.2). After the fill cycle, the cure cycle begins, during which the mold is heated and resin polymerizes to become a rigid plastic. The process is suitable for molding both simple and complex structural components for use in the automotive and aerospace industries. Typical product sizes can range from a few square centimeters up to 20 m^2.

A typical RTM process involves the sequence of operations shown in Figure 16.3. A wide range of resin systems, including polyester, vinyl, epoxy, phenolic, methyl methacrylates, and others, are used for the process. These ingredients are combined with pigments and fillers, including aluminum trihydrates and calcium carbonates, to form the molding compound. The fiber pack can be either glass, carbon, arimid, or a combination of these. A large variety of weights and styles are commonly available.

RTM has been used in structural applications in other industries. It gained popularity in the automotive industry on APX-produced composite body panels for many passenger vehicles. The major interest in RTM composite parts compared to steel lies in part consolidation and reduction in the number of tools involved, thus reducing capital cost and lead time for the production of new components. A major issue in this

---

[*]Review experiment file *BKCS-161.Q4W* for detail design and analysis.

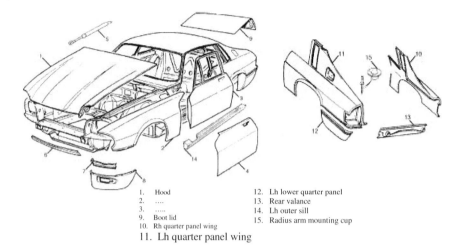

1.	Hood	12.	Lh lower quarter panel
2.	.....	13.	Rear valance
3.	.....	14.	Lh outer sill
9.	Boot lid	15.	Radius arm mounting cup
10.	Rh quarter panel wing		
11.	Lh quarter panel wing		

Figure 16.1   Automobile body panels.

new manufacturing process is the elimination of void spaces in the resin fill operation, so that products better conform to the geometry and specifications.

Objectives and Method of Evaluation.   Traditionally, the parts made by the RTM process (mold and material) are inherently overdesigned for strength. The higher strength generally comes with increased weight of the part. The objective of this study was to determine the process parameters that produce the part with minimum weight. The results of the experiments were measured in terms of part weight, and the *smaller is better* QC was used for analysis.

Factors.   A team approach was used to select the controllable factors included in the study. The team consisted of the supplier of the material, the area manager, the release engineer, and the corporate supplier quality specialist. It was determined that the facilities, equipment, and material used, although they may actually vary, were to be

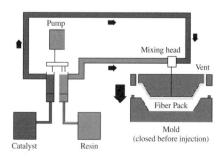

Figure 16.2   Resin transfer molding process.

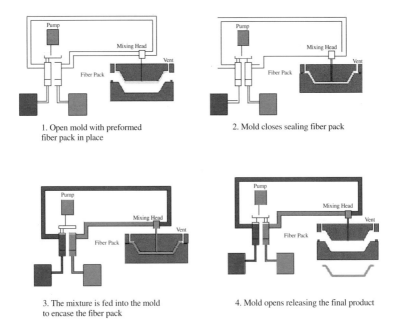

Figure 16.3 RTM process sequence.

held fixed. The variables in the study would be time (in the mold after injection), temperature (of the mold and material), and pressure (of the mold) (Table 16.1). From past history, these factors were considered to be linear, and all factors were to be studied at two levels. The interactions among the factors selected were also found to be of interest and were considered for the study.

Experimental Design. An L-8 array was used to design the experiment with 3 two-level factors and three interactions among them as shown in Figure 16.4. The eight experiments were randomly run during the same shift, the same day, and using the same material batch.

TABLE 16.1 Factors, Levels, and Interactions for Case Study 16.1

	Level	
	1	2
A: time (sec)	65	45
B: temperature (°F)	500	350
C: pressure (lb/in^2)	10	6

(Interactions: $A \times B$, $B \times C$, and $C \times A$)

	Factors	Level 1	Level 2
1	A: Hold Time	65 sec.	45 sec.
2	B: Mold Temperatu	500 deg.	350 deg. F
3	INTER COLS 1 x 2	*INTER*	------------
4	C: Pressure	10 psi	6 psi
5	INTER COLS 1 x 4	*INTER*	------------
6	INTER COLS 2 x 4	*INTER*	------------
7	COLUMN UNUSED	*UNUSED*	------------

Figure 16.4   Factor description and column assignments.

Collection of Data.   Three samples were molded in each of the eight trial conditions and the thickness (inches) of the panel was measured selecting 10 points along the deck-lid interface. The average of the 10 thickness readings was considered to be representative performance for the sample (see Figure 16.5).

Analysis of Results.   The results were analyzed using S/N ratios of the trial results, and the main effects of all the factors were obtained. Analysis of variance (ANOVA) yields the percent contribution of each factor, degrees of freedom, and F-ratio as shown in Figure 16.6. ANOVA showed that the hold time factor had the least influence on the variation. Since the interaction between mold temperature and pressure has the most influence (59.283%), their levels for the optimum design were selected based on the interaction between them (review the interaction plot in QT4).

Optimum Condition.   The expected performance at the optimum condition in terms of S/N is predicted to be -12.824, which transforms to 4.377 mm of thickness. The levels of the factors for the optimum condition are second levels of all factors. Since the time factor does not have much influence on the results, it could be set at

	1	2	3	4	5	6	7		1	2	3
1	1	1	1	1	1	1	0		4.2	5.4	4.8
2	1	1	1	2	2	2	0		6	5.4	6
3	1	2	2	1	1	2	0		6	6	4.8
4	1	2	2	2	2	1	0		5.4	4.8	4.8
5	2	1	2	1	2	1	0		4.8	5.4	6
6	2	1	2	2	1	2	0		5.4	6	5.4
7	2	2	1	1	2	2	0		4.8	5.4	6
8	2	2	1	2	1	1	0		4.8	4.2	4.2
			Inner Array								Results

Figure 16.5   L-8 orthogonal array and the results.

Expt. File: BKCS-161 Q4W	Data Type:	S/N Ratio			Print	Ok
	QC Type:	Smaller is Better			Help	Cance

Col # / Factor	DOF (f)	Sum of Sqrs. (S)	Variance (V)	F - Ratio (F)	Pure Sum (S')	Percent P(%)
1 A: Hold Time	1	.065	.065	1.829	.029	.638
2 B: Mold Temperatu	1	.517	.517	14.569	.482	10.439
3 INTER COLS 1 x 2	1	.572	.572	16.108	.537	11.623
4 C: Pressure	1	.103	.103	2.914	.068	1.472
5 INTER COLS 1 x 4	1	.551	.551	15.502	.515	11.157
6 INTER COLS 2 x 4	1	2.774	2.774	78.057	2.739	59.283
Other/Error	1	.034	.034			5.388
Total:	7	4.62				100.00%

Figure 16.6 Pooled ANOVA.

any of the levels tested. For productivity improvement, however, the least time (45 seconds) is preferred.

Variation Reduction. The improvement achieved from the new design as indicated by a gain of 1.574 in *S/N* (see Figure 16.7) was also expressed in terms of the expected savings of 29.4% of the current loss, as shown in the variation diagram of Figure 16.8.

Expt. File: BKCS-161 Q4W	Data Type	S/N Ratio		Print
	QC	Smaller is Better		Help

Column # / Factor	Level Description	Level	Contribution
1 A: Hold Time	45 sec.	2	.09
2 B: Mold Temperatu	350 deg. F	2	.254
3 INTER COLS 1 x 2	*INTER*	1	.267
4 C: Pressure	6 psi	2	.113
5 INTER COLS 1 x 4	*INTER*	1	.262
6 INTER COLS 2 x 4	*INTER*	1	.588
Total Contribution From All Factors...			1.574
Current Grand Average Of Performance...			-14.398
Expected Result At Optimum Condition...			-12.824

Figure 16.7 Optimum condition and performance.

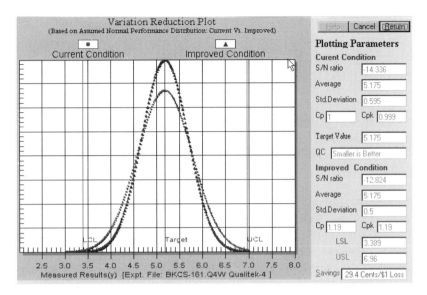

Figure 16.8   Variation reduction plot for improved body panel preform.

Conclusions and Recommendations.   The interaction between temperature and pressure was found to have the largest impact on the process. Because the error term was low (see ANOVA in Figure 16.6), it can be concluded that the factors selected for study are indeed the important ones. The factor time was found to be insignificant at the levels studied. To further improve the cycle time of the RTM process, another DOE may be performed by considering the lower time ranges. Also, any future DOE should consider the effect of these factors on the other important quality observations, such as void spaces in the resin fill operation. A confirmation test was performed and the thickness data were found to come within the predicted range.

CASE STUDY 16.2: WINDOW-CRANKING EFFORT REDUCTION STUDY[*]

It is customary to check how the component performs at the beginning of builds (assembly operations) for any new automotive product line. A group of production engineers at a local automobile manufacturer found that window-cranking efforts for one of their vehicles was in excess of the desired level (24 in.-lb.). Company performance standards specify a maximum of 18 in.-lb. As different door assemblies were tested, window efforts remained high. Realizing that the window design did not meet company objectives, a team of engineers was assigned to determine what improvements could be made to the system to reduce the torque down to the target specification of 15 in.-lb.

---

[*] Review experiment file *BKCS-162.Q4W* for detail design and analysis.

Objectives and Evaluation Criteria. The project team first undertook the task of understanding how various components combine to produce the cranking torque. A functional schematic of crank-up torque measurements was developed as the various components of the window system were installed. Initial efforts did not reveal how the components interacted with each other. Therefore, a design of experiments (DOE) team was formed to study the problem and determine the design parameters required in reducing the window-cranking effort. Everyone in the team agreed that window cranking would be measured in inch-pounds of torque and that the design parameter for minimum torque will be sought.

Experimental Design. Five design factors shown in Figure 16.9, were selected for study. Based on the available design options, two levels were considered for each factor. An L-8 orthogonal array was used to design the experiment, which allowed studies of two interactions and five factors. The team chose interactions between the window regulator and the glass runner seals, and between the window regulator and the glass lift plates (see Figure 16.9). Variations in noise factors were not considered in the experiment, due to time and money constraints. Given the opportunity, the experiment would include variations in temperature, humidity, and test operators.

Component parts for window-cranking mechanisms were fabricated to produce one sample for each of the eight design conditions prescribed by the L-8 design. The test samples were then assembled following these assembly guidelines:

- The window regulator mechanism winds the cable onto a drum. The new window regulator design changed the drum diameter from 28 mm to 23 mm and modified the cable routing.
- The new glass runner design had changes in the hinge points and lip thickness.
- The glass lift plate mounts to the window using attaching holes. The new glass lift plates increased the distance between the holes by 2 inches, therefore increasing the hole spacing from 150 mm to 210 mm.

	Factors	Level 1	Level 2	
1	A: Window Regulat	28 mm Dru	23 mm Drum	-
2	B: Runner Seals	Old Brack	New Hinge pt.	-
3	INTER COLS 1 x 2	*INTER*	------------	-
4	C: Lift Plates	150 mm Sp	210 mm Spc.	-
5	INTER COLS 1 x 4	*INTER*	------------	-
6	D: Inner Belt	3Lips	1 Lip	-
7	E: Outer Belt	4 Lips	2Lip	-

Figure 16.9 Factor and level descriptions.

	1	2	3	4	5	6	7		1	2	3
1	1	1	1	1	1	1	1		23.1	23.1	23.1
2	1	1	1	2	2	2	2		26.4	26.4	26.4
3	1	2	2	1	1	2	2		18.7	18.7	18.7
4	1	2	2	2	2	1	1		16.5	16.5	16.5
5	2	1	2	1	2	1	2		18.7	18.7	18.7
6	2	1	2	2	1	2	1		17.6	17.6	17.6
7	2	2	1	1	2	2	1		13.2	13.2	13.2
8	2	2	1	2	1	1	2		15.4	15.4	15.4
			Inner Array								Results

Figure 16.10 L-8 array and trial results.

- Both the inner and outer belts have lips that wipe against the window. For both the inner and outer belts, the new design changed from two lips to one lip.

Data Collection. The test was performed after setting up the sample door on a test stand using the window system hardware listed previously. Data were gathered by inserting a transducer between the window regulator and the crank handle. This transducer recorded the amount of torque being applied to the regulator versus regulator handle revolutions. Computer software was used to record these data from the transducer. To ensure that all the tests were performed at the same cranking rate, a small variable stepper motor was used. Each measurement was taken three times to estimate and control measurement error. The results recorded are entered in QT4 software for analysis, as shown in Figure 16.10.

Analysis. The results of this experiment were analyzed using the *Standard analysis* option of QT4 (experiment file *BKCS-162.Q4W*) and by selecting the *smaller is better* quality characteristic. ANOVA and optimum performance from analysis are as shown in Figures 16.11 and 16.12. The ANOVA calculation showed that factors *C* and *D*

Col # / Factor	DOF (f)	Sum of Sqrs. (S)	Variance (V)	F - Ratio (F)	Pure Sum (S')	Percent P(%)
1 A: Window Regulat	1	147.014	147.014	768.996	146.823	38.891
2 B: Runner Seals	1	181.499	181.499	949.376	181.308	48.026
3 INTER COLS 1 x 2	1	16.334	16.334	85.443	16.143	4.276
4 C: Lift Plates	(1)	(1.815)		POOLED	(CL=99.96%)	
5 INTER COLS 1 x 4	(1)	(0)		POOLED	(CL=96.51%)	
6 D: Inner Belt	(1)	(1.813)		POOLED	(CL=99.99%)	
7 E: Outer Belt	1	29.038	29.038	151.895	28.847	7.641
Other/Error	19	3.632	.191			1.166
Total:	23	377.52				100.00%

Figure 16.11 ANOVA from experimental results.

| Expt. File:BKCS-162.Q4W | Data Type Average Value | | Print |
| | QC Smaller is Better | | Help |

Column # / Factor	Level Description	Level	Contribution
1 A: Window Regulat	23 mm Drum	2	-2.475
2 B: Runner Seals	New Hinge pt.	2	-2.751
3 INTER COLS 1 x 2	------------	2	-.826
7 E: Outer Belt	4 Lips	1	-1.101

Total Contribution From All Factors...	-7.153
Current Grand Average Of Performance...	18.699
Expected Result At Optimum Condition...	11.546

Figure 16.12   Optimum condition and performance.

and interaction $A \times C$ are not significant. It also indicated that the factors A and B were most influential to the variation in cranking torque. Based solely on factor averages, the *optimum condition* is $A_2B_2E_1$. Since factors C and D were found to be insignificant (shown pooled in ANOVA), they can be set at any convenient level. Levels C_1 and D_1 for these factors were chosen for the new design condition. Cranking torque at the optimum condition is estimated to be 11.546 in.-lb. The new design is being incorporated in the production vehicle.

CASE STUDY 16.3: REDUCTION OF HYDROGEN EMBRITTLEMENT IN ELECTROPLATING[*]

Hydrogen embrittlement (HE) is a potentially dangerous condition that occurs only with tensile loading in high-strength steel. For this reason, high-strength bolts used as fasteners are extremely susceptible to HE which can occur in various processes in the manufacture of high-strength bolts. One common way to reduce HE is by electroplating (process).

Hydrogen embrittlement is a common and troublesome form of stress cracking. Several theories have been proposed to explain it, but the exact mechanism is still unknown. What is known is that if hydrogen is trapped in a bolt by poor electroplating practices, it can encourage stress cracking. Bolts can fail, suddenly and unexpectedly, under a normal load. Experience has shown that by manipulating process parameters, HE can be minimized. Several experiments have been run in the recent past to study the effects of one variable at a time. To determine a more robust design, a study using a designed experiment was undertaken.

To evaluate how well the electroplating process reduces HE and increases functional life, the stress limit to which the bolts can be subjected without failure was con-

---

[*]Review experiment file *BKCS-163.Q4W* for detail design and analysis.

sidered as the criterion of evaluation. The failure stress, expressed as a percent of the ultimate stress of the material, was considered as the result. A higher percentage value was desired from the optimum design (QC = *bigger is better* for analysis). After brainstorming, the group identified a long list of factors, from which 2 two-level factors (*D* and *G*) and 5 three-level factors (*A*, *B*, *C*, *E*, and *F*) were selected for the study:

A: temperature

B: current density per square foot of surface area

C: pH

D: hydrochloric acid (HCl) concentration

E: time in hydrochloric acid (minutes)

F: time of bake (hours)

G: tank size

Several noise factors were identified to have an influence on the result, but none were considered feasible to control for experimental purposes. The experiment was designed using a modified L-18 array (levels of the second column downgraded from three to two), as shown in Figure 16.13. One sample in each of the 18 trial conditions was tested for HE. The recorded failure stress was then expressed as a percentage of the known ultimate stress for the material. Following are the results: 48, 65, 64, 32, 55, 48, 40, 62, 64, 62, 53, 58, 57, 61, 76, 60, 75, and 52%.

Observations and Conclusions. The experiment showed that hydrochloric concentration, temperature, pH, and time of bake were the only important factors. When these factors are adjusted in the new design according to the optimum condition, an improvement of 17.885 units of strength is expected. (See Figure 16.14; review experiment file *BKCS-163.Q4W* for details of experiment design and analysis of results.)

	Factors	Level 1	Level 2	Level 3
1	D:Hcl Conc.	30 %	60 %	-----------
2	G:Tank Size	Rectangul	Circular	-----------
3	A:Temperature	60 deg.F	75 deg.F	90 deg.F
4	B:Current Density	2.5 amps	5 amps	10 amps
5	C:pH	6 pH	6.5 pH	7 pH
6	E:Time in HCl	5 minutes	10 minutes	30 minutes
7	F:Bake Time	2 hours	4 hours	8 hours
8	COLUMN UNUSED	*UNUSED*	-------------	-------------

Inner Array Design

Array Type: L-18

Use <ctrl> + <arrows> to move cursor.

Figure 16.13 Factor and level descriptions.

Expt. File:BKCS-163.RES	Data Type Average Value			Print
	QC Bigger is Better			Help

Column # / Factor	Level Description	Level	Contribution
1 D:Hcl Conc.	60 %	2	4.222
3 A:Temperature	75 deg.F	2	4.499
5 C:pH	7 pH	3	5.499
7 F:Bake Time	4 hours	2	3.666

Total Contribution From All Factors...	17.885
Current Grand Average Of Performance...	57.333
Expected Result At Optimum Condition...	75.219

Figure 16.14   Optimum performance and factor levels.

CASE STUDY 16.4: WATER-JET CUTTING PROCESS STUDY[*]

Water-jet cutting is used in the automotive industry for many applications, such as headliners, interior trim, carpet, and instrument panels. A typical water-jet cutting machine (robot) is shown in Figure 16.15. It can also be used to cut metal parts such as aluminum wheels and automotive tooling fixtures. Figure 16.16 shows a specimen metal plate undergoing water-jet cutting operation. When cutting metal parts, the surface roughness can vary greatly. Often, surface roughness is critical in meeting automotive specifications. Cycle time can also become a key issue when productivity adjustment is desired. A DOE project was undertaken to determine the process parameters for improvement in surface roughness.

Method of Evaluation.   The criterion for evaluation of the cutting process was the surface roughness of the cut metal. The surface roughness was measured with a profilometer in terms of micrometers (μm). The expected reading of the roughness was found to range between 0.05 and 5 μm. *Smaller is better* was selected as the quality characteristic for analysis of the results since surface roughness is smoother with lower numbers.

Factors and Levels.   Several factors were considered to influence the surface roughness and cycle time of a water-jet cutting process. Since the surface roughness depends on the material being cut, only aluminum material was used for the test. Nozzle angle, nozzle speed, water pressure, and stream diameter are all known to affect the surface roughness and were tested at three levels each. These factors and their levels are described in Table 16.2.

Noise factors that are considered to affect the process include the vibration of the robot arm, the quality of the water supply (i.e., soft vs. hard water), and whether or

[*]Review experiment file *BKCS-164.Q4W* for detail design and analysis.

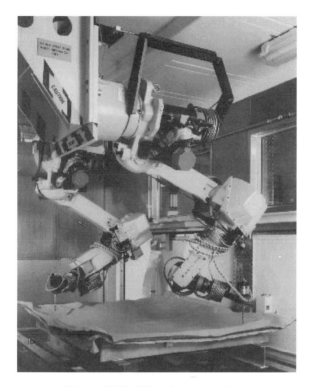

Figure 16.15   Water-jet cutting robot.

Figure 16.16   Water-jet cutting of metal plate.

TABLE 16.2   Factors and Levels Affecting Surface Roughness

	Level		
Factor	1	2	3
A: water pressure (MPa)	200	300	400
B: nozzle speed (mm/s)	250	500	750
C: nozzle angle (deg)	+5	Normal	−5
D: stream diameter (mm)	0.15	0.25	0.35

not abrasives (such as fine sand) are added to the water stream. Vibration is known to cause the water stream to vary from the tool path and cause roughness. The water quality can cause differences in the water flow through the system, and can affect the cutting capability of the stream and the longevity of the machine parts. Abrasives in the water stream can affect the cutting capability of the water stream. Water quality was simulated using softened and nonsoftened water. Robot arm vibration was controlled either by manipulating the robot parameters, or was left uncontrolled (random vibration). A set amount of abrasives was deliberately added to the water to influence the cutting process.

Array Selection and Design Strategy.   The experiment used a standard L-9 array (4 three-level factors) to study the 4 three-level factors. The effects of the three noise factors were included in the experiment by means of an L-4 outer array. Four samples in each of the nine trial conditions prescribed by the outer array were tested and results were evaluated. The trial conditions were randomly selected and all samples in the same trial were tested in sequence for expediency. The experimental design and the results obtained are shown in Table 16.3.

Observations and Conclusions

- Water pressure is found to be most influential among the factors. Nozzle angle had the least effect on surface roughness (see Figure 16.17).

- The arm vibration noise factor is most likely to cause variation in the surface finish.

- Based on *S/N* analysis (QC = S), the roughness number could be reduced to 1.8 μm from a current average of about 3 μm (see Figures 16.18 and 16.19).

The experiment provided valuable insight into the process and the rationale for several similar optimization studies.

TABLE 16.3 Experimental Design and Results

				Z: Arm Vibration	1	2	2	1
				Y: Water Quality	1	2	1	2
				X: Abrasives	1	1	2	2
Trial	A: Water Pressure	B: Nozzle Speed	C: Nozzle Angle	D: Stream Diameter	R1	R2	R3	R4
1	1	1	1	1	1.2	1.9	3.7	2.9
2	1	2	2	2	3.1	4.3	2.1	0.3
3	1	3	3	3	0.6	2.1	4.8	0.6
4	2	1	2	3	0.5	2.1	3.3	3.2
5	2	2	3	1	2.2	1.7	3.4	1.6
6	2	3	1	2	2.9	0.2	3.3	.09
7	3	1	3	2	4.6	4.7	1.7	4.7
8	3	2	1	3	0.05	4.9	2	4.2
9	3	3	2	1	0.8	2	0.7	4.4

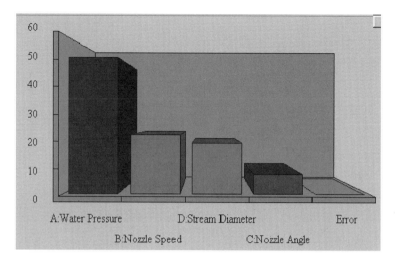

Figure 16.17   Relative influence of factors.

Optimum Conditions and Performance				Qualite

Expt. File:BKCS-164.RES   Data Type S/N Ratio   Print
                           QC   Smaller is Better   Help

Column # / Factor	Level Description	Level	Contribution
1 A:Water Pressure	300 Mpa	2	1.298
2 B:Nozzle Speed	750 mm/sec	3	1.005
3 C:Nozzle Angle	Normal	2	.437
4 D:Stream Diameter	0.15 mm	1	.941

Total Contribution From All Factors...	3.68
Current Grand Average Of Performance...	-8.796
Expected Result At Optimum Condition...	-5.115

Figure 16.18   Optimum condition and performance.

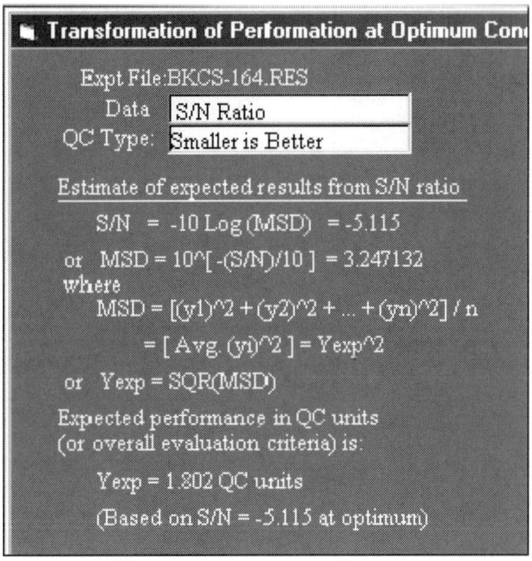

Figure 16.19   Results expected at the optimum condition.

CASE STUDY 16.5: PERFORMANCE OPTIMIZATION OF AN AIRBAG INFLATOR [*]

An automotive passenger airbag system comprises a bag, deployment door (or cover), reaction can, and inflator. The inflator is the component that generates the gas needed to fill the bag, which provides restraint to the colliding passenger body. Conventional inflators use a solid propellant, which is expensive and hazardous to produce. In addition, many propellants are toxic and require special handling and disposal. In an effort to overcome these issues, manufacturers have developed an "all-gas" family of inflators (heated gas inflator). This study was undertaken to optimize the performance of the heated gas inflator (HGI) design.

The HGI inflator is filled with a compressed mixture of air and hydrogen. A cross section of the inflator is shown in Figure 16.20. When the igniter is fired, a needle is propelled forward, removing the burst disk to both release and ignite the stored gas mix. An understanding of the significant factors and their interaction effects on inflator performance is essential to optimize the crash performance of a vehicle under accidental collisions. The knowledge gathered from this study helped preparation of an analytical simulation, which is expected to optimize HGI designs for other vehicle platforms without actually running additional tests.

Method of Evaluation.   Inflator performance was measured using ballistic tank tests. For these tests, inflators were placed in a sealed chamber of known volume and fired. Curves of resulting tank pressure vs. time and internal inflator pressure vs. time were generated for every test, using transducers. Inflator performance was evaluated

[*]Review experiment file *BKCS-165.Q4W* for detail design and analysis.

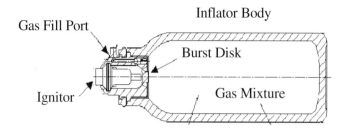

Figure 16.20 HGI inflator cross section.

using three criteria: peak tank pressure [kilopascal (kPa)], aggressiveness (or slope of inflation pressure, kPa/ms), and peak internal inflator pressure (kPa). For peak tank pressure, the highest output is desired for any given inflator size. If a particular configuration provides more output than needed, a smaller inflator can be substituted, saving cost and weight. For inflator slope, smaller is better. The less aggressive the slope (or onset) of the inflation, the less the risk of injury to small children or "out-of-position" occupants. The lower limit of slope (often not achievable) is determined by the need to deploy and fill the bag fast enough to provide restraint in a high-speed crash. For internal inflator pressure, the upper limit is determined by safety factor requirements for the inflator. If pressure exceeds the maximum allowable pressure, the inflator body may rupture. If lower pressures can be achieved, a thinner wall or weaker material may be used for the inflator body construction, reducing the cost of inflator design.

An overall evaluation criterion (OEC) was used for this experiment to create a weighted average of all three evaluation criteria. The criteria descriptions, their relative weight, and sample evaluations are shown in Table 16.4. For the sample evaluation in the table, the OEC would be calculated as follows:

$$OEC = \left| \frac{450-400}{600-400} \right| \times 50 + \left(1 - \left| \frac{12-8}{25-8} \right| \right) \times 30 + \left(1 - \left| \frac{52-45}{70.5-45} \right| \right) \times 20 = 49.95$$

TABLE 16.4 Overall Evaluation Criteria

Criterion	Worst Value	Best Value	QC	Relative Weight (%)	Sample Evaluation
1: peak tank output (kPa)	400	600	Bigger is better	50	450
2: peak tank slope (kPa/ms)	25	8	Smaller is better	30	12
3: inflator pressure (MPa)	70.5	45	Smaller is better	20	52

This OEC value takes into account the relative importance of each of the three criteria and considers their individual quality characteristics. The configuration that yields the highest OEC was chosen as the optimum design.

Factors and Levels. The 4 two-level factors included in the study are gas mix fuel percentage, gas fill pressure, gas exit nozzle area, and bottle size (volume). Table 16.5 summarizes the factors and levels to be considered. In an effort to reduce test costs, all factors are being evaluated at two levels only. This is acceptable because engineering judgment predicts that response to the factors will be approximately linear in the areas of interest. The differences between the levels have been selected to be as large as possible while keeping the practical considerations of functionality and safety in mind. Prior to running the designed experiments, a few samples were built with a few extreme conditions and tested to ensure that the factor levels selected were functional.

Three interactions—fuel% × fill pressure $(A \times B)$, fuel% nozzle area $(A \times C)$, and fill pressure × nozzle area $(B \times C)$—were also deemed likely to be significant by the engineering design team. Both fuel percentage and fill pressure increase are known to increase the inflation pressure independently. However, too high a fill pressure mixed with a higher fuel % was expected to produce some unknown inflation pressure. Reduction in nozzle size can also increase backpressure, resulting in higher burn efficiencies. Therefore, nozzle size can complement the effects of higher fuel percentage and fill pressure.

The airbag deployment process is also considered to be influenced by several manufacturing variables (noise factors) that are not possible to control in real life. Three such noise factors and their expected influence are described in Table 16.6. Although these factors are not controllable in field applications, special care was taken to adjust them for the sake of running experiments for robustness. The influence of these factors was included in the experiment by designing an outer array.

Experimental Array. The experiment was designed using an L-8 inner array to include the four factors and the three interactions described above. The 3 two-level noise factors were also included in the experiment by using an L-4 outer array. The experimental configuration, including the control factors, interactions, and noise

TABLE 16.5 Factors and Levels for Case Study 16.5

	Level	
Factor	1	2
A: fuel percentage ($\%H_2$)	12.5	14.0
B: fill pressure (lb/in^2)	2500	3200
C: nozzle area (mm^2)	120	50
D: bottle size (cm^3)	400	600

TABLE 16.6 Noise Factors

Noise Factor	Description	Level	
		1	2
X: line warm-up	The gas fill system's solenoids and valves performance may be affected by warm-up time.	<5 min run time	>1 hr run time
Y: burst disk pressure	Actual burst pressure variations will occur based on machining tolerances and variations in material properties; maximum and minimum design allowables will be considered.	6500 lb/in^2	7000 lb/in^2
Z: igniter charge weight	Pyrotechnic charge of the igniter will vary during mass production; $\pm 3\sigma$ charge weights based on vendor capability studies determine levels.	170 µg	200 µg

factors, is shown in Table 16.7. The presence of the noise factors required that four samples be tested for each of the eight trial conditions. The tests were carried out by selecting trial conditions at random, and the results of all 32 samples were evaluated under the three criteria of evaluation (see Table 16.4). The sample evaluations, which form the results (R11, R12, etc. in Table 16.7) are shown in Table 16.8. The OEC capability of the QT4 program was utilized to combine the sample evaluations (see Figure 16.21) and the results (calculated OEC) were analyzed using the S/N ratios and *bigger is better* quality characteristic.

TABLE 16.7 Experimental Configuration

							Z	1	2	2	1
							Y	1	2	1	2
							X	1	1	2	2
	A	B	$A \times B$	C	$C \times D$	$B \times C$	D	Results			
Trial	1	2	3	4	5	6	7	1	2	3	4
1	1	1	1	1	1	1	1	R11	R12		
2	1	1	1	2	2	2	2	R21	etc.		
3	1	2	2	1	1	2	2				
4	1	2	2	2	2	1	1				
5	2	1	2	1	2	1	2				
6	2	1	2	2	1	2	1				
7	2	2	1	1	2	2	1				
8	2	2	1	2	1	1	2				

TABLE 16.8 Evaluations of Experimental Samples[a]

Trial	Sample 1			Sample 2			Sample 3			Sample 4		
	EC 1	EC 2	EC 3	EC 1	EC 2	EC 3	EC 1	EC 2	EC 3	EC 1	EC 2	EC 3
1	450	12	52	445	14	57	470	11	49	411	16	51
2	462	14	68	596	20	58	511	14	47	460	9	46
3	463	12	57	489	17	61	509	22	66	435	9	50
4	519	24	49	527	24	46	422	16	54	432	14	47
5	440	8	50	493	8	48	482	11	51	489	19	54
6	404	15	49	445	14	58	445	15	56	491	18	68
7	537	12	65	597	14	64	545	13	66	565	8	58
8	486	15	47	556	16	59	508	18	52	532	12	66

[a]EC 1, evaluation criterion 1 (peak tank output); EC 2, evaluation criterion 2; etc.

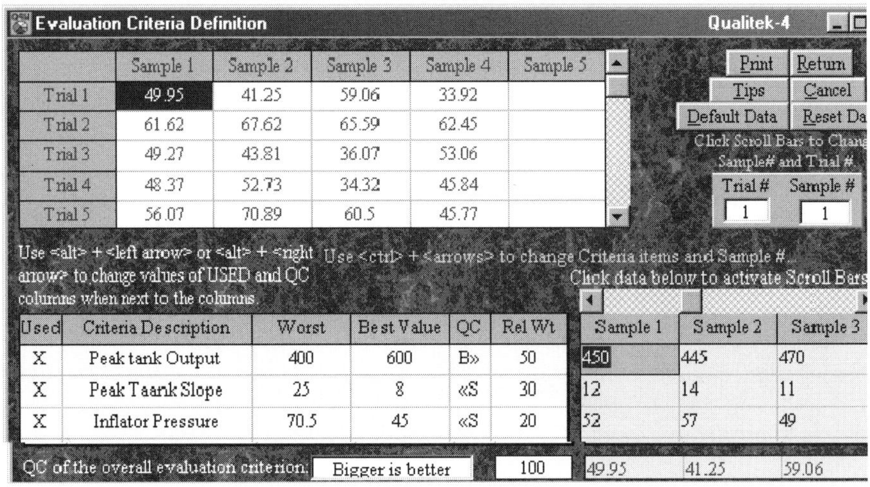

Figure 16.21 Description of evaluation criteria and evaluations for three samples of trial 1.

Observations and Conclusions

- All but one factor (*C*: nozzle area) were found to be significant.
- The experiment identified an optimum design that is expected to be insensitive to variations in the noise factors (see Figure 16.22). The analysis indicated a gain of 3.5 in *S/N* ratios, based on the OEC value of performance expected.

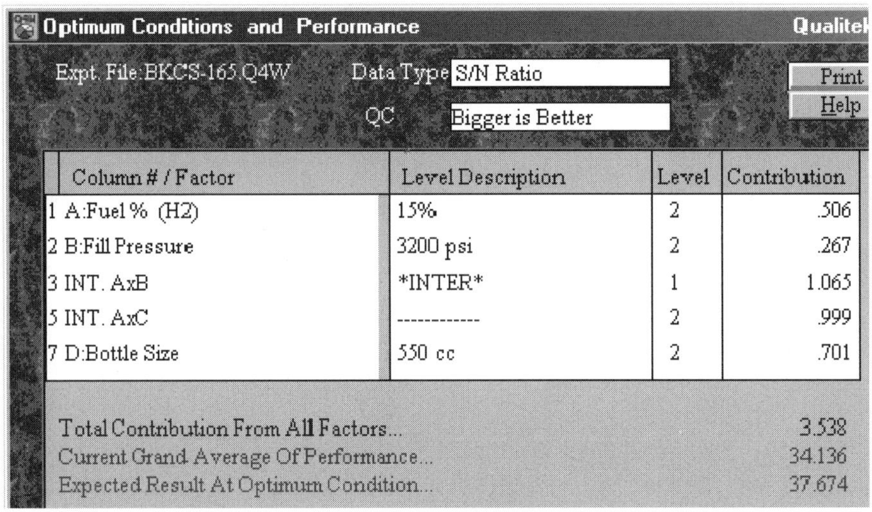

Figure 16.22 Optimum condition and the expected performance.

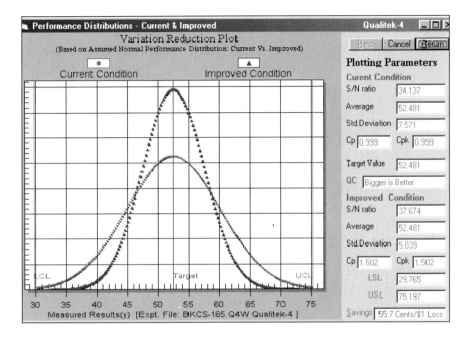

Figure 16.23   Estimates of cost savings and performance distribution expected.

- Assuming that the gain in *S/N* value is due to a reduction in variability around the target, the analysis also predicted a narrower performance distribution and 55.7% cost savings (see Figure 16.23).

CASE STUDY 16.6: SUMMARY OF LASER WELDING PROCESS STUDY

Automotive manufacturers are continuously searching for cost reduction and quality improvement programs to aid in the success of their business. This study was conducted to evaluate the feasibility of replacing the traditional resistance spot weld process with a laser welding process (see Figure 16.24). The experiments were conducted as a beta-site experiment in the assembly side of a stamping plant. The process is used to laser-weld three brackets to the cowl side of a sport utility vehicle. The ultimate goal is to determine process capability and laser process robustness in a production environment. As the beta-site experiment began the in-plant production phase, the laser-welded parts were found to contain unacceptable porosity on the surface of the weld. Porosity above a certain level is considered unacceptable in the assembly plant. Since the part is located between a wet and a dry area of the automobile, it is feared that excessive weld porosity might lead to leaking and corrosion. The primary purpose of the study was to minimize porosity by finding a suitable combination of welding process variables.

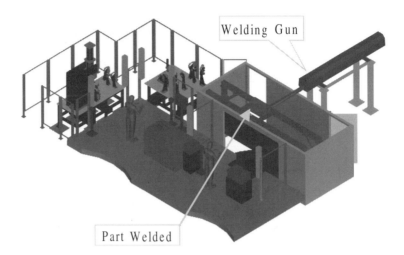

Figure 16.24 Laser welding station.

The three laser welding process factors to be tested are weld shape, laser welding speed, and laser focus. The weld shape alters porosity by allowing more gas to escape during some configurations and more weld blow-through holes in others. The laser welding speed variation and laser focus level both affect the penetration and porosity of the weld between the two panels. A higher speed is known to decrease porosity. By defocusing the laser, the penetration and porosity will also be decreased. Each of the three variables will be tested at three different levels to achieve the desired level of porosity.

Method of Evaluation. The quality of the weld was evaluated by the level of weld porosity. The porosity is visually inspected by the presence of porous holes on the front side of the weld. The level of porosity was evaluated by counting the number of visible holes in the welded part. For each test sample, several welds were inspected and the total number of porous holes counted.

Factors and Levels. The 3 three-level factors are studied to find their optimum levels to produce a nonporous weld (Table 16.9). Figure 16.25 shows a process diagram for laser welding with a brief description of the operation's input and output (results). The laser welding process also has a number of uncontrollable (noise) factors. The noise factors, shown in Figure 16.25, are factors that can change during the welding process without operator control. The plant environment and part surface temperature are suspected to have an influence on the weld quality. The part surface temperature is expected to vary when the helium gas blowing onto the surface varies in pressure and rate of flow.

TABLE 16.9   Laser Welding Factor and Level Description

Factor	Level		
	1	2	3
Laser focus level	100% original focus	90% focus (defocused)	80% focus (larger defocus level)
Feed rate (mm/s)	1750	2400	3000
Weld shape	Sine wave shape	Racetrack shape	Staple shape

Experimental Design and Results.   The experiment was designed using an L-9 inner array for the 3 three-level factors and an L-4 outer array to study the 3 two-level noise factors. The layout of the experiment called for four separate test conditions (called *cells*) in each of the nine trial conditions. Three samples were tested in each of the cells ($9 \times 4 = 36$). The cycle time of the process (including loading time and laser welding time) was approximately 30 seconds.

The optimum condition determined from this experiment reduced the rejects due to porosity by 60%. Future vehicle programs are also expected to benefit from knowledge gathered in this study regarding the behavior of laser welding process parameters.

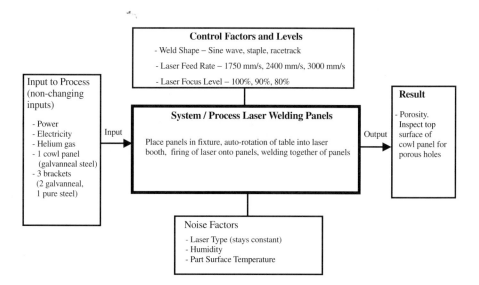

Figure 16.25   Laser welding process diagram.

CASE STUDY 16.7: SUMMARY OF PAINT AND URETHANE BOND STRENGTH STUDY

Engineering testing of paint durability is critical in determining the feasibility of a new color applied to an automobile component. Federal motor vehicle safety standards (FMVSS) require that a new compound comply to strength requirements before it can be approved for production. This study was undertaken to determine the bond strength of adhesive urethane to a painted windshield flange (Figure 16.26).

Method of Evaluation. An adhesive urethane bead is used to bond a glass windshield to a vehicle. It is applied to the painted flange of the windshield. Federal safety regulations require that the pull force needed to rip the urethane off the painted flange exceed 250 pounds per square inch. For successful bonding of the urethane bead to the windshield flange, a smooth layer of paint must first be applied to the flange. The flange is painted by the supplier [the original equipment manufacturer (OEM)]. If the painted flange has an uneven coating, the urethane does not bond completely to the area. This increases the likelihood of reduced bond strength. Past experience has shown that a single layer of paint on the flange produces a better result.

Occasionally, a cloud of paint (basecoat or clearcoat) is found on the OEM layer, usually caused by an overspray paint. In a reprocessing operation it is possible for overspray to land on a flange unit. The current assembly process requires the windshield flange to be covered or "masked" completely before a vehicle is reprocessed. This prevents overspray from damaging the flange and creating a potential uneven bonding site for the urethane. However, the masking process is very time consuming

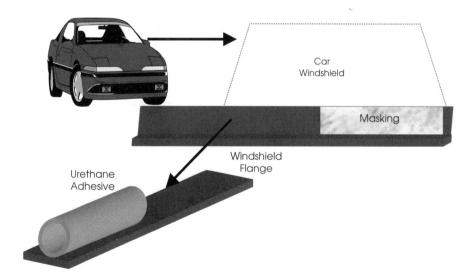

Figure 16.26 Adhesive urethane windshield application.

TABLE 16.10   Factors in Initial Experiment

Factor	Level 1	Level 2
Process	Reprocessed panels	OEM panels
Topcoat bake	350°F	250°F
Number of free rides	6	0
Basecoat thickness	High film	Overspray
Clearcoat thickness	High (3.0 mils)	Low (0.5 mil)
Paint technology	Metallic	Straight shade
Color	Dark	Light
Surface preparation	Sanded	No sanding
Timing of free rides	Before reprocessing	After reprocessing

and expensive, due to additional labor and materials. The assembly group requested that an investigation be conducted to determine if the masking requirement could be eliminated. There was considerable interest in determining if overspray, alone or in combination with other factors, could lead to a potential adhesion failure.

Factors and Experimental Design.   Nine two-level factors were identified as critical in this initial experiment (Table 16.10). An L-12 array was used to design the experiment, and three samples in each of the 12 trial conditions were tested.

Results and Conclusions.   Although the optimum design obtained from this experiment showed a significant increase in pull strength, it did not eliminate the need to mask the flange. A more comprehensive DOE study, including the noise factors, was recommended.

CASE STUDY 16.8: SUMMARY OF FINISH TURNING PROCESS OPTIMIZATION STUDY

The warranty results for an automotive supplier of a transmission shaft revealed poor quality of the finishing process as the basic cause of failure. An investigation led to the conclusion that tighter control of the diameter of the roll-formed shaft is a potential solution. As part of the finishing process, the roll-formed diameter can be either ground or turned; however, the cost of controlling this dimension through grinding is much more than that of turning. This study was launched to find a combination of turning process parameters that would minimize the variation.

Method of Evaluation.   For evaluation purposes, the machined diameters of a sample were recorded at several locations along the length. The average of the diameter readings was considered to be the representative sample diameter. To analyze the experimental results, *S/N* analysis using the *nominal is best* quality characteristic was performed.

Factors Affecting Variations. The planning session produced a long list of factors suspected to have an influence on diameter variation. The list included some factors that were easier to control economically than others.

Potential Factors

- Coolant pressure
- Coolant volume (flow rate)
- Coolant type (brand)
- Coolant concentration
- Toolholder type (key is how the toolholder locks into the turret)
- Toolholder geometry (controls the orientation of the insert)
- Rake angle
- Insert geometry
- Insert coating
- Tool nose radius
- Chip breaker design
- Material type
- Material consistency
- Cutting speed (ft/min)
- Feed rate (in./rev or mm/rev)
- Depth of cut
- Temperature
- Vibration (vibration from other machines)
- Machine (type of lathe and brand)
- Chuck design (including collet/jaw design)
- Chuck clamp pressure
- Machine foundation
- Machine leveling pads (antivibratory)

For the initial experiment, seven control factors (Table 16.11) and three noise factors were identified.

1. Vibration from environment (low to high).
2. Coolant volume (50 to 100 gal/min). The coolant volume is affected by the central system pumps and by the use of other machines on the same central system.
3. Ambient temperature (67°F) [high temperature (73°F)]. The facility temperature is known to vary widely from day to day.

TABLE 16.11 Control Factors for the Study

Factor	Level	
	1	2
Insert geometry	55° diamond	35° diamond
Tool nose radius (in.)	0.019	0.031
Rake angle	Posative	Negative
Speed (standard ft/min)	450	600
Feed rate (in./rev)	0.010	0.015
Depth of cut (in.)	0.005	0.010
Clamp pressure	Minimum	Maximum

Experimental Design and Benefits.   The experiment for this study involved an L-8 array for the 7 two-level control factors and an L-4 outer array for the 3 two-level noise factors. Analysis of the experimental results identified a new set of process parameters that significantly reduced the variation (gain in S/N of 4 points).

CASE STUDY 16.9: DRIVER COMFORT SIMULATION STUDY

Interior vehicle packaging plays an important role in customer satisfaction. Pedal location, steering wheel location, seat track travel, and seat location can all affect the driver's comfort. Vehicle design engineers are continually looking for ways to make the interior more driver-friendly in future vehicle designs. For a new vehicle that is targeted mostly for high school graduates, improving driver comfort level of people of smaller stature while maintaining the same for people of larger stature was aimed. A computer simulation tool, RAMSIS, was used to evaluate driver comfort. RAMSIS is a three-dimensional analytical model (representations) of human drivers of various statures. The software makes use of a database of driver data (gender, age, length, size, weight, etc.) from regions around the world. The required inputs for the software include three key dimensional categories: gender, age group, and country or region. In this experiment, the North American database, female gender, and 18 to 30-year-old age group will be used. In addition to these inputs, vehicle interior components (steering wheel, brake, etc.) need to be located with respect to the position of the three-dimensional model of the driver. Once positioned, the software provides a rating for overall discomfort and discomfort for specific body parts of the simulated driver, as shown in Figure 16.27. The computer model reports its evaluation such that the larger the number, the greater the discomfort. This output is used to analyze the specific design changes being considered.

Method of Evaluation.   The result was measured by using the built-in discomfort assessment functionality available in the software. No specific units are associated with the discomfort scale, which ranges from zero, the least amount of discomfort, to

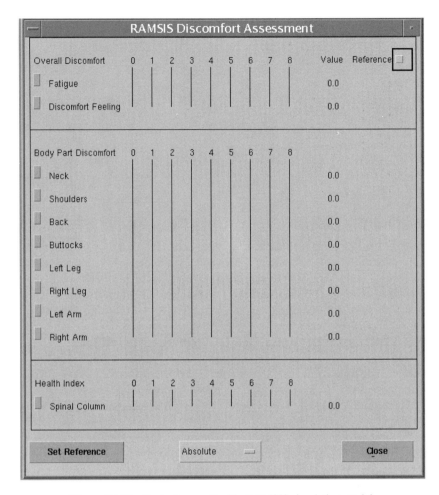

Figure 16.27   Evaluation method in RAMSIS simulation model.

eight, the greatest discomfort. The software provides a discomfort rating for a specific body part and for the overall driver. In this study, only overall comfort, which includes two categories, comfort feeling and fatigue was analyzed. Description of these criteria and their relative weights are given in Table 16.12.

TABLE 16.12   Criteria for Case Study 16.9

Criterion	Worst Value	Best Value	QC	Relative Weight (%)
Fatigue	8	0	Smaller	60
Comfort feeling	8	0	Smaller	40

Factors and Levels. Four two-level factors and three interactions among them were considered for the study (Table 16.12).

Experimental Design and Benefits. The study was carried out by setting up the computer runs based on an experiment designed using an L-8 array for the 4 two-level factors (A, B, C, and D) and the three dependent interactions ($A \times B$, $B \times C$, and $A \times C$). The results of the eight trial runs were evaluated in terms of the two criteria of evaluations described above. The evaluations were first combined into OEC values using the QT4 program, then analyzed to determine the component location for minimum driver discomfort. The results of this study provided valuable guidance to the vehicle design group.

CASE STUDY 16.10: OPTIMIZATION OF TENSILE STRENGTH OF AN AIRBAG STITCH SEAM

The integrity of an inflatable-restraint module is critical to the performance of an air-bag system when a vehicle is involved in a collision. In particular, the airbag's cushion, which is made of a soft nylon fabric sewn to form panels and chambers, is subjected to a high degree of pressure and temperature. During the deployment process, an airbag is subjected to nearly instantaneous extreme pressures and temperatures. The sewn seams are considered the weakest point in the bag, and efforts are made to assure that they are strong enough to withstand the bursting force. If the seams were to fail (i.e., tear, break, separate, etc.), the deployed airbag would not be able to provide cushioning during an accident. Reports indicate that several manufacturers' airbag programs are experiencing marginal performance and integrity with airbag seams. This study was undertaken to understand the influence of various sewing parameters and to identify the optimal seam configuration.

Method of Evaluation. In this study the tensile strength of the stitched joints was evaluated in terms of maximum force the sample was able to withstand. To evaluate strength, test samples were created from sewn fabric 4 inches wide by 5 inches long,

TABLE 16.13 Factors, Levels, and Interactions for Case Study 16.9

Factor	Level	
	1	2
A: steering wheel position	Current position	Rearward 10 mm
B: pedal position	Current position	Rearward 15 mm
C: seat height	Current	Moved down 10 mm
D: seatback angle	Current (24°)	12°

Interactions: $A \times B$, $B \times C$, and $A \times C$

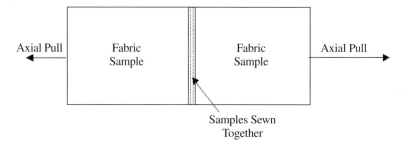

Figure 16.28 Sewn airbag fabric sample.

as shown in Figure 16.28. The test samples were then tested using an Instron Tensile Test Machine. The ultimate tensile strength data were recorded in pounds-force, and the Instron machine was used to generate a graphical output of the tensile loading, along with a summary sheet containing the sample's maximum tensile strength. The results recorded (ultimate strength) ranged between 100 and 500 pounds-force.

Factors and Levels. Five factors were selected for investigation in this study. The first factor, *uncoated fabric type*, is measured in denier, a measurement of the nylon fabric fiber's weight per unit length. Denier is a direct numbering system in which the lower numbers represent the finer sizes and the higher numbers the coarser (or heavier, thicker) sizes of fibers in the fabric. The fabric type was expected to have a significant influence on the tensile strength of the stitched seam. Four popular fabrics were chosen (four-level array) for the study.

The second factor, *stitches per inch*, indicates the number of thread stitches per inch of fabric. This factor was assigned four levels commonly used to sew the various bags produced within the company. The stitch levels used vary from 4 stitches per inch (very large, dispersed stitches) to 16 stitches per inch (very small gap between stitches).

The third factor, *thread type,* classifies the threads by the weight (number of winds) of the thread. Two commonly used thread types are "92" and "138".

The fourth factor is the *type of stitch*. There are two types of stitches. In a *chain stitch* seam, two fabric panels are joined together using a chain-link type of stitch, which is believed to be slightly weaker than a lock stitch. A chain stitch has the ability to stretch (much like a rubber band) as the seam becomes stressed. However, once it begins to tear, it opens like a zipper. A good example of a chain stitch is the type of stitching used on dog food bags. In a *lock stitch* seam, each stitch interlocks with itself. A *lock stitch* is attractive because it is believed to be stronger (against tearing), but it does not have the "give" that a chain stitch has.

The last factor that is being investigated is the *number of needle passes* to create the stitch seam. In a single needle stitch, one single-pass line of stitching comprises the seam, whereas in a double needle stitch, two passes of stitching comprise the stitched seam.

TABLE 16.14   Factors and Levels for Case Study 16.10

Main Factors	Level			
	1	2	3	4
Uncoated fabric type (denier)	210	350	525	630
Stitches per inch	4	8	12	16
Thread type (weight)	92	138		
Type of stitch	Chain	Lock		
Number of needle passes	Single	Double		

The 2 four-level factors and 3 two-level factors and level descriptions are given in Table 16.14.

Noise Factors.   A number of noise factors were identified for possible inclusion in the study: sewing machine operator, the type of sewing machine used (two different types used), and the orientation of the sample as it is clamped in the Instron tensile machine. The third noise factor (clamping orientation) applies more to test method inaccuracies rather than test results; however, it will show evidence of how the stitching reacts to *nonaxial loading* (applying force nonperpendicular to the stitched seam).

Experimental Design and Results.   An L-16 array was modified to accommodate the 2 four-level factors and the 3 two-level factors. Five samples in each of the 16 trial conditions were tested under random levels of the noise factors. The results, measured in terms of ultimate strengths, were analyzed using *S/N* ratios. The experiment identified an optimum design of the sewing process that increased the strength by 30%.

CASE STUDY 16.11: SUMMARY OF RESISTANCE SPOT WELDING STUDY

The use of aluminum in the automotive industry continues to grow, due to the demand for weight savings. Many automotive components, such as doors, contain a number of aluminum parts that are spot-welded to join with each other. The joining of inner and outer panels in an automotive door system is a typical example. Resistance spot welding is a simple and cost-effective process that can be used in high-speed automated production operation. This study investigated the resistance spot welding of an aluminum bracket used in an automobile body.

Method of Evaluation.   The overall evaluation criterion (OEC) used to evaluate the test samples combined three separate criteria of evaluation. The quality of a

TABLE 16.15   Criteria for Case Study 16.11

Criterion	Worst Value	Best Value	Quality Characteristic	Relative Weighting (%)
Shear strength (lb)	500	800	Bigger	60
Nugget diameter (mm)	3	7	Nominal	25
Appearence	0	10	Bigger	15

resistance spot weld was evaluated by examining its shear strength, nugget diameter, and appearance. *Shear strength* is a direct measure of the structural integrity of the bond. *Nugget diameter* (the width of the weld bead) is an important indicator of the quality of the weld. The *appearance* of the weld is also important when the weld is highly visible to the customer. For visible locations, the spot weld should have symmetry, no discoloration, and minimum surface indentation. The *appearance* is a subjective measure and was evaluated on an arbitrary scale from 0 to 10 (10 being the best appearance). Descriptions of the three evaluation criteria and their respective characteristics are given in Table 16.15.

Factors and Interactions.   There are a number of different factors that can affect the quality of a resistance spot weld. For this experiment, 5 two-level factors were selected, with two interactions (Table 16.16).

Experimental Design and Results.   An L-8 orthogonal array was used to design the experiment. Six brackets were welded in each of the eight trial conditions and evaluated by the three criteria of evaluation. The combined evaluation (OEC) was analyzed to determine the optimum weld condition employed in vehicle body production at present.

TABLE 16.16   Factors and Levels for Case Study 16.11

	Level	
Factors/Variables	1	2
A: Current (A)	12,000	14,000
B: Tip pressure (lb)	600	650
C: Cycles	3	5
D: Electrode material	Copper	Elkonite
E: Electrode shape	Pointed	Radius

Interactions: $A \times B$ and $A \times D$

SUMMARY

In this step you have seen a number of projects that were carried out by experimenters in manufacturing industries. These case studies show you typical products and processes in which the DOE technique can be utilized to optimize design. By reviewing the completed studies (and the experiment files, where provided), you should be well prepared to apply the technique to your own project and prepare reports on the findings.

◼◼◼ APPENDIX

GLOSSARY

ANOVA Acronym for *analysis of variance*, a statistical data treatment for sorting out the relative influence of factors to variation of results. ANOVA statistics are needed for the calculation of *confidence limits*.

C.I. *Confidence interval*. Defines limits within which the mean performance of samples tested at the *optimum condition* is expected to fall.

Controllable factors Design variable that is considered to influence the response and is included in the experiment. Its levels can be controlled at the will of the experimenter.

DOE Acronym for *design of experiments*.

Error Amount of variation in the response caused by factors other than controllable factors included in the experiment.

Evaluation criteria Measures or characteristics that are used to determine how well the objective is met.

Factors Input variables (or parameters) to the project.

Histogram Graphical representation of the sample data using classes on the horizontal axis and frequency on the vertical axis.

Interaction Activity between two factors in which the influence of each depends on the value of the other.

Levels Values or conditions of the factors utilized in carrying out an experiment.

Linear graph Graphical representation of relative column locations of factors and their interactions.

Loss function Mathematical expression proposed by Dr. Taguchi to determine quantitatively the harm caused by the lack of quality in a product.

Main effect Trend of influence of the factor to the objective/characteristics of measure. The numbers listed as *main effects* from DOE analyses have only relative, not absolute, meaning.

Noise factors Factors that have influence over the response but cannot be controlled in actual applications. They are of three kinds:

Outer noise Environmental conditions vs. humidity, temperature, operators, etc.

Inner noise Deterioration of machines, tools, and parts.

Between product noise Variation from piece to piece.

514

OEC Acronym for *overall evaluation criterion.*

Off-line quality control Quality enhancement efforts in activities before production. These are activities such as upstream planning, R&D, systems design, parameter design, tolerance design and loss function, etc.

Optimum condition Represents the combination/setting of factor levels that is expected to produce the best performance. Depending on the quality characteristic, the best performance may be the nominal, smallest, or largest value.

Optimum performance Estimate of performance at the *optimum condition.* If a number of samples are tested at the optimum condition, their average performance is expected to be close to this value.

Orthogonal array Set of tables containing information on how to determine the least number of experiments and their conditions. The word *orthogonal* means balanced.

Quality characteristic (QC) Yardstick that measures the performance of a product or process under study.

Response Quantitative value of a measured quality characteristic (e.g., stiffness, weight, flatness, etc.).

Results Numerical data representing the evaluation of performance. When multiple criteria are involved, the OECs become the results.

Robustness Condition in which a product/process is least influenced by a variation of individual factors. To become robust is to become less sensitive to variation.

Signal factor Factor that influences the average value but not the variability in response.

S/N ratio (S/N) *Signal-to-noise ratio,* the ratio of the power of the signals to the power of noise (error). A high *S/N* ratio will mean that there is high sensitivity with the least error of measurement. In Taguchi analysis using *S/N* ratios, a higher value is always desirable regardless of the quality characteristic.

System design Design of a product or a process using special Taguchi techniques.

Target value Value that a product is expected to possess. Most often this value is different from what a single unit actually does. For a 9-volt transistor battery, the target value is 9 volts.

Tolerance design Sophisticated version of parameter design used to optimize tolerance, reduce cost, and increase customer satisfaction.

Variables, factors, or parameters Words used synonymously to indicate the controllable factors in an experiment. In a plastic molding experiment, molding temperature, injection pressure, and set time are a few of the factors.

F-TABLES [*]

F-Table (90%)

				n_1			
n_2	1	2	3	4	5	6	7
1	39.864	49.3500	53.593	55.833	57.241	58.204	58.906
2	8.5263	9.0000	9.1618	9.2434	9.2926	9.3255	9.3491
3	5.5393	5.4624	5.3908	5.3427	5.3092	5.2847	5.2662
4	4.5448	4.3246	4.1908	4.1073	4.0506	4.0098	3.9790
5	4.0604	3.7797	3.6195	3.5202	3.4530	3.4045	3.3679
6	3.7760	3.4633	3.2888	3.1808	3.1075	3.0546	3.0145
7	3.5894	3.2574	3.0741	2.9605	2.8833	2.8274	2.7849
8	3.4579	3.1131	2.6238	2.8064	2.7265	2.6683	2.6241
9	3.3603	3.0065	2.8129	2.6927	2.6106	2.5509	2.5053
10	3.2850	2.9245	2.7277	2.6053	2.5216	2.4606	2.4140
11	3.2252	2.8595	2.6602	2.5362	2.4512	2.3891	2.3416
12	3.1765	2.8068	2.6055	2.4801	2.3940	2.3310	2.2828
13	3.1362	2.7632	2.5603	2.4337	2.3467	2.2830	2.2341
14	3.1022	2.7265	2.5222	2.3947	2.3069	2.2426	2.1931
15	3.0732	2.6952	2.4898	2.3614	2.2730	2.2081	2.1582
etc.							

F-Table (95%)

				n_1			
n_2	1	2	3	4	5	6	7
1	161.45	199.50	215.71	224.58	230.16	233.99	236.77
2	18.513	19.000	19.164	19.247	19.296	19.330	19.353
3	10.128	9.5521	9.2766	9.1172	9.0135	8.9406	8.8868
4	7.7086	6.9443	6.5914	6.3883	6.2560	6.1631	6.0942
5	6.6079	5.7862	5.4095	5.1922	4.3874	4.2839	4.2066
6	5.9874	5.1433	4.7571	4.5337	4.3874	4.2839	4.2066
7	5.5914	4.7374	4.3468	4.1203	3.9725	3.8660	3.7870
8	5.3277	4.4590	4.0661	3.8378	3.6875	3.5806	3.5005
9	5.1174	4.2565	3.7626	3.6331	3.4817	3.3738	3.2927
10	4.9646	4.1028	3.7083	3.4780	3.3258	3.2172	3.1355
11	4.8443	3.9823	3.5874	3.3567	3.2039	3.0946	3.0123
12	4.7472	3.8853	3.4903	3.2592	3.1059	2.9961	2.9134
13	4.6672	3.8056	3.4105	3.1791	3.0254	2.9153	2.8321
14	4.6001	3.7389	3.3439	3.1122	2.9582	2.8477	2.7642
15	4.5431	3.6823	3.2847	3.0556	2.9013	2.7905	2.7066
etc.							

[*]For complete *F*-table information, please consult any text on the subject. A list of texts can be found in the reference.

_F_-Table (97.5%)

				n_1			
n_2	1	2	3	4	5	6	7
1	647.79	799.50	864.16	899.58	921.85	937.11	948.22
2	38.506	39.000	39.165	39.248	39.298	39.331	39.355
3	17.443	16.044	15.439	15.101	14.885	14.735	14.624
4	12.218	10.649	9.9792	9.6045	9.3645	9.1973	9.0741
5	10.007	8.4336	7.7636	7.3879	7.1464	6.9777	6.8531
6	8.8131	7.2598	6.5988	6.2272	5.9876	5.8197	5.6955
7	8.0727	6.5415	5.8898	5.5226	5.2852	5.1186	4.9949
8	7.5709	6.0595	5.4160	5.0526	4.8173	4.6517	4.5286
9	7.2093	5.7147	5.0781	4.7181	4.4844	4.3197	4.1971
10	6.9367	5.4564	4.8256	4.4683	4.2361	4.0721	3.9498
11	6.7241	5.2559	4.6300	4.2751	4.0440	3.8807	3.7586
12	6.5538	5.0959	4.4742	4.1212	3.8911	3.7283	3.6065
13	6.4143	4.9653	4.3472	3.9959	3.7667	3.6043	3.4827
14	6.2979	4.8567	4.2417	3.8919	3.6634	3.5014	3.3799
15	6.1995	4.7650	4.1528	3.8043	3.5764	3.4147	3.2194
etc.							

_F_-Table (99%)

				n_1			
n_2	1	2	3	4	5	6	7
1	4052.2	4999.5	5403.3	5624.6	5763.7	5859.0	5928.3
2	98.503	99.000	99.166	99.249	99.299	99.332	99.356
3	34.116	30.817	29.457	28.710	28.237	27.911	27.672
4	21.198	18.000	16.694	15.977	15.522	15.207	14.976
5	16.258	13.274	12.060	11.392	10.967	10.672	10.456
6	13.745	10.925	9.7795	9.1483	8.7459	8.4661	8.2600
7	12.246	9.5466	8.4513	7.8467	7.4604	7.1914	6.9928
8	11.259	8.6491	7.5910	7.0060	6.6318	6.3707	6.1776
9	10.561	8.0215	6.9919	6.4221	6.0569	5.8018	5.6129
10	10.044	7.5594	6.5523	5.9943	5.6363	5.3858	5.2001
11	9.6460	7.2057	6.2167	5.6683	5.3160	5.0692	4.8861
12	9.3302	6.9266	5.9526	5.4119	5.0643	4.8206	4.6395
13	9.0738	6.7010	5.7394	5.2053	4.8616	4.6204	4.4410
14	8.8616	6.5149	5.5639	5.0354	4.6950	4.4558	4.2779
15	8.6831	6.3589	5.4170	4.8932	4.5556	4.3183	4.1415
etc.							

COMMON ORTHOGONAL ARRAYS[*]

Type of Array	Number of Factors	Levels
L-4(2^3)	3	2
L-8(2^7)	7	2
L-12(2^{11})	11	2
L-16(2^{15})	15	2
L-32(2^{31})	31	2
L-9(2^4)	4	3
L-18($2^1,3^7$)	1	2
	7	3
L-27(3^{13})	13	3
L-16(4^5) modified	5	4
L-32($2^1,4^9$) modified	1	2
	9	4
etc.		

Two-Level Orthogonal Arrays and Interactions (Linear Graphs)

L-4 (2^3) Array

Trial	Column		
	1	2	3
1	1	1	1
2	1	2	2
3	2	1	2
4	2	2	1

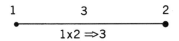

$$1x2 \Rightarrow 3$$

L-8 (2^7) Array

Trial	Column						
	1	2	3	4	5	6	7
1	1	1	1	1	1	1	1
2	1	1	1	2	2	2	2
3	1	2	2	1	1	2	2
4	1	2	2	2	2	1	1
5	2	1	2	1	2	1	2
6	2	1	2	2	1	2	1
7	2	2	1	1	2	2	1
8	2	2	1	2	1	1	2

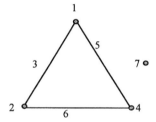 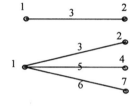

L-12 (2^{11}) Array[a]

Trial	Column										
	1	2	3	4	5	6	7	8	9	10	11
1	1	1	1	1	1	1	1	1	1	1	1
2	1	1	1	1	1	2	2	2	2	2	2
3	1	1	2	2	2	1	1	1	2	2	2
4	1	2	1	2	2	1	2	2	1	1	2
5	1	2	2	1	2	2	1	2	1	2	1
6	1	2	2	2	1	2	2	1	2	1	1
7	2	1	2	2	1	1	2	2	1	2	1
8	2	1	2	1	2	2	2	1	1	1	2
9	2	1	1	2	2	2	1	2	2	1	1
10	2	2	2	1	1	1	1	2	2	1	2
11	2	2	1	2	1	2	1	1	1	2	2
12	2	2	1	1	2	1	2	1	2	2	1

[a]The L-12 is a special array designed to investigate main effects of 11 two-level factors.

L-16 (2^{15}) Array

Trial								Column							
	1	2	3	4	5	6	7	8	9	10	11	12	13	14	15
1	1	1	1	1	1	1	1	1	1	1	1	1	1	1	1
2	1	1	1	1	1	1	1	2	2	2	2	2	2	2	2
3	1	1	1	2	2	2	2	1	1	1	1	2	2	2	2
4	1	1	1	2	2	2	2	2	2	2	2	1	1	1	1
5	1	2	2	1	1	2	2	1	1	2	2	1	1	2	2
6	1	2	2	1	1	2	2	2	2	1	1	2	2	1	1
7	1	2	2	2	2	1	1	1	1	2	2	2	2	1	1
8	1	2	2	2	2	1	1	2	2	1	1	1	1	2	2
9	2	1	2	1	2	1	2	1	2	1	2	1	2	1	2
10	2	1	2	1	2	1	2	2	1	2	1	2	1	2	1
11	2	1	2	2	1	2	1	1	2	1	2	2	1	2	1
12	2	1	2	2	1	2	1	2	1	2	1	1	2	1	2
13	2	2	1	1	2	2	1	1	2	2	1	1	2	2	1
14	2	2	1	1	2	2	1	2	1	1	2	2	1	1	2
15	2	2	1	2	1	1	2	1	2	2	1	2	1	1	2
16	2	2	1	2	1	1	2	2	1	1	2	1	2	2	1

L-32 (2^{31}) Array

Column

Trial	1	2	3	4	5	6	7	8	9	10	11	12	13	14	15	16	17	18	19	20	21	22	23	24	25	26	27	28	29	30	31
1	1	1	1	1	1	1	1	1	1	1	1	1	1	1	1	1	1	1	1	1	1	1	1	1	1	1	1	1	1	1	1
2	1	1	1	1	1	1	1	1	1	1	1	1	1	1	1	2	2	2	2	2	2	2	2	2	2	2	2	2	2	2	2
3	1	1	1	1	1	1	1	2	2	2	2	2	2	2	2	1	1	1	1	1	1	1	1	2	2	2	2	2	2	2	2
4	1	1	1	1	1	1	1	2	2	2	2	2	2	2	2	2	2	2	2	2	2	2	2	1	1	1	1	1	1	1	1
5	1	1	1	2	2	2	2	1	1	1	1	2	2	2	2	1	1	1	1	2	2	2	2	1	1	1	1	2	2	2	2
6	1	1	1	2	2	2	2	1	1	1	1	2	2	2	2	2	2	2	2	1	1	1	1	2	2	2	2	1	1	1	1
7	1	1	1	2	2	2	2	2	2	2	2	1	1	1	1	1	1	1	1	2	2	2	2	2	2	2	2	1	1	1	1
8	1	1	1	2	2	2	2	2	2	2	2	1	1	1	1	2	2	2	2	1	1	1	1	1	1	1	1	2	2	2	2
9	1	2	2	1	1	2	2	1	1	2	2	1	1	2	2	1	1	2	2	1	1	2	2	1	1	2	2	1	1	2	2
10	1	2	2	1	1	2	2	1	1	2	2	1	1	2	2	2	2	1	1	2	2	1	1	2	2	1	1	2	2	1	1
11	1	2	2	1	1	2	2	2	2	1	1	2	2	1	1	1	1	2	2	1	1	2	2	2	2	1	1	2	2	1	1
12	1	2	2	1	1	2	2	2	2	1	1	2	2	1	1	2	2	1	1	2	2	1	1	1	1	2	2	1	1	2	2
13	1	2	2	2	2	1	1	1	1	2	2	2	2	1	1	1	1	2	2	2	2	1	1	1	1	2	2	2	2	1	1
14	1	2	2	2	2	1	1	1	1	2	2	2	2	1	1	2	2	1	1	1	1	2	2	2	2	1	1	1	1	2	2
15	1	2	2	2	2	1	1	2	2	1	1	1	1	2	2	1	1	2	2	2	2	1	1	2	2	1	1	1	1	2	2
16	1	2	2	2	2	1	1	2	2	1	1	1	1	2	2	2	2	1	1	1	1	2	2	1	1	2	2	2	2	1	1
17	2	1	2	1	2	1	2	1	2	1	2	1	2	1	2	1	2	1	2	1	2	1	2	1	2	1	2	1	2	1	2
18	2	1	2	1	2	1	2	1	2	1	2	1	2	1	2	2	1	2	1	2	1	2	1	2	1	2	1	2	1	2	1
19	2	1	2	1	2	1	2	2	1	2	1	2	1	2	1	1	2	1	2	1	2	1	2	2	1	2	1	2	1	2	1
20	2	1	2	1	2	1	2	2	1	2	1	2	1	2	1	2	1	2	1	2	1	2	1	1	2	1	2	1	2	1	2
21	2	1	2	2	1	2	1	1	2	1	2	2	1	2	1	1	2	1	2	2	1	2	1	1	2	1	2	2	1	2	1
22	2	1	2	2	1	2	1	1	2	1	2	2	1	2	1	2	1	2	1	1	2	1	2	2	1	2	1	1	2	1	2
23	2	1	2	2	1	2	1	2	1	2	1	1	2	1	2	1	2	1	2	2	1	2	1	2	1	2	1	1	2	1	2
24	2	1	2	2	1	2	1	2	1	2	1	1	2	1	2	2	1	2	1	1	2	1	2	1	2	1	2	2	1	2	1
25	2	2	1	1	2	2	1	1	2	2	1	1	2	2	1	1	2	2	1	1	2	2	1	1	2	2	1	1	2	2	1
26	2	2	1	1	2	2	1	1	2	2	1	1	2	2	1	2	1	1	2	2	1	1	2	2	1	1	2	2	1	1	2
27	2	2	1	1	2	2	1	2	1	1	2	2	1	1	2	1	2	2	1	1	2	2	1	2	1	1	2	2	1	1	2
28	2	2	1	1	2	2	1	2	1	1	2	2	1	1	2	2	1	1	2	2	1	1	2	1	2	2	1	1	2	2	1
29	2	2	1	2	1	1	2	1	2	2	1	2	1	1	2	1	2	2	1	2	1	1	2	1	2	2	1	2	1	1	2
30	2	2	1	2	1	1	2	1	2	2	1	2	1	1	2	2	1	1	2	1	2	2	1	2	1	1	2	1	2	2	1
31	2	2	1	2	1	1	2	2	1	1	2	1	2	2	1	1	2	2	1	2	1	1	2	2	1	1	2	1	2	2	1
32	2	2	1	2	1	1	2	2	1	1	2	1	2	2	1	2	1	1	2	1	2	2	1	1	2	2	1	2	1	1	2

Triangular Table for Two-Level Orthogonal Arrays[a]

1	2	3	4	5	6	7	8	9	10	11	12	13	14	15
(1)	3	2	5	4	7	6	9	8	11	10	13	12	15	14
	(2)	1	6	7	4	5	10	11	8	9	14	15	12	13
		(3)	7	6	5	4	11	10	9	8	15	14	13	12
			(4)	1	2	3	12	13	14	15	8	9	10	11
				(5)	3	2	13	12	15	14	9	8	11	10
					(6)	1	14	15	12	13	10	11	8	9
						(7)	15	14	13	12	11	10	9	8
							(8)	1	2	3	4	5	6	7
								(9)	3	2	5	4	7	6
									(10)	1	6	7	4	5
										(11)	7	6	5	4
											(12)	1	2	3
												(13)	3	2
													(14)	1
														(15)
														etc...

[a]*How to read the triangular table*: To find the interaction column between factor *A* placed in column 4 and factor *B* placed in column 7, look for the number at the intersection of the horizontal line through (4) and the vertical line through (7), which is 3.

$$4 \times 7 \Rightarrow 3$$

Similarly,

$$1 \times 2 \Rightarrow 3$$

$$3 \times 5 \Rightarrow 6$$

The set of three columns (4, 7, 3), (1, 2, 3), etc. are called *interacting groups of columns*.

Three-Level Orthogonal Arrays

L-9 (3^4) Array

Trial	Column			
	1	2	3	4
1	1	1	1	1
2	1	2	2	2
3	1	3	3	3
4	2	1	2	3
5	2	2	3	1
6	2	3	1	2
7	3	1	3	2
8	3	2	1	3
9	3	3	2	1

L-18 ($2^1 3^{7)}$) Array

Trial	Column							
	1	2	3	4	5	6	7	8
1	1	1	1	1	1	1	1	1
2	1	1	2	2	2	2	2	2
3	1	1	3	3	3	3	3	3
4	1	2	1	1	2	2	3	3
5	1	2	2	2	3	3	1	1
6	1	2	3	3	1	1	2	2
7	1	3	1	2	1	3	2	3
8	1	3	2	3	2	1	3	1
9	1	3	3	1	3	2	1	2
10	2	1	1	3	3	2	2	1
11	2	1	2	1	1	3	3	2
12	2	1	3	2	2	1	1	3
13	2	2	1	2	3	1	3	2
14	2	2	2	3	1	2	1	3
15	2	2	3	1	2	3	2	1
16	2	3	1	3	2	3	1	2
17	2	3	2	1	3	1	2	3
18	2	3	3	2	1	2	3	1

L-27 (3^{13}) Array

Trial	Column												
	1	2	3	4	5	6	7	8	9	10	11	12	13
1	1	1	1	1	1	1	1	1	1	1	1	1	1
2	1	1	1	1	2	2	2	2	2	2	2	2	2
3	1	1	1	1	3	3	3	3	3	3	3	3	3
4	1	2	2	2	1	1	1	2	2	2	3	3	3
5	1	2	2	2	2	2	2	3	3	3	1	1	1
6	1	2	2	2	3	3	3	1	1	1	2	2	2
7	1	3	3	3	1	1	1	3	3	3	2	2	2
8	1	3	3	3	2	2	2	1	1	1	3	3	3
9	1	3	3	3	3	3	3	2	2	2	1	1	1
10	2	1	2	3	1	2	3	1	2	3	1	2	3
11	2	1	2	3	2	3	1	2	3	1	2	3	1
12	2	1	2	3	3	1	2	3	1	2	3	1	2
13	2	2	3	1	1	2	3	2	3	1	3	1	2
14	2	2	3	1	2	3	1	3	1	2	1	2	3
15	2	2	3	1	3	1	2	1	2	3	2	3	1
16	2	3	1	2	1	2	3	3	1	2	2	3	1
17	2	3	1	2	2	3	1	1	2	3	3	1	2
18	2	3	1	2	3	1	2	2	3	1	1	2	3
19	3	1	3	2	1	3	2	1	3	2	1	3	2
20	3	1	3	2	2	1	3	2	1	3	2	1	3
21	3	1	3	2	3	2	1	3	2	1	3	2	1
22	3	2	1	3	1	3	2	2	1	3	3	2	1
23	3	2	1	3	2	1	3	3	2	1	1	3	2
24	3	2	1	3	3	2	1	1	3	2	2	1	3
25	3	3	2	1	1	3	2	3	2	1	2	1	3
26	3	3	2	1	2	1	3	1	3	2	3	2	1
27	3	3	2	1	3	2	1	2	1	3	1	3	2

Four-Level Orthogonal Arrays

This array is called the *modified L-16 array* and is made by combining the five interacting groups of columns of the original 16 two-level columns.

L-16 (4^5) Array

Trial	1	2	3	4	5
1	1	1	1	1	1
2	1	2	2	2	2
3	1	3	3	3	3
4	1	4	4	4	4
5	2	1	2	3	4
6	2	2	1	4	3
7	2	3	4	1	2
8	2	4	3	2	1

L-16 (4^5) Array (Continued)

Trial	1	2	3	4	5
9	3	1	3	4	2
10	3	2	4	3	1
11	3	3	1	2	4
12	3	4	2	1	3
13	4	1	4	2	3
14	4	2	3	1	4
15	4	3	2	4	1
16	4	4	1	3	2

L-32 ($2^1 \times 4^9$) Array

Trial	Column									
	1	2	3	4	5	6	7	8	9	10
1	1	1	1	1	1	1	1	1	1	1
2	1	1	2	2	2	2	2	2	2	2
3	1	1	3	3	3	3	3	3	3	3
4	1	1	4	4	4	4	4	4	4	4
5	1	2	1	1	2	2	3	3	4	4
6	1	2	2	2	1	1	4	4	3	3
7	1	2	3	3	4	4	1	1	2	2
8	1	2	4	4	3	3	2	2	1	1
9	1	3	1	2	3	4	1	2	3	4
10	1	3	2	1	4	3	2	1	4	3
11	1	3	3	4	1	2	3	4	1	2
12	1	3	4	3	2	1	4	3	2	1
13	1	4	1	2	4	3	3	4	2	1
14	1	4	2	1	3	4	4	3	1	2
15	1	4	3	4	2	1	1	2	4	3
16	1	4	4	3	1	2	2	1	3	4
17	2	1	1	4	1	4	2	3	2	3
18	2	1	2	3	2	3	1	4	1	4
19	2	1	3	2	3	2	4	1	4	1
20	2	1	4	1	4	1	3	2	3	2
21	2	2	1	4	2	3	4	1	3	2
22	2	2	2	3	1	4	3	2	4	1
23	2	2	3	2	4	1	2	3	1	4
24	2	2	4	1	3	2	1	4	2	3
25	2	3	1	3	3	1	2	4	4	2
26	2	3	2	4	4	2	1	3	3	1
27	2	3	3	1	1	3	4	2	2	4
28	2	3	4	2	2	4	3	1	1	3
29	2	4	1	3	4	2	4	2	1	3
30	2	4	2	4	3	1	3	1	2	4
31	2	4	3	1	2	4	2	4	3	1
32	2	4	4	2	1	3	1	3	4	2

Linear Graphs for Two-Level Orthogonal Arrays

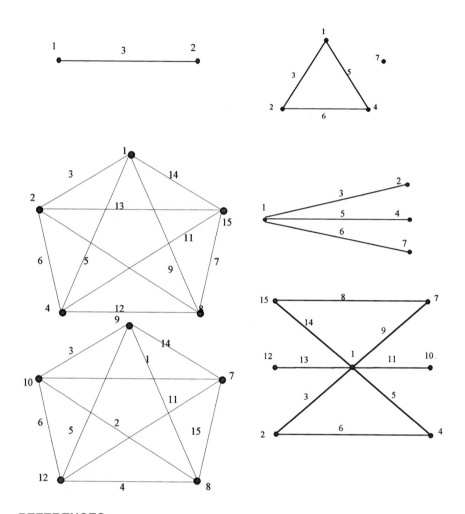

REFERENCES

American Supplier Institute. 1998. *Robust Design Using Taguchi Methods: Workshop Manual.* Livonia, MI: ASI.

Baker, Thomas B., and Don P. Causing. March 1984. Quality engineering by design—the Taguchi method. *Proc. 40th Annual ASQC Conference.*

Burgman, Patrick M. 1985. Design of experiments—the Taguchi way. *Manufacturing Engineering*, May, pp. 44–46.

Gunter, Burton. 1987. A perspective on the Taguchi methods. *Quality Progress.*

Iman, Ronald L., and W. J. Conover. 1983. *A Modern Approach to Statistics.* New York: Wiley.

Peace, Glenn Stuart. 1993. *Taguchi Methods.* Reading, MA: Addison-Wesley.

Phadke, Madhav S. 1989. *Quality Engineering Using Robust Design*. Upper Saddle River, NJ: Prentice Hall.

Quinlan, Jim. 1985. *Product Improvement by Application of Taguchi Methods*. Midvale, OH: Flex Products. Winner of Taguchi Applications Award by American Supplier Institute.

Ross, Philip J. 1988. *Taguchi Techniques for Quality Engineering*. New York: McGraw-Hill.

Roy, Ranjit K. 1990. *A Primer on the Taguchi Method*. Dearborn, MI: Society of Manufacturing Engineers.

Roy, Ranjit K. 1996. *Qualitek-4 (for Windows): Software for Automatic Design of Experiment Using Taguchi Approach, IBM or Compatible Computer*. Bloomfield Hills, MI: NUTEK, Inc. Download free DEMO from: *http://www.rkroy.com*

Sullivan, Lawrence P. June 1987. A power of the Taguchi methods. *Quality Progress*.

Taguchi, Genichi. 1987. *System of Experimental Design; UNIPUB*. New York: Kraus International Publications.

Taguchi, Genechi and Yuin Wu. 1989. *Taguchi Methods: Case Studies from U.S. and Europe*, Vol. 6. Livonia, MI: ASI Press.

Wu, Yuin. 1986. *Orthogonal Arrays and Linear Graphs*. Dearborn, MI:. American Supplier Institute.

Wu, Yuin, and Willie Hobbs Moore. 1986. *Quality Engineering Product and Process Optimization*. Dearborn, MI: American Supplier Institute.

PRACTICE SESSION USING QUALITEK-4 SOFTWARE

1. Run the Qualitek-4 (QT4) program on your computer and click *OK* past the registration screen to the experiment configuration screen.
2. Notice that the experiment file *PISTON.Q4W* is already loaded in memory. If not, open file *PISTON.Q4W* (or any of over 60 example experiment files).

Analyze Results of PISTON and Other Example Experiments

3. Click on the *Analysis* menu and select *Standard analysis*. Since PISTON has three samples/results per trial, QT4 will remind you to perform *S/N* analysis. Click *OK* to proceed. In the next screen, check the *Bigger is better* quality characteristic and click *OK*.
4. Review the results and click on *Graph* if you wish to view it. Click *OK* when done.
5. While in the *Main effects* screen, click *Plot* to view plots of the main effects. Once you are in the graphics screen, click the "<<" or ">>" buttons at the bottom of the graph to display other factor plots. Click *OK* to return to the *Main effects* screen.
6. Click on the *Interaction* button to display interaction plots between any two factors. Select the *Automatic* option if you want QT4 to calculate interactions between all possible factors. Even though you may not have thought about all the interactions, or included them in your study, QT4 can calculate the strength (severity) of their presence. QT4 calculates $N \times (N - 1)/2$ possible pairs of

interactions for *N* factors and ranks them in order of the *severity* of their presence (severity Index, 0 to 100%). Review the list to note the important pairs of interaction for future reference. Click *OK* or *Cancel* to return to the *Main effects* screen and click *OK* to proceed to the *ANOVA* screen.

7. In the *ANOVA* screen, try pooling the factor with the least amount of influence (the last column in ANOVA). To pool, double-click on the factor description. At the prompt, review the percent confidence levels and click *OK* to proceed with pooling. Notice how the ANOVA table is updated automatically. While in the ANOVA screen, you may also try one or more of the following tasks.

   a. Pool another factor with a smaller influence value.

   b. Reset ANOVA.

   c. Pool all factors with less than a 98.6% confidence level.

   d. Plot a bar graph, pie chart, and restart ANOVA.

   e. Review the pie diagram.

   Click *OK* or the *ANOVA* button to move to the optimum screen.

8. At the optimum screen:

   a. Examine the response expected.

   b. Convert this result to a real number when the performance expected is expressed in terms of the *S/N* ratio.

   c. Plot the graph of factor contribution.

   d. Find the response for all factors at level 2.

   e. Go back to ANOVA and pool all factors below 10% influence.

   f. Determine the 90% confidence interval of the optimum response.

   g. Go to the optimum screen from ANOVA.

   Click *OK* to return to the main screen. From the file menu select *Open* and select any other experiment file from the list of over 50 files. Analyze results by following steps 3 to 8.

Experiment Designs

9. Prepare a list of factors and levels for a practice experiment (say 5 factors each at the two-level). Click on the *Design* menu and select *Manual design*. Check the L-8 array for your design. When in the design screen, describe factors and their levels by assigning to any rows (which are columns of the array). Since you will be using five of the seven available columns, click on the *Unused* button to designate the two columns as unused. Click *OK* to move to the *Orthogonal array* screen. Click *OK* to the experiment file screen. Supply only the first eight characters of your file name. The file extension .Q4W is added automatically.

10. After the experiment design is completed, you may review the trial condition by clicking on the *Review* menu and selecting the *Trial Condition* option.

Should you want to carry out these experiments, you may print some or all of the trial conditions.

Result Entry

11. Once the experiments are carried out, you will need to enter the results by clicking on the *Result* menu and then selecting the *Enter result* option. If your result includes multiple evaluation criteria, you will see the *Multiple criterion (OEC)* option. Enter results/evaluation criteria as applicable. When done entering results, perform analysis following steps 3 to 8 as described above.

SPECIAL TASKS: CAPTURING, PASTING, AND CROPPING QT4 SCREENS FOR REPORTS AND PRESENTATIONS

In addition to the standard print output option in QT4, you can capture any screens and conveniently size and paste it into your *Word* (by Microsoft) document through the following tasks:

1. Be in the screen of your choice
2. Press *Alt* and *Print screen* together.
3. Open the *Word* document.
4. Select *Edit* and *Paste* to place the screen on the *Word* document.
5. Select the picture (screen), then from the *Insert* menu, select *Frame*.

To crop the picture you pasted, click on the picture. Make sure that the cropping tool is available and is displayed. Crop and size the screen to your satisfaction following the procedure necessary for your version of the *Word* program. (Refer to the *Practice tips* menu item in the QT4 main screen.)

What's on the Disk

The companion CD-ROM attached to the back cover of this book contains **Qualitek-4** software developed by the author. The CD also contains a large number of example experiments that you will need to study this book, plus many additional case studies. This version of the software has restricted use for personal projects but provides opportunities to review example experiments of all sizes. You will be asked to install the software while studying Step 2, just before you are ready to start experiment design and analysis. Installation requires only a few clicks of a mouse. Here is what you do:

Windows 95 Installation Instructions
1. Remove the CD-ROM from the back cover of the book.
2. Turn your computer on and make sure there are no other programs are running in the background. If there are, close all programs. Now insert the Qualitek-4 CD-ROM in the CD drive.
3. From the *Start* menu of your Windows (left lower corner of your monitor), select the *Run* option and type *D* or *?:\Setup.exe* (the ? mark indicates any other character designations appropriate for your computer). Now click *OK* to proceed with the set up.
4. Follow the prompts and type *Demo* when asked for a registration number. (Click *OK* at all prompts when you accept the default descriptions.)

How You Will Be Able to Use the Software (Qualitek-4) on the Disk

You *will be able to* review all examples and exercise experiments provided on the disk and with the program. You will also be able to use this version of the software to design your own experiments using the L-8 orthogonal array.

You *will not be able to* design your experiment using any arrays other than the L-8 array.

The limited-capability (DEMO) version of the Qualitek-4 software can also be downloaded by visiting www.rkroy.com.

More detailed installation instructions are provided in Step 2 of this book. You may also review the *Reference Manual for Qualitek-4* (Microsoft Word Version 7.0 document) provided in the CD-ROM.

To use this CD-ROM, your system must meet the following requirements:

Platform/Processor/Operating System. Windows 3.1 or higher.

RAM. 32 MB.

Hard Drive Space. 10 MB minmum.

Peripherals. CD-ROM drive, Microsoft PowerPoint, Microsoft Word to view additional documents.

To install Qualitek-4, insert the CD-ROM and launch the setup.exe file your CD-ROM drive (i.e., click Start, Run, and type d:/setup.exe where "d" indicates your CD-ROM directory). Follow the instructions on screen.

For technical support, please e-mail techhelp@wiley.com and indicate the title, ISBN, and the nature of the problem.